Revolt

Revolt

The Worldwide Uprising Against Globalization

Nadav Eyal

TRANSLATED FROM THE HEBREW BY HAIM WATZMAN

ecco

An Imprint of Harper Collins*Publishers*

 For information, address HarperCollins Publishers, 195 Broadway, New York, NY 10007.

HarperCollins books may be purchased for educational, business, or sales promotional use. For information, please email the Special Markets Department at SPsales@harpercollins.com.

Originally published as *The Revolt Against Globalization* in Israel in 2018 by Yediot Ahronot Books.

FIRST EDITION

Designed by Paula Russell Szafranski

Frontispiece © NWM/Shutterstock

Library of Congress Cataloging-in-Publication Data has been applied for.

ISBN 978-0-06-297335-1

21 22 23 24 25 LSC 10 9 8 7 6 5 4 3 2 1

To Tamar

". . . And they seemed unto him but a few days"

If there is no struggle, there is no progress.

—FREDERICK DOUGLASS,
WEST INDIA EMANCIPATION SPEECH, 1857

Contents

Introduction: The Death of an Age

The building looks like a typical office tower that can be found in every prosperous city center, from Manhattan to London to Tel Aviv. The VIPs are led through a back corridor to a small service elevator, completely inappropriate to the occasion, adding to the mysterious ambience. The elevator descends and its door opens, revealing the venue for the night's event—a family wine cellar, a secret one, our host says. At one end of the room, a famous chef is preparing our dinner. All along the walls, behind glass, lie bottles of wine flown in from vineyards all over the world. The guests—high-tech entrepreneurs, a former prime minister, a former senior army officer who is now a social entrepreneur, the CEOs of leading corporations—are impressed, and they are not easily impressed. Everyone there—indeed, most everyone everywhere—knows the generous host's name.

As we seat ourselves around a table, I look around and count the uber-wealthy. I'm pretty sure I am the only guest who drove here in a Toyota Corolla with a loose bumper.

I've been invited to speak about the international situation, about globalization and the revolt against it. My audience in the expertly lit wine cellar listens attentively to my accounts of populations bypassed by the prosperity generated by the present world order, and of how giant technology companies have avoided responsibility for the ills of the connected world they have created. I argue that liberal values are being challenged by a resurgence of enemies of progress, and suggest that young people have become less inclined to fight for democracy and instead are calling for radical solutions. The numbers, I note, show that humanity is doing well in general. Why then, do so many people feel so trapped?

I should have known what the reaction would be. The 1 percenters feel, for the most part, that the crisis of 2008 was but a passing cloud; that Trump's election was but a onetime historical fluke; and that progress—that is the aristocratic version of progress to which they subscribe—is unstoppable. Our munificent host and one or two of his guests grasp the thrust of the analysis, even if they do not accept it. The others balk. "It's overblown pessimism," one of them suddenly says, and the others begin to chant "pe-ssi-mi-sim." They quickly counter me with the common wisdom: It's a "wave of populism," a brief backlash that will pass without causing significant damage. The conversation degrades into the kind of anachronistic discourse characteristic of people born in the 1950s and '60s, including clichés such as "confidence breeds success," "fortune favors the bold," "the young will grow up," and "we can't go back to the Dark Ages." Most of them have no interest in listening to what I have to say. Instead, they want to instruct me—and through me, my generation—that all will be well if we just think positively. Dessert is served, elegantly ending the debate, such as it was. It's easy to disagree politely when your children's future is ensured by low-risk bonds.

The dinner somehow reminded me of a much more dramatic event I attended as a journalist two years previously. Anxiety was ubiquitous at both gatherings. It's just that when the super-rich are anxious, they wrap themselves in cellophane-like packaging that crackles with optimism. The middle class adopts a much simpler tactic: outrage.

The evening of November 8, 2016, was festive and crisp in Manhattan. A cloudless sky was visible through the glass ceiling of the Javits Center, ready for the crowning of the new leader of the free world. Hawkers outside were doing a brisk business—President Hillary t-shirts, showing her in a Superwoman costume; First Husband Bill Clinton t-shirts; campaign buttons of all colors, souvenirs of the historic day. Hundreds of policemen and security personnel were deployed outside, along with an army of broadcast vehicles and an entire field of satellite dishes. The media presence was many times larger than that assigned to the more spare headquarters of the Trump campaign, less than half a mile away as the crow flies. "She means to rise," wrote the poet Maya Angelou about Clinton back in 2008; now she was about to break free of those rusty chains and become the most powerful person in the world.

Representatives of America, of all the colors of the rainbow, were placed on the stage. There were straights and gays, Hispanics and blacks and whites, women and children. They were to serve as models of the new age that Clinton's election heralded. With infinite patience, they sat there for long hours, waiting for those few seconds their children would see on television and forever cherish, that image of them with the first woman to be elected president of the United States of America. Even when the skies grew dark over the Javits Center, they did not budge from their seats.

In the end, of course, Clinton never showed up. She did not see the celebration prepared for her. Night fell and swept it all away.

There's something brutal about the journalist's gaze. He

sees the image as it develops, from a distance that gives him perspective. He observes the disappointment as it spreads through the crowd, the shocked gasps, the tears and the heartbreak, the banality of human reaction—denial, disappointment, the desperate hope that continues to percolate among the believers.

When the results began to come in, the eyes of the Clintonistas were glued to their smartphones, murmuring in disbelief. That was exactly the point. They couldn't believe it, they couldn't understand how this could be happening. Many wept. One told me that, as a Jew and a homosexual, he feared a new Holocaust.

I asked him if that was just a figure of speech.

"No," he sobbed, "I'm really scared."

On the face of it there seems to be no connection between the distraught and panicked Clinton campaigners on that autumn night and the self-assured rich whom I met in the wine cellar. The latter were resolutely optimistic, determined to explain just how the world order that is so good for them, personally, is so great for everyone else. Clinton's supporters sensed that democracy was endangered and that they had been robbed of their future. But the point is that both shared a deep, unspoken fear. The 1 percenters dealt with it by euphorically hiding their heads in the sand; the Clinton campaigners coped by covering the floor of the Javits Center with the tears they shed.

They were not only frightened by the prospect that Trump, the advocates of Brexit, European nationalists, or Islamist fundamentalists would propel the world toward catastrophe. After all, if such a catastrophe were to come, it would demonstrate just how correct they had been in their faithfulness to liberal values or to the market economy. No, what they feared was not a cataclysm but the opposite—that the other side, that Trump, might be successful. His success would mean a world with an enduring anti-liberal order and severely constrained global cooperation.

It would be a world in which bedrock beliefs—in the victory of good over evil in World War II, in freedom as a precondition for prosperity, in the rejection of bigotry, in the principle of women's rights over their bodies, and most of all the fervent faith in the overarching value of progress—would turn out to be ephemeral. For them, history would stop, and then reverse. For many, the years since have proved that the shift has already begun.

I AM NEITHER AMERICAN NOR EUROPEAN. I LIVE IN A DISTANT province that shelters under the wings of the American empire. From here I can be an observer, with the luxury of some emotional detachment from the coming storm. In 2016, some months before Election Day, I set out on a journey through the United States, seeking an answer to a simple question: If Trump were to win, how would it happen? The polls said it was nearly impossible, but I was skeptical. In Pennsylvania, one of the cornerstones of the Industrial Revolution, I sat in the living room of a coal mining family as rain fell outside and the wind shrieked. The family was as grim and despondent as the weather, lacking a trace of the American optimism that I so put my trust in. Black activists in Philadelphia told me that President Obama was simply one more mask worn by whites who were killing innocent residents of their neighborhoods. They vowed not to vote for "that Hillary person." A little girl in Charlotte, North Carolina, told me with tears in her eyes that a classmate had stopped inviting her to her birthday parties because her mothers are transgender women. In her story I could feel the burgeoning animosity toward the new America. In the same state, I attended Sunday services in a church whose preacher maintains that the United States would be punished with a plague worse than Ebola for condoning homosexual sodomy. I asked him if his America was not passing from this world; his response was, "Hey, don't bury us yet!"

What has taken place in the United States under Trump is no routine political change; nor is it a revolution based on a new and coherent political idea. Neither is there a coherent political idea behind Brexit. The rise of populism and nationalism, from Brazil to Italy to Hungary, constitutes an attack, albeit diffused, on today's globalization, growing out of an echo chamber of injustices that have plagued the middle class throughout the industrialized world. Those who are overly focused on what is going on in the Americas, Europe, Africa, or Asia miss the most important social, cultural, and political phenomenon of our time. As in a pointillist painting, the little dots come together to form a picture, of revolt. Large numbers of people are rejecting globalization as an economic, cultural, and universal value system. The revolt is worldwide, unorchestrated, and fluid. It is more about the rejection of current power structures than about the fine details of building new ones.

Fundamental opposition to globalization began at opposing poles—anarchist-radical on one side and fundamentalist-religious on the other. Spurred by growing social disquiet, radical and reactionary ideas began to make their way into the middle class. The revolt is manifested in the British decision to leave the European Union, the rise of the extreme right in Europe, the growth of fundamentalism, as well as increasing support for the radical left and burgeoning resentment of the rich and of the concentration of wealth. Politicians are desperately trying to ride the tiger. After his election, the president of the United States has inundated the American and international discourse with relentless provocations. The tap of his keyboard as he was tweeting was so deafening that we have forgotten what we all realized when he won: Trump is a manifestation of a much broader phenomenon, which preceded the 2016 and 2020 elections. Now, a few years down the line, we can do what is required and look back on recent decades as one section of the political and historical mosaic that is our current world. The era of the revolt is too momentous, too consequential, to be defined by Trump or by the media's addiction to him.

The rebels are a disparate coalition of rejects. Some claim that globalization, the liberal values to which it is tied, and the technology it has both spawned and fed upon, have been toxic to their lives, their communities, and their deeply held values and beliefs. Others are up in arms, sometimes literally, against a political class that promised that global solutions would bring prosperity to all while at the same time becoming bedmates of the upper 1 percent. They are in revolt because they were told that globalization makes the world flat—everything lies before you, everything is immediate, everything is within reach, all you need to do is take it. That, it goes without saying, is a hollow notion, because the international economy is built more on inequality than on equality. The rebels see their children forsaking their culture, and the demand for political correctness spreading and preventing them from giving voice to their understandable frustrations. They are rising up because their security, identity, and livelihood are all endangered. Terrorism could strike at any moment, immigrants want to go everywhere, and employers are constantly thinking about terminating them. The COVID-19 pandemic that swept across the globe in 2020 revealed the degeneration of twentieth-century-style politics and its incapacity for coping with contemporary challenges, like the spread of a new pathogen in a world that was highly interconnected. As a matter of course, both political systems and their leaders routinely put on a facade of control, certainty, and security to the public. But during the course of history, epidemics have shattered that illusion. They also expose which rulers are effective and able and which are feckless and dangerous. Luchino Visconti, who ruled Milan in the fourteenth century, imposed quarantine on homes in which the Black Death had broken out, saving many lives in his city during the epidemic's first wave. Other rulers fled to their summer palaces as their subjects died, in a way not dissimilar to Donald Trump playing golf as the coronavirus raged. "In a dark time, the eye begins to see," wrote the American poet Theodore Roethke. It is no coincidence that widespread protests broke out in many countries as the virus spread. COVID-19 further catalyzed the uprising against a fractured world order.

This outpouring of grievances, this surge of resentment, is changing the world. Contrary to the picture often painted by the media, the protests against international trade or, on a different plane, against universal values, are far more than flare-ups of hatred and ignorance or passing phenomena. Protesting the increase in immigration in Western societies is not always hypernationalist, jingoistic propaganda. Globalization has bettered the human condition, but it has also decimated communities and ravaged ecosystems, sowing the seeds of insurrection. The revolt erupted at the end of the age of responsibility.

AFTER WORLD WAR II, THE WORLD ENTERED AN ERA OF relative stability, guided by caution and a sense of duty. This was the age of responsibility. It was in a very profound sense molded by the horrifying personal experiences of both voters and the representatives they elected. Before them lay a devastated and burned world, a planet in a state of shock. They saw the horrifying consequences of racism, hypernationalist vengeance, economic decline, trade wars, and addiction to ideological extremism, and they rejected it all. For a brief time after the war ended, civilization was awash in optimism, like rain after a drought. President Franklin Roosevelt gave voice to these feelings as early as 1943, two years before the war ended: "We have faith that future generations will know that here, in the middle of the twentieth century, there came a time when men of good will found a way to unite, and produce, and fight to destroy the forces of ignorance, and intolerance, and slavery, and war."[1]

The simple goal he articulated was achieved. The Soviets, Americans, Chinese, British, and French all agreed that it had been a just war, and grasped the significance of the horrors they had witnessed. But that's as far as the consensus went. Roosevelt spoke of future generations, but his generation saw Hiroshima and Nagasaki, and were shortly afterward terrified by the first Soviet nuclear test in 1949. A new world had been born, but one that was faced with the prospect of its own extinction.

The trembling world's greatest fear was that another world war was on the way, one that would be set off by the dangerous antagonisms of the Cold War. Optimism was soon overwhelmed by profound pessimism. If, immediately after the end of World War II, Americans thought that the Soviet Union would cooperate to achieve world peace, just a year later few Americans believed that the Soviets could be trusted, and 65 percent predicted another global conflagration within no more than a quarter century. At the same time, according to one survey, six out of every ten Americans wanted a stronger United Nations, or even a single world government.[2]

Anxieties and fears are sometimes advantageous, especially for rulers. One advantage is that they can compel caution. And caution begets responsibility.

The age of responsibility was virtually defined in 1947 by William A. Lydgate, the editor of the Gallup Poll, in a lengthy analysis. "Let's-drop-a-few-atom-bombs-on-Moscow extremism doesn't appeal to our people . . . The very fact that the situation appears so gloomy may, however, be a healthy sign. Instead of idealistically supposing, as many did after 1918, that the world was safe for democracy, the nation today soberly realizes that you have to work to keep peace."[3]

Nostalgia is as deceptive as it is dangerous. The Cold War didn't *feel* like the age of responsibility. The West shed its colonies in the developing world grudgingly and often violently. The world heard the drums of war in the Cuban missile crisis, in the tensions over Berlin, and in the Korean and Vietnam conflicts. The two superpowers sparred through a slew of proxy wars, in which people of the so-called Third World were sacrificed on the altar of preventing a nuclear war between the West and the East.

Yet it was nonetheless a responsible world, and recognition of that fact, even if in retrospect, is useful now. It is difficult to discern the good in the present, and even harder to follow the rapid trajectory of evil. After World War II, the world's leaders lived in steady anxiety about a new and truly calamitous conflict. It was their anxiety that held them back, in most occasions, from walking down the road of militaristic adventurism. Even more significantly, public opinion limited them. In both Soviet propaganda and the pronouncements of American generals, peace was the highest value, or at least leaders wanted the public to believe that it was peace they were pursuing. Even the bellicose General Douglas MacArthur spoke a lot about peace. "The soldier above all others prays for peace," he said, and spoke of the need "to preserve in peace what we won in war." He even said that honor should be sacrificed for the sake of peace.[4] Was it ideologies that restrained or constrained leaders with the bonds of responsibility? Not really. It was a much more profound force—the personal and collective memory of the horrors of the war, and the moral lessons learned from them. "All wars start from stupidity," President John F. Kennedy said during the Berlin crisis of 1961.[5] During the Cuban missile crisis, when the military leadership presented Kennedy with a plan to stage a nuclear first strike that would destroy the entire Soviet bloc (the plan included dropping 170 atomic and hydrogen bombs on Moscow alone), Kennedy left the room, appalled. "And we call ourselves the human race," he bitterly remarked to Secretary of State Dean Rusk on his way to the Oval Office.

The leaders of that world—Nikita Khrushchev and Kennedy, as well as Yugoslavia's Josip Broz Tito, Konrad Adenauer in West Germany, Israel's David Ben-Gurion, Britain's Clement Attlee, Leonid Brezhnev of the Soviet Union, and France's François Mitterrand—had lived through a great and destructive war, or even both world wars. They were not naive pacifists. Rather, they had pragmatic goals, which were consistent with their particular national interests—stability, international institutions, avoiding the next great war.

In the West, responsibility also took the form of a decline of extremist forces on both the right and the left and increasing support of democracy. Political scientists Roberto S. Foa and Yascha Mounk have shown that more than 70 percent of Americans born in the 1930s felt that it was "essential" for them to live in a democracy. Almost as many British subjects born in that decade—65 percent—felt the same way.

Democracy was an essential value for those born in the 1940s and 1950s as well.[6] The people who built the West shared a single, terrible formative experience—the awful destruction of war. These parents and grandparents of the current generation shared an ethos that cut across national borders. They exhibited an almost religious diligence and scrupulousness, and sanctified the present rather than harboring fantasies about the future. They demanded a more or less mainstream responsible politics, and that is what they got.

Slowly and painfully, the age of responsibility led to relative stability and peace. The two superpowers maintained an adversarial, competitive relationship that was fundamentally rational and responsible. They eschewed populism and focused on science and technology to win the Cold War and as the means for improving the material conditions of societies. Each in their separate spheres of influence, the superpowers idealized international cooperation within their blocs.

Indeed, following World War II, with the exception of a temporary spike in conflict following the fall of Communism, the number of interstate wars declined.[7] The last time complete armored regiments fought battles was in the Second Gulf War in 2003. The number of fatalities in conflicts around the world is in steep decline, as is the number of people living on less than $2 a day. Child mortality is in decline. In 1950, less than half of the world's inhabitants could read and write; today the figure is 86 percent.[8] Between 2003 and 2013, the world's median income per capita nearly doubled.[9] None of this happened by chance. The scarred societies and apprehensive leaders of the postwar period planted a tree of stability. These are its fruits.

Two things about the age of responsibility need to be kept in mind. First, it was an exception in the turbulent and war-torn modern age. World War II struck extremism and populism speechless. The silence lasted for a moment in history, but it was during that period that most of this book's readers were born. Then the memory of the war began to fade. Unlike the generation born in the 1930s, people born in the 1980s in Britain and the United States do not tend to believe that democracy is vital. Only 30 percent think it is.[10] Their grandfathers may have made the ultimate sacrifice on the beaches of Normandy to defend democracy, but they themselves think that the term has lost its meaning.

The second thing you need to know about the age of responsibility is what you already sense: that it's over.

THE AGE OF RESPONSIBILITY ENDED WHEN THE WORLD Trade Center towers came crashing down. We are living in the initial aftermath of 9/11. Al-Qaeda's attacks on American soil were an act of war by fundamentalists against the universalist vision that the US represented. The terrorists sought a global war between Christianity and Islam, and in the process they unleashed demons previously kept in check, many of which had nothing to do with the two faiths. It was the start of a battle to determine the fate of the world, an engagement fought not between religions but between ideas. On one side are those who believe that the world is moving slowly toward political and cultural integration, and on the other are those for whom such a prospect is a nightmare, and who are willing to fight to ensure that it never happens. In the middle is the world's, and especially the West's, middle class, wavering uncertainly between the nation-state and globalization, between particular identity and universal values.

Today's globalization is not sustainable; the relative peace of the post–World War II era is under threat, and the signs of instability are multiplying. The most serious of these is the climate crisis. The prosperity of the industrial age was paid for by abuse of the present and future natural world.

This book is a journey through the trenches of the revolt, both its visible contours and its dark corners. In northern Sri Lanka I saw the last elephant herds that have been pushed into patches of forest that are slowly being destroyed by indigent farmers who are themselves trying to cope with the consequences of international trade. Teenage Syrian refugees spoke to me about their future as we trod along a railroad track on their long trek from Greece to Germany. In Japan, which is facing an unprecedented demographic crisis, an elderly woman told me, in a deserted school, about her longing for the lost sounds of children playing. I saw the Greeks riot, protesting the severe recession there, and I was in London at the outbreak of the great financial crisis of 2008, the most severe since the Great Depression of the 1930s. I spoke with starry-eyed racists and nationalists about their hopes for the future.

It is a story that offers conversations with and observations of particular people coping with local problems at specific places and specific times, but much larger issues emerge from them. It tells of the advent of a global consciousness that crosses geographical and cultural borders, and the way in which globalization has changed moral sensibilities.

We are living in a time in which an era of relative peace has impelled a huge wave of refugees to flee their homes in centers of catastrophe in search of sanctuary in the West; in which a great economic crisis has passed but nevertheless continues to fracture the middle class and to threaten globalization and its institutions; in which cooperation between the world's people and institutions and states is declining just as the world needs to address the greatest global crisis ever, that of the climate. Fundamentalism is flourishing in an era of rapidly declining poverty and increasing education, ever-improving health services and ever-growing incomes—but as people are producing fewer and fewer children, with all the implications that ensue. An international community founded on a liberal vision accepted by consensus is turning more to the extremes.

These tensions have spawned a crusade against the very idea of progress. Progress in the sense of Enlightenment values depends on trust in facts and reason, acceptance of science as essential to bettering the human condition, and an open society in which tradition has no absolute veto over critical thinking. The energy of the revolt against globalization is being harnessed by both old and new opponents of progress. Their ambition is not to address the grievances stemming from an unsustainable global system, but only to use them as a decoy. Populist-racist politicians, anti-science charlatans, Bakuninite anarchists, fundamentalists, virtual communities on social networks, totalitarian ideologues, neo-Luddites, and the votaries of conspiracy theories—they are all on the march.

The revolt and the politics it engenders can lead to a more just, and thus stronger, international system, one that will balance the local and the global, require more equality of opportunity, and facilitate the environmental cooperation that is crucial for our survival. But this optimistic scenario is neither obvious nor inevitable. If there is something we have learned in the last twenty years, it is that nothing is preordained, and that no progress is irreversible.

Progress affects to be muscular, but it is actually quite fragile. It is wholly dependent on the readiness of communities to fight for it, and on the determination of leaders to avoid folly. People around the globe are living through a radical moment. This book is an attempt to listen to them.

CHAPTER 1

An Attack on a Newspaper

I once had a hand in an attack on a Pakistani newspaper by more than two dozen armed men. I could hardly have anticipated it and certainly did not want it. I knew neither the attackers nor the victims; indeed, I had never visited the newspaper's offices. Pakistan and Israel, where I live, do not have diplomatic relations. But in a globalized world, things a person does in one country can have dire, occasionally overwhelming consequences for people living far away. Sometimes it is more ominous than anything you expected.

I met Ammara Durrani, then a senior editor for Pakistan's Jang Media Group and a writer for the country's largest English-language newspaper, the *News International*, in 2004. We were members of a group of journalists who had come to the United States for a lengthy professional program funded by the State Department, at the invitation of one of the country's best-known public radio stations, WBUR of Boston. The organizers from the station had what they thought was a brilliant idea. They'd bring together hostile tribes, Israelis and Palestinians, Indians and Pakistanis. The program focused on the media's role in conflicts, a polite way of saying that journalists feed the fires of conflict and inflame public opinion, and perhaps it would be better if they didn't. The Bush administration was interested in projects of this sort because, in the midst of its war on terror and the occupation of Iraq, it needed the fig leaf of promoting dialogue between hostile peoples as a demonstration of its commitment to resolving international conflicts by peaceful means. The organizers may have believed that Israelis and Palestinians might be able, thousands of miles from home, and in the presence of a parallel conflict on the Indian subcontinent, to find a common language. It was a vain hope. With foreigners in the room, they entrenched themselves in their traditional positions. So did the Pakistanis and Indians. Nevertheless, some exceptional and culture-crossing friendships emerged. Everyone got along with Ammara. She was the quintessential Oxfordian, speaking eloquently serious and polished English. All the Middle Easterners, whether Israeli or Palestinian, envied her.

Her passport, like all those issued by her country, specified that it was valid for travel to all countries except Israel. There is a long tradition of cold hostility between the Jewish state and the Islamic Republic of Pakistan. It dates back to the birth of both countries, within a year of each other, as Britain divested itself of its empire. Despite and in fact because of this, Durrani and I remained in contact by email after the American seminar. In 2005 she began work on an in-depth article on the unofficial relations between the two countries, and the possibility that these might be upgraded to full diplomatic recognition. She wrote to me that she'd be delighted to interview Prime Minister Ariel Sharon for the piece. My guess was that it would not be easy to get him to grant an interview. But if she wanted, I suggested, I could probably land her an interview with Vice Prime Minister Shimon Peres, whom

I knew well. Durrani seized the opportunity. Peres, a former prime minister and Nobel laureate, was no less an international figure than Sharon—in fact, he was probably better known. But there was a problem. She told me that, because of the hostility between the two countries, she could not place a telephone call from Karachi to Jerusalem. In 2005, Skype and other such services were not available. I thus suggested that she send her questions by email. I would arrange an interview through Peres's press spokesman. I would ask him her questions just as she wrote them, tape-record his answers, and then transcribe and send them to her.

Peres's office was only too enthusiastic to have him interviewed by a prominent Pakistani newspaper, and Peres himself was always more than happy to broadcast his indefatigable political optimism. So it happened that, one day in mid-January 2005, I sat across the table from Peres in the Knesset's cafeteria, and instead of chatting him up, as usual, about the possibility that he might seek to recapture the leadership of the Labor Party—a routine issue of the type I dealt with on a daily basis on my politics beat—I interviewed him for a Pakistani newspaper, adding some questions of my own. I typed up his answers and sent them to a very pleased Ammara Durrani, who wrote them up for the *News International.*

Fourteen years later the two countries still had no official relations, but by this time Ammara Durrani and I could place video calls between Karachi and Tel Aviv and reminisce about that interview and its aftermath. Ammara told me that, at the time, she had not been entirely frank about her feelings.

"I was afraid," she told me. "This was the first time that a top Israeli official had given a statement to a Pakistani media group. This was unprecedented. So I was terribly afraid and expected a negative impact, and a big one. What really gave me the confidence was support from my editors—it was an immediate 'Yes, let's do it.'" And they sure did. The interview appeared on the front page, following Durrani's four-page article on the relationship between the two countries, citing officials in Israel, the US, and Pakistan.

The headline was "Peres: If Pakistan and India Can Do It, So Can Israel and Pakistan." The subhead: "Says There Is No Shame in Peace; If Pakistan Wants to Be a Part of the ME Peace Process, It Cannot Do So with 'Remote Control.'"

The piece led to neither peace nor diplomatic relations. A day after it appeared, in the dark of the night, a group of about thirty armed men on motorcycles arrived at the main offices of the Jang Media Group. They fired shots in the air, overwhelmed and beat the security guards, broke into the editorial offices, trashed the newsroom, and tried to set it on fire. Fortunately, no one was killed. They left shouting "Allahu Akbar!" It was clear to everyone in Pakistan that the attack was a direct response to the interview. A reaction not necessarily to what Peres had said but simply to the precedent that had been set, that a large and well-known Pakistani media outlet could publish an interview with a senior Israeli official calling for peace between the countries. The attack was reported by international news agencies, such as Reuters, largely because of this context. The Pakistani government condemned the attack, as did Reporters Without Borders. Closing the circle, the attack was also reported in Israel, where the interview that set off the incident took place. It was news creating news.

Let's take a close look at what happened here.

Two journalists who had grown up on the far corners of a huge continent met in a class sponsored by the government of a country on a continent on the opposite side of the world, a superpower seeking to bolster its position by ongoing mediation of conflicts around the globe—at the same time that it itself occupied a large swath of the Middle East. The journalists' countries were enemies, but the two of them could communicate freely, thanks to technology that collapses the huge distance and breaches the diplomatic and political barriers between them. Extremists responded to an interview signaling the possibility of peace and conciliation—with violence. The attack was reported all over the world, returning to Israel as a news item.

This entire incident, from beginning to end, took place over just a few days. It is a story of human connections, the viral nature of ideas, the technological challenge to hidebound politics, fundamentalism, media involvement. It's also a story, of course, of capitalist interests, in this case the need for a newsworthy headline so as to sell newspapers. This latter factor is the prime generator of the whole sequence of events. The violent end of the story demonstrates how these supranational interactions pose an increasing threat to local power structures, traditions, and beliefs. Opponents do not, and will not, sit idly by. They are rebelling.

Just three years later it became clear that this doesn't happen only in a country like Pakistan. It is happening everywhere, in different ways, and by different means. I saw that when, during a stay in London, the entire world slid into its most severe financial crisis since the Great Depression.

A WALKER IN LONDON FINDS HIMSELF OUTSIDE TIME AND gradually oblivious to his schedule. The eyes drink in the street, its intensity, the sediments of humanity laid down and mineralized there over centuries. Human diversity is so typical of London today and so much a part of British history that one might think that all these people accept it as a matter of course. Not true. Many people on the street feel a profound sense of alienation, of being strangers among themselves.

It is a feeling that both disconcerts and stimulates the city. Nearly 40 percent of Londoners were born outside Britain, most of them outside the European Union. Three hundred languages are spoken in the metropolis. Alienation is at the root of its current identity.

I was a stranger among those mutual strangers. My wife and I needed a break from the steeplechases of our local Israeli careers. We wanted to experience life elsewhere, so we decided to pursue graduate degrees far from home. New York, London, Paris, Washington—the truth is that it didn't really matter to us where we might land. We came from a distant province and, as far as we were concerned, each of those places was the center of the universe, wonderfully foreign and tantalizing for us.

My route to the university was a fixed one. I strode along the streets bordering Bloomsbury to Theobalds Road and then to my favorite spot. It was in a sort of alleyway, narrow and ancient-looking, heading off from the main thoroughfare. Reeking of fried food, the alley was adorned with an old pub and a few cheap cafés offering tasteless sandwiches. I imagined it teeming with rats bearing the Black Death and people emptying excrement into the street. The alley's filthy walls and congestion exuded that. The modern city had transformed this little walkway, making it almost exotic. It bustled with human traffic, the hurried strides of suits in the morning rush hour.

At the end of the alley, past a small park, I reached the clutch of buildings that form the urban campus of the London School of Economics and Political Science (LSE), not far from Holborn station and the British Museum. It's not Oxford or Cambridge—instead of green spaces and bike paths, there is the bustle of an ambitious city preoccupied with its own affairs.

It was September 2007, and the world was more or less coherent, even if deeply polarized between the ideology of the Bush administration and the international community. Those with sensitive ears could hear, as the bullet train of change shot forward, that the ties in the tracks laid by the previous era were groaning. But few yet had grasped the deep meaning of the terror attacks of September 11, 2001, and their aftermath. My fellow-students and I in the LSE program were slated to study global politics, comprising global governance, the challenges being faced by economic institutions like the World Bank, international trade, interest rate policy, post-imperialism, equality and the growing international income gap, and immigration policy. As I came from a small country in the Middle East, and most of my time was devoted to its turbulent politics, I was less expert than my classmates about matters like international trade policy or foreign direct investment. However, unlike the rest of them, I was a journalist. I had covered election campaigns, seeing prime ministers go ballistic when they were asked probing questions. I covered the second Lebanon war, running for cover when rockets rained onto northern Israel, and went to the Oval Office to cover official visits. That was the baggage I arrived with. In other words, like every reporter in distress, I could make up for insufficient knowledge with anecdotes—like the story of the Pakistani newspaper. But my baggage, like that of the other students, would soon prove itself to be of very limited relevance. Just a few months later, in the midst of our studies, globalization would face its worst crisis since the Great Depression, and international politics would begin to change and challenge the assumptions an entire world order was built upon.

This tectonic shift in international economics and politics was, of course, not included in our weighty textbooks or in the lectures we heard, which had been written and delivered before the crisis. Only the most radical approaches in the syllabus addressed, in some way, the earthshaking turn of events that swept away the complacence of the experts.

At the end of 2007, the Federal Reserve, the central bank of the United States, realized that a liquidity crisis was impending because of defaults on subprime housing mortgages, which led to a collapse in the speculative derivatives market based on those mortgages. The United States soon faced a large-scale financial crisis. At the beginning of 2008, the Bush administration tried to counter it with a stimulus package, but that didn't work. Then, between the spring and autumn of that year, giant American firms like Bear Stearns and Lehman Brothers began to fail. These were the very same firms that my classmates had been hoping to get jobs at.

It was one of those instances in which our textbooks became obsolete as we read them, their theories proven invalid as soon as they were put to the test. As the crisis smashed models and refuted the pronouncements of pundits, we were forced to question much of what we thought was certain. Born in the 1980s or at the beginning of the 1990s, my classmates and I had grown up in a world of expanding interconnectedness, changing at an exponential pace. It had seemed obvious that the entire globe would become more integrated into a single economy and order, and that this would bring us and everyone else more prosperity. But then the false premise of globalization's inevitability collapsed.

A Constant Revolution

During the last ten years, globalization has lost a great deal of its luster. The data itself points to the shrinking or stagnation of international trade, cross-border investment, and bank loans relative to world GDP, a phenomenon *The Economist* calls "slowbalisation." The great economic crisis undoubtedly undermined globalization's fundamental assumptions. Perhaps people simply tired of the optimistic prophecies of a globalized world that dangerously downplayed the dark side of the force.

But the fickle fashions of public discourse cannot change the stark truth that globalization is a constant revolution. I use the word "constant" to denote the aggressive way in which globalization is changing, in an ongoing and intensive way, how people have lived from time immemorial. It has created a climate in which human beings must cope with the world, materially and conceptually, as a single and integrated place. The minute such a matrix is in place, the circumstances of our lives change constantly and radically. It is a political perpetual motion machine fueled by the energy produced out of the ever-growing tension between the local and the global.

The ebb and flow of globalization shapes the international milieu, and will continue to do so for the foreseeable future. Globalization expresses a fundamental uncertainty that has permeated history from the imperial ages of China and Rome to the present day. Is the world melding into a single whole, or remaining a collection of separate communities?

As global challenges arise, globalization in its widest definition has become the central issue of our time. History did not end with the unchallenged reign of liberal democracy, as predicted by Francis Fukuyama in *The End of History and the Last Man*; nor did it deteriorate into a permanent clash of civilizations, as Samuel P. Huntington had it in his book of that name. But we are now locked in fierce battle over a more ancient question: To what extent are human beings destined to ultimately live in a consolidated world, a cosmos in which basic values are held in common and local communities fuse to a supranational economy? That is the real question, and always has been. Both globalization and resistance to it are responses to the question. Fittingly, a growing number of today's leaders, from Recep Tayyip Erdoğan of Turkey through Emmanuel Macron of France and Donald Trump of the United States, have based critical parts of their policies on their hostility toward or sympathy for globalization and values attributed to it.

For mainstream economists, globalization holds the promise of ending poverty; for French farmers it is a malignant infection that threatens to destroy communities and even livelihoods. Neither flu epidemics nor the competitive market for smartphones in Asia can be understood without understanding how globalization functions. It has become so pervasive that it is either everything or nothing, just a meaningless cliché. But it is fundamentally clear as a concept—it refers to an ever-tightening network of interrelations between everything and everyone.

The result is increasing integration, the inevitable result of international commerce that requires and creates flows of capital, labor, knowledge, culture, and technology among industrialized nations. Human beings are now virtual walking atlases, adorned as they are with clothing and accessories that bear the imprint of countries all over the globe.

Think of the hutch, that item of furniture that some associate with their grandparents. It displayed, behind thick glass doors, the family's most precious objects, including, often, chinaware, some of which might have actually come from China. Perhaps there was a painted lion figurine from Iran. A pair of silver candlesticks that may have been fashioned in England or Germany. People often liked to own items made far away. The better-off they were, the more such objects they had. Long-distance transport and trade was always risky, whether the goods traveled by land or by sea. For that reason, the cost of items from distant places took that risk into account, meaning that prices were high. Such goods, from tea leaves to fabrics to porcelain to certain spices, often from the Far East, were termed "exotic" and thus especially valued. An exotic item was also an emblem of the tenuous ties between cultures. Today could hardly be more different. The family hutch has been dismantled and abandoned. If we had such an item today, it would be more appropriate, in the global North, to use it to display knickknacks produced locally, which today are usually more expensive than imports. Relations between distant places are no longer tenuous—they are broad, deep, and intensive.

Each of us wears or uses products consisting of components and designs coming from dozens of countries on different continents, from eyeglass lenses to jewelry to pacemakers. We bear on our bodies the dramas and opportunities of places distant from us and people we will never know.

Emancipating Revolution

Globalization is not only self-perpetuating—it also offers opportunities that make it emancipatory. The paramount advance of our lifetime is that, since 1990, more than a billion people have emerged from abject poverty.[1] Never before have so many moved so rapidly from constantly battling for their very survival to a life of opportunities, as modest as those might be. In 2000 the UN set itself a "millennium development goal" of reducing extreme poverty, defined by the World Bank as those living on less than $1.25 a day, by half. This was to be achieved by 2015; in fact, it happened five years ahead of schedule. Most of those who have escaped desperate poverty live in India and China, but other countries have benefited as well—Vietnam, Ethiopia, Rwanda, and Bangladesh are all prime examples. Extreme poverty is generally measured by daily income or consumption per capita, but other indicators illustrate the improvement in material existence around the world—the plummeting infant mortality rate, the rise in life expectancy, and the leap in literacy. Every place that has enjoyed economic growth and rising incomes also exhibits the dramatic influence of technological progress, followed by participation in international commerce.[2] Such indicators, seen in a broad historical context, show without much ambiguity a direct continuation of the improvement in the human condition that began with the Industrial Revolution and the global interrelationships that followed in its wake.

Up until about two centuries ago, life expectancy at birth stood, throughout the world, at about thirty to forty years.[3] In mid-nineteenth-century Great Britain, children who made it to the age of five could expect to die in their early fifties.[4] People survived on the equivalent of about $400 a year or less, in today's dollars. The population at large was illiterate, unwell, and indigent. Many people lived in slavery of one sort or another—not just nonwhite slaves, whose servitude was a product of racism, but also European and Asian peasants, serfs, and indentured workers who were, in one sense or another, the property of aristocrats and capitalists.

Those who were legally free, to the extent it was possible in a world devoid of democracy and equal rights for women, were slaves to unremitting poverty. Economists estimate that at least 84 percent of the public in previous centuries lived in grinding destitution of the type in which every ounce of a person's effort is, on a daily basis, devoted to survival,[5] creating appalling inefficiencies in exploitation of resources. Imagine a serf who could earn money by cutting and selling firewood, except that he lacks an ax. And even if he had an ax, he would lack a wagon to transport the wood to market.

The most agonizing experience in the life of average people was helplessly watching their children die. At the beginning of the nineteenth century, some 40 percent of a family's progeny died before the age of five. In most places, high death rates for babies and children continued into the 1920s and 1930s.[6] The human condition, for most of humankind, most of the time, was wretched—nearly intolerable, sometimes to the point of insensibility.

The fierce faith in an unchanging, cyclical world, and a hierarchy that sanctified what were believed to be everlasting beliefs and values, dictated distorted and distorting ideas. For most of history, poverty was seen as a natural and necessary part of human society, and elites sought to justify it. Martin Ravallion of Georgetown University has collected some of these views and analyzed how the world came to recognize a need to reduce poverty.[7] A British writer declared in 1771 that "everyone but an idiot knows that the lower classes must be kept poor or they will never be industrious,"[8] while an eighteenth-century economist declared that "to make the Society happy and people easy under the meanest circumstances, it is requisite that great numbers of them should be ignorant as well as poor."[9] For these people, poverty seemed a necessity and a natural characteristic of a healthy society because, as Philippe Hecquet put it, "The poor . . . are like the shadows in a painting: they provide the necessary contrast."[10]

The human condition did not improve because of some cosmic event or gift of the gods. *Ideas* brought about the change—the ideas at the foundation of the scientific revolution and the Enlightenment. The deliverance of humanity from the horrifying misery of previous generations was engendered by free thinking, liberation from superstition, the shattering of the Catholic Church's monopoly on knowledge, and a recognition of the need to respect individual autonomy. Beginning in the fifteenth century, political competition in Europe created incentives for advancements in technology, military science, and other fields, which in turn led to the need for new economic arrangements. Enlightenment values provided a foundation for building social institutions and protecting private property, facilitating these reforms and simultaneously being promoted by them. "Enlightenment is man's emergence from his self-imposed immaturity," wrote Immanuel Kant. "Immaturity is the inability to use one's understanding without guidance from another. This immaturity is self-imposed when its cause lies not in lack of understanding, but in lack of resolve and courage to use it without guidance from another. *Sapere Aude*! 'Have courage to use your own understanding!'—that is the motto of enlightenment."[11] The values of the Enlightenment were the armor that protected the achievements of the scientific revolution and, in doing so, made the Industrial Revolution possible. In turn, industry and capitalism required globalization in order to survive, distributing its products around the world.

Horribly Efficient

Imagine the owner of a textile mill in Manchester, England, in the nineteenth century. The revolutionary adoption of the spinning jenny, invented in 1764, and the power loom, invented twenty years later, made it possible to produce fabric more quickly and in much larger quantities than the local market demanded. Innovations in transport and communications technologies enabled the mill owner to turn this increased productivity into profit. Considering his hefty investment, and local markets already flooded by competitors' products, the owner needed to raise revenues as quickly as possible, so he sought to market his merchandise wherever he could, from London to Asia. Furthermore, constant technological advances required him to preserve his competitiveness and expand by buying new machines and keeping up to date, which often meant that he had to raise capital from creditors. At this stage, if our mill owner can't find those new markets, he will go broke.

Here come the politicians to the rescue. If Britain has to force its colonies to buy its own products rather than local ones, or to send its navy to force other nations to open their markets to these new tycoons, so be it. Marx and Engels got it right in 1848: "The need of a constantly expanding market for its products chases the bourgeoisie over the entire surface of the globe. It must nestle everywhere, settle everywhere, establish connections everywhere."[12]

Not long after *The Communist Manifesto* was written, Britain was already producing half the world's cotton cloth—even though it grew no cotton.[13] It did so not because of the political power held by the bourgeoisie or the violence employed by capitalists to defend their control of the means of production. It was simply a matter of the efficiency of the Industrial Revolution, which made it possible to produce products more cheaply and ship them vast distances, and the huge temptation of the prosperity that this brought.

Globalization does not ask—it commands, and its command is efficiency. This efficiency is judged only through the lens of business, focused on revenue. Local concerns are relevant only to the extent that they serve or interfere with profit-making. Therefore, by its very nature, globalization creates phenomena like today's Indonesian textile sweatshops, or the dumping of massive amounts of toxic waste in the countries of the global South. When operating with no moral codes or meaningful regulation, it is blind, no more than a simple engine of supply and demand fueled by efficiency.

It is a sweeping and powerful process. The conversation about it has revolved around predictions of technology-based and irreversible globalization, or global prophecies of wrath. The reality is more complex, with both light and dark sides. One positive result of the Industrial Revolution and the way it went global was the emergence of stronger social institutions, first and foremost education.

As the Industrial Revolution gained pace, it required a workforce with at least a basic education, so that factories could be staffed.[14] Schools, whether publicly funded or so-called factory schools for adolescent workers, supplied two essential services for capitalists: workers acquired basic technical experience and literacy, both needed in a society that required the use of bills of exchange, the writing of letters, the reading of notices, and apprenticeships. A second service was behavioral: workers employed in a large factory, as opposed to a farm or family cottage industry as in the past, needed to learn how to obey orders and be punctual, and to understand communal responsibility. Education for the masses was a capitalist need.

In time, however, public education detached itself from the instrumental circumstances of its birth and became a value in its own right, one linked to equality. During the nineteenth century alone, throughout the world, the percentage of people aged fifteen and above with a basic education almost doubled, from 17 to 33 percent. By the middle of the twentieth century it reached 50 percent, and by the year 2000 it was at 80 percent.[15]

This leap was a historical result of the need of the moneyed class for human capital. But, in parallel with its exploitative character, public education empowered entire publics that had previously been suppressed and gave them tools for improving their lives on the personal and political levels, in part by demolishing class structures and reinforcing democracy and the rights of workers.[16]

The Inequality Revolution

Globalization is a constant, emancipating, and terrifyingly efficient revolution. It's not a village in which members of different nations and races sit in a circle and sing "Kumbaya." The "flat world" is a mirage obscuring bumps and twists that globalization requires for it to sustain itself. Indeed, the worst-case scenario for the current model of globalization is that the world might become a communal egalitarian village. The global economy is fueled by inequality. International production and commerce require differentials and arbitrage gaps in the cost of labor, purchasing power, the prices of commodities and raw materials, and currency rates.

Entrepreneurs have taken advantage of these disparities to build profitable companies in an export-import global economy, a process that accelerated after the fall of the Berlin Wall. Simultaneously, approximately 128,000 people escaped poverty, every day, in the twenty-five years that followed.[17] Inequality and the attempt to capitalize it on a global level have been essential drivers of the improvement of incomes and standards of living.

The end point is of critical importance. Today's version of globalization is wholly without precedent. There was always international commerce moving goods across the globe, at changing intensities. But in contrast with the present, in the past it did not raise the global standard of living and certainly did not reduce penury. The exploited and the oppressed were players in a zero-sum game in which they always lost. The world was in any case caught in the Malthusian trap—slow technological advancements and increased food production led to population growth, requiring resources to be distributed over a larger population, until eventually the standard of living returned to its initial dire condition.

In the eighteenth century, the Enlightenment philosopher Voltaire defended luxury and attacked the hypocrisy of critics who, he charged, pontificated against consumer culture while enjoying the good things in life, including a cup of coffee. "Does it not have to be ravished by human industry from the fields of Arabia?" Voltaire said of the beverage. "The porcelain and the fragile beauty of this enamel coated in China, was made for you by a thousand hands, baked and re-baked, and painted and decorated. This fine silver, chased and fluted, whether flat or made into vessels or saucers, was torn from the deep earth, in Potosa, from the heart of the new world. The whole universe has worked for you, so that in your complacent rage with pious acrimony, you can insult the whole world, exhausted to give you pleasure."[18]

Voltaire provides an early version of the trickle-down economics argument. Luxury, or what we would call consumerism, unites the world because it provides employment, which leads to trade and industry. It was certainly a false claim when he made it. The economic historian Gregory Clark puts it succinctly: "The average person in the world of 1800 was no better off than the average person of 100,000 BC. Indeed, in 1800 the bulk of the world population was poorer than their remote ancestors."[19]

It was not the universe that labored for the sake of the hedonists of eighteenth-century Paris, as Voltaire argued. It was human beings who were racially enslaved, and sometimes worked to death, without any chance of improving their material welfare. Those who enjoyed the luxury products coming from foreign shores were a thin stratum of aristocrats and wealthy bourgeoisie, like Voltaire himself. Not only were the masses left in poverty, but the economy did not prosper, either—the average growth rate of output per capita in Western Europe was 0.14 percent in the years 1500–1820.[20]

The Industrial Revolution, followed by contemporary globalization, changed all this profoundly. The industrialized, massive, and liberal nature of these phenomena sharply redirected human history, for the first time creating opportunities for most human beings. Globalization is both an enabler of exploitation *and* a proven remedy for global poverty.

The process is so forceful that we tend to forget that it is not a natural phenomenon, not a march of progress or a global village. It is a political-economic creation that forces all of us, for better or worse, to be part of the same story. Sometimes the story writes itself in London, or in Karachi. More and more, though, it is happening in Beijing.

CHAPTER 2

Showering Twice a Month

Michael Wong comes from the first generation that grew up in China's globalized economy. He and I have been friends for years, and from time to time we speak, he from blaring Shanghai, his home city, and I from sweltering Tel Aviv. Both of us were born in 1979, exactly on the cusp between the analog and digital ages. For China, it was a time of unprecedented reforms. When Michael was born, the country's gross domestic product (GDP) per capita (in current US dollars) was less than $200. Israel's was thirty times that. Since then the gap has closed dramatically. Michael, hardworking and serious, always with a ready smile, is an expert on both Western and Chinese hip-hop and performs it himself. I like the mathematical precision with which he decodes such cultural phenomena. He's effortlessly cool.

A few years ago we met with a group of friends on a wintry California night, not far from San Francisco. The chat took us back to our childhoods. The best day of the week for me, I told him, had been the day of my after-school computer class. I had no interest in programming, but we didn't have a computer at home at first, and at the end of the class they gave us a few minutes to play those grainy 1980s computer games like Montezuma's Revenge.

Michael's best day of the week was entirely different—it was the day he and his parents could afford to shower at a public bathhouse.

Michael grew up with parents who were first-generation residents of Shanghai. During the Cultural Revolution, his parents' high schools were shut down and many members of their families were sent to work on farms, cogs in the Maoist program to restructure Chinese society. Because of the upheaval, his father "self-learned high school," as Michael put it. Only afterward, when the country's universities were reopened, did he begin to study mechanical engineering; later, he taught himself programming.

"When I was in kindergarten and primary school, life was really a struggle," he said.

"It was very tough. We lived with our grandparents and cousins, all in the same apartment. I slept in a tiny room with no windows, and didn't even have a table or bed. My dad had to use some wood to make a bed for me. We needed food coupons, since food was limited, so we usually didn't eat any meat, mostly rice and vegetables. On holidays and special occasions, when we sat at a big festive table, only the children and the grandparents would have meat. All the parents at the table left the meat for the children—each of us was an only child."

His family had no refrigerator; they used the cool water of a small well in the yard to chill their food, "and we ate well-salted food, because it kept better in the summer months." He grew up in an apartment building without showers; the toilets were in an outhouse in the back. They washed themselves from a metal pail. "There was a tiny small space under the stairs and people put a curtain on it and you just used your own water to shower yourself. Then, once a month or every couple of weeks, we went to the public showers with our parents. There you could actually thoroughly clean yourself. You could not go there often—it cost money."

Living this way was not at all unusual in China, or Asia in general, at the beginning of the 1980s. Michael's family was not poor in Chinese terms—the rural poor were much worse off.

Then, at the end of the 1980s, conditions began improving in ways that few could have imagined. "First, stores were no longer always out of what you needed," Michael recalled. "Suddenly there were goods to buy. Second, there were private markets. You could actually buy and sell on your own. Starting then, there were a lot of people who started to do business on their own, and we began to have a free market." In short, "life was getting better and better. It changed every year."

His father and mother worked at factories that manufactured electronic equipment, and they assembled their own black-and-white television from parts they picked up at different places. Michael won an important math competition for Shanghai schoolchildren, which set him on the road to success.

He told me that, at the beginning of the 1990s, he began downloading files via BBS, an early technology that linked computers via a dial-up connection, a kind of preliminary version of the internet. A funny thing happened to the conversation at this juncture: our experiences suddenly converged and we have a shared childhood memory. My life was nothing like Michael's up to this point. I grew up in an Israeli middle-class family that could afford overseas trips from time to time. We even had two cars. While we lived in incredibly different circumstances, the newly emerging internet was something we had in common, and it started to make our worlds a bit more alike. Both of us, the same age, were sharing files and communicating with dial-up connection, classic children of the 1980s, the first generation to grow up with the internet as an integral part of our lives.

Michael's story is not just about open markets and their effects. As a child, he benefited from government investment in a school system that identified his talent. Traditional Chinese values, which made education a top priority for families, and his parents, who were exceptionally technically adept, also played a role. But he is the first to admit that he was very lucky. Today, Michael, a kid from Shanghai whose parents saved up money to shower every few weeks at a public bathhouse, is an entrepreneur, one of the founders of a company whose shares are traded on the New York Stock Exchange. There is a good chance that many readers of this book use the application that his company developed. "My generation feels so grateful," he said. "It is a treasure for me because nothing is to be taken for granted. Those crises in our childhood actually make us more appreciative of many things. We are grateful to our parents, to progress, to the government, because we ourselves experienced the change."

The change was brought about by Deng Xiaoping and his allies. China embarked on major reforms in 1978, under the leadership of a resolute Deng. He permitted limited trade in private markets and created special economic zones to foster manufacturing and export. Private enterprise quickly transformed everyday life in China. At the same time, foreign investors flocked there to take advantage of its low labor costs. The country's economy began to grow almost immediately, at an average rate of 10 percent annually, in some years reaching 15 percent. In 1980 China's GDP per capita stood at $195. As of 2018 it had reached $9,770.[1] Between 1980 and 1990 the number of Chinese living in extreme poverty went down by 167 million.[2] By 2013, more than 850 million had escaped that deadly trap.[3] Globalization is accelerating interdependent relations, and its significance in the Chinese case was lightning-fast change. These were not economic policies packaged as cumbersome development plans that require decades to show results—they brought about practical improvements in all areas of life in a very short time. In 1990, two out of every ten Chinese citizens—hundreds of millions of people—were illiterate. Two decades later, 95 percent could read and write. At the beginning of the 1990s, only 68 percent of the women could do so; by 2010 there was virtually no gender gap.[4] Between 1990 and 2017 the infant and child mortality rate, up to the age of five, plummeted by 83 percent.[5] By every possible criterion, life in China improved in a profound way. In fact, it happened all over Asia, at varying rates, except for in North Korea, the world's last Stalinist dictatorship.

Industrialization is key. There is no correlation more fateful to humans than that between industrialization and rising standards of living. The Chinese were late arrivals to the Industrial Revolution. The train came into the station in the nineteenth century, but they boarded only in the twentieth. Yet that is only the blink of an eye in human history. Seven out of every ten Chinese worked in agriculture or related fields in 1978. By 2018, the situation had reversed—seven to eight out of every ten Chinese now work in nonagricultural fields, in trade, industry, and services. I sometimes ask audiences at my lectures to name the most important leader of the twentieth century. The usual replies are Churchill, Hitler, and Stalin. Perhaps look further east, I suggest. Stalin thought he was building a Soviet superpower that would last forever and become the future of mankind. Churchill hoped to save the British Empire, and Hitler dreamed of a thousand-year

Reich. All three of them failed, although Churchill saved Western civilization in the process. Only one twentieth-century leader inherited a backward and poor country and gave back a superpower-in-waiting—Deng Xiaoping. He was able to do so because, in his case only, globalization was his close ally.

Becoming Avatars

Michael and many others like him illustrate the rapid pace of global change. Not vague hopes that our children will live better lives than we did, but the potential for an immediate change in the way we live our lives. Millions moved directly from life without running water to work in export-oriented companies or software and application development.

In and of itself, trade between nations and cultures is nothing new. In Rome, Pliny the Elder protested the global nature of the market for luxuries in his time. "We have come to see . . . journeys made to Seres [China] to obtain cloth, the abysses of the Red Sea explored for pearls, and the depths of the earth scoured for emeralds," he wrote. "At the lowest computation, India and Seres and the [Arabian] Peninsula together drain our empire of one hundred million *sesterces* every year. That is the price that our luxuries and our womankind cost us."[6] That was written more than two thousand years ago, and may be the first (male chauvinist) screed against a trade deficit—that is, when a country pays more for imports than it receives from exports. Pliny confined himself to luxury items enjoyed by the empire's tiny upper crust. But for most of the world's inhabitants, such goods did not become affordable until about two hundred years ago. Those people did not buy vanilla beans or silk cloth. Most of their time was spent obtaining food for today and tomorrow.

Reciprocal global trade was confined. It took place between tiny aristocratic and wealthy classes. There was little commerce that truly crossed huge territories. The global nature of the Silk Road was a myth that began to spread during the nineteenth century. It depicted an ancient world of abundance and variety, open and functioning commerce, intercontinental transport and intercultural dialogue. Today we know that the picture of a heavily traveled Silk Road, packed with caravans crisscrossing Asia, each with trains of camels laden with silk to be sold in exchange for Roman coins, is an exaggerated romantic illusion. Goods moved at no more than ten to thirteen miles a day, largely between rural and agricultural centers that provided for their own needs through local trade. It was carried out by what Valerie Hansen, in her book *The Silk Road: A New History*, terms "peddlers."[7]

The broadening of such trade in today's world is manifested in the fact that *most* of the goods in the home of a person in an industrialized country were not produced close by. Indeed, what do "close" and "far" even mean in a world in which goods can be transported by air from one hemisphere to the other in a day, and money and information by optical fiber in less than a second?

In 1881, the Royal Geographical Society of Britain published a large map of a kind that would never again appear. It was painted green, yellow, orange, and blue, the colors indicating travel times from London. In the days of travel by horse-drawn carriage and boat, such a map was essential for planning long and arduous journeys. All of Europe appeared in dark green on the map, meaning that a traveler from London could expect to arrive at his destination within ten days. In the United States, the Eastern Seaboard was yellow, meaning twenty days of travel from the British capital—which was the time it took for a relatively fast boat to cross the Atlantic. Really distant destinations—East Asia, for example—appeared in brown, as they required a trip of at least six weeks.

This disconnected world that took so long to traverse, in which the news that a war had ended depended on wind speed, wave heights, and the size and strength of sails, has been replaced by the instant world, in which information and merchandise move at enormous speeds, and in which deals are closed and implemented immediately. Even more important, change is accelerating. After the invention of the telephone, it took fifty years for half of Americans to have one in their homes. Thirty-eight years passed from the moment the radio was invented until it had an audience of 50 million listeners in the United States. It took thirteen years for television to achieve the same number.[8] Facebook, in contrast, had 6 million users in its very first year, and that increased a hundredfold within five years.[9]

These developments are a product not only of trade and technology but also—perhaps largely—of the relative political stability achieved in 1945, solidified after the fall of the Berlin Wall. The surge in the flow of information, capital, and goods was enabled thanks to the cautious and meticulous decision makers and voters of the age of responsibility. International tariff and taxation standards were established, transport costs declined, and investors in international markets felt more secure. Just as economies do not prosper without strong institutions, so globalization cannot expand without an international order that proceeds with moderation.

IT IS A LESSON THE WORLD HAS LEARNED THE HARD WAY. The belief that technology, science, and profit would power an irresistible march of progress was widespread among political elites during the first decade of the twentieth century. It shattered on the shoals of World War I. That early version of globalization, lasting from the end of the Franco-Prussian War in 1871 to the roar of the guns of August 1914, is often referred to as the Belle Époque, the beautiful age. It was a time of incredible human flourishing. The world experienced one of the largest waves of peacetime human migration, much of it with North America as its destination. Italians, Irish, Jews, Dutch, Germans, Czechs, Englishmen, Scots, Poles, and many others left the old world in search of a new future. They frequently found it. Scientific discoveries and technologies appeared one on the heels of another. Marie and Pierre Curie investigated the secrets of radioactivity; Louis Pasteur and Robert Koch uncovered the way bacteria cause fermentation and disease. Henry Ford pioneered the mass production of automobiles; Alexander Graham Bell designed the first useful telephone, Thomas Edison the first incandescent lightbulb. The Lumière brothers held the first public screening of motion pictures. Any one of these advances alone would have changed the way people lived in major ways; coming together, over the space of just a few decades, they transformed the world.

The Belle Époque was also an age of cultural efflorescence, producing, among other things, some of the art most loved to this day—the work of the impressionists, post-impressionists, cubists, and expressionists. It was the great era of literary realism and plumbing of the human psyche by innovative modernist writers such as Thomas Mann and Marcel Proust. But one data point supports the claim that the globalization of today is merely a replay (with more advanced technology) of that previous era—international trade as a percentage of leading countries' GDP and of the world's GDP. International trade accounted for 44 percent of British GDP in 1913, a level that would not be reached again until sixty years later.[10] The value of exported goods as a share of world GDP stood at 14 percent on the eve of World War I and would not reach that level again until the 1980s.[11]

The Great War smashed it all to pieces. "The lamps are going out all over Europe, we shall not see them lit again in our life-time," said British foreign secretary Edward Grey on the eve of that bloody conflict. The gory trenches of 1914–1918 were followed by the turbulent 1920s and 1930s, culminating in another world war. That was followed by the world of the Western and Soviet blocs, a world of walls, tariffs, and barbed wire.

A Japanese friend once told me that the Cold War did for the world what snow does for the cherry trees in Japan. The colder the weather in the winter, the more vivid the spring blossoms. The infrastructure laid down in the age of responsibility proved itself when the cold passed. The fall of the Berlin Wall and the Eastern bloc led to a revival of international trade of exceptional scope. The new globalization broke all the records of the Belle Époque.

Something else happened. It was not just an acceleration and expansion of interdependent relations between countries; these relations also became deep and profound for individuals. The livelihood of an industrial laborer in Indonesia now depends on the supply and demand on American websites. This laborer uses a cell phone manufactured in China based on American patents, and whether he remains employed or gets fired is affected by the interest rates set by the Federal Reserve in the United States. A German citizen can reside in Berlin at the same time that the center of his life is on another continent. His business, friends, and hobbies do not have to be in the city where his bed is. He reads professional journals written on a third continent, on his tablet, via the internet. He makes his purchases on international websites, invests his savings in companies headquartered elsewhere, and might choose to adopt the values, spirituality, exercise routine, and diet of a foreign culture (or more than one) from another continent.

This choice, to live as a global avatar of one's physical presence, is becoming more common. It is a possibility that raises questions and dilemmas humans have never yet faced. Globalization has penetrated deep into our veins, our blood, into the genetic tests we do before bringing children into the world, and in the way we bring them up.

Global Consciousness

An ongoing survey conducted for the BBC on a periodic basis over many years asked people whether they identify with the statement "I see myself more as a global citizen than as a citizen of my country." It showed that, in 2016, the concept of global citizen reached its zenith—for the first time, half of the citizens of the countries included in the survey viewed themselves as citizens of the world.[12] Similar findings can be found in an American study of 2017, in which about half the respondents said they felt a commitment to the values of a "global human community." There was no significant difference among different demographic groups.[13] A person who says that he feels more a citizen of the world than a citizen of his country feels, or wants to feel, that his life is not entirely nailed down to anything local. It is a feeling that "There's the whole world at your feet," to quote Bert from the film *Mary Poppins*, and that it shouldn't be only "the birds, the stars, and the chimney sweeps" who enjoy it.

For nearly all of human history, the opposite was true. Far-off experiences or events, no matter how potent or momentous, had very little material impact on the lives of most people. A good illustration of this is the Great Fire of London. London in 1666 was already the capital of a growing maritime empire, with vast lands beyond the seas. The September fire destroyed a large part of the enormous and important capital of this expanding realm. Three-quarters of the original medieval city burned down—more than 13,000 houses, 87 churches, and much more. The fire had cultural, architectural, literary, social, and even religious impact. But who knew about it? Who heard that it happened?

Of course, Londoners did. Almost certainly people throughout England did, and probably many people throughout Great Britain. Following a parliamentary inquiry, the English latched on to a scapegoat, accusing the "Popish faction"—that is, Catholics—of setting the fire. The advocates of religious intolerance and xenophobia used it as an excuse to persecute foreigners and Catholics for a time. Only in 1830 was the inscription blaming the Catholics for the fire expunged from the monument to the tragedy in London.

For most of humanity, and most Europeans, the fire never happened, for all intents and purposes. They did not hear of it; nor did they have any interest in knowing about it or any particular incentive to take an interest. The world they lived in was extremely local. As always, rumors and narratives spread and were disseminated in social situations—in the church sermon on Sunday or at the local pub. But these were glints of knowledge, hints of a larger world beyond the village. A person defined himself in relation to his community or the district in which he was born. Clergymen, the aristocracy, and a small class of wealthy merchants were part of a global elite that possessed knowledge, leisure time, and money, all of which allowed them to know more about the world. An understanding of the world—what it was, and what was happening in it—was limited to small and privileged classes.

It is easy enough to imagine a scenario in which the Great Fire of London impinges on the average Englishman. The forest near his home is cut down to provide the wood needed for reconstruction of the capital. Cutting down the forest affected the plebian Englishman in many ways, but he played no role in the drama. He was a small and mute pawn on an arbitrary chessboard. Workers hired by the local lord come to cut down the forest, and he can only watch. Bits and scraps of information about reasons for the cutting might reach him—perhaps he would hear about a fire in a distant place. But more likely he did not. Even if he did, would being in possession of that information make any difference in his life? The only place he could influence decisions and control his life was in his own home, a structure that was sometimes not even his own property.

Compare that to the reach of a modern catastrophe—the collapse of the twin towers in al-Qaeda's attack on the United States on September 11, 2001. More than two billion people saw the second tower crumble.[14] More than half of humanity, by conservative estimate, were exposed to the crash of airplanes into the towers, which destroyed the World Trade Center and killed 2,606 people in the towers and immediate area. The attack was an event of huge importance, with broad geopolitical implications. Yet what happened in Manhattan was less consequential than what happened in Auschwitz, or in Hiroshima and Nagasaki at the end of World War II.

But it was filmed and aired live. That's the point. The fall of the towers was an ultimate image that crossed national borders and entered the international public consciousness. A large part of the human species took part in the trauma by watching it, although people in Pakistan and America construed the image in disparate ways. People drew opposite inferences from it and felt entirely different feelings about it.

But they all knew about it, and the image was everywhere, leading to thousands and then millions of individual decisions around the world.

When the interrelations between places and people are as dense and intensive as they are today, events in distant places can have a powerful local impact. Individuals therefore have an incentive to create a common foundation of ideas, facts, and images. Yet what stands out is not that people know more about what affects their lives—after all, such knowledge is in their self-interest. It is that they know so much about what ostensibly has no immediate bearing on their lives. About 2.5 billion people watched Princess Diana's funeral in 1997. The opening of the World Cup games of 2018 was watched by 3.5 billion. A billion human beings listened to or watched the rescue of the miners who were caught when the tunnels they worked in collapsed in Chile in 2010. Any person who does not live in abject poverty, who is not struggling for subsistence, can now form a global outlook. A privilege enjoyed a few centuries ago by the likes of a Benedictine monk bent over books in his isolated priory is now available to almost all.

Knowing how to read and write, having access to running water, electricity, and the internet, are binary states. Either one has them or one does not. Having them changes the human condition and enables a broad view to those who seek it. The continually connected world creates a common consciousness. A child can talk with another child about an online video game; adults remember exactly where they were when the World Trade Center collapsed. Two strangers can snigger cynically about a clownish political leader they know by sight and by reputation. As interrelationships grow stronger, people share more ways of thinking. Each piece of additional knowledge, each image or paradigm, augments their shared view of the world. People need not love or accept pornography, fast food, Hollywood entertainment, the power of the dollar, fear of terror, smartphones, religious fundamentalism, or the empowerment of women, but all these constitute a growing part of a common human consciousness. This expanding consciousness nurtures both common aspirations and common fears, which influence and disrupt social conventions everywhere, from consumer demand to domestic politics. And technology is an enabler and an accelerator of this process.

A good example is a study of education and computer literacy, the "Hole in the Wall" experiment conducted by Sugata Mitra of Newcastle, England. It was the inspiration for Vikas Swarup's novel *Q & A* and its film adaptation, *Slumdog Millionaire*. In 1999, Mitra placed a computer monitor in a sealed wall in a slum neighborhood in New Delhi. Next to it was a mouse that could be used to surf the internet. As the name implies, it was simply a hole in the wall to which a computer had been bolted. It was not guarded and there were no responsible adults overseeing it. Mitra used a hidden camera to record the reactions of children, many of whom were thus exposed for the first time to surfing the internet. The camera recorded how the children taught themselves, in groups, how to use the computer without any formal instruction, how to access websites and to download software, games, and music. He expanded the experiment to other cities, always in poor neighborhoods, including remote places where there was no internet. In such places he placed a library of disks with games and educational software—all in English, a language none of the children spoke. When he revisited one of these places, he heard this from the children: "We need a better processor and mouse." They also said, "You've given us machines that work only in English, so we learned English."

Mitra's research showed how access to the internet, without any adult oversight, enables groups of children to gain abilities, education, and knowledge that they would have had no access to without the computer in their neighborhood. This included basic knowledge of how to operate a computer but also how to search for information, gain facility with mathematics, learn a language, develop critical thinking skills, and more. The interactivity that is intrinsic to the internet and to computers themselves led to a process in which children independently acquired knowledge.[15] "Tomorrow's illiterate," said the psychologist Herbert Gerjuoy, "will not be the man who can't read; he will be the man who has not learned how to learn."[16] Children who share a smartphone in Mumbai (or who, in the past, spent time in internet cafés) learn on their own how to learn. They face other obstacles, often formidable, but they hold these children back less than the old world's chains of ignorance. In today's world, facts are just a click away. But, as we know, lies are just as easily available.

WHEN THE TUNISIAN REVOLUTION BROKE OUT IN 2010, Western journalists needed a name for it. They settled on the Jasmine Revolution, in keeping with the country's national symbol. (More meaningfully, the Tunisians themselves called it Thawrat al-Karamah, the Dignity Rebellion.) It took just a few weeks for the uprising there to spread across North Africa and the Middle East in a wave that came to be called the Arab Spring.

Yet it also reverberated in the Far East, in China in particular. In February 2011 protests broke out in Beijing and other cities. There was a demand for political reforms. The demonstrators used the jasmine blossom, which has deep cultural roots in Chinese tradition, as a code for political change. They distributed flowers while singing a familiar

Chinese song, "Such a Beautiful Jasmine." As the Chinese public was well aware of what had happened in Tunisia, there was no need for any other slogan—the context was clear. The government responded by censoring the word "jasmine." It blocked searches for the word, and for the phrase "Jasmine Revolution," on social media and apps. When the protests spread to Egypt, some Chinese websites also blocked the word "Egypt."[17] Censorship was pervasive. The country's previous president, Hu Jintao, had once been recorded singing the jasmine song; suddenly it could no longer be accessed on the internet. China hosts an annual international jasmine festival, but that year it was suddenly postponed. In some places, the police actually forbade the sale of jasmine flowers, causing losses for the growers of ornamental jasmine in the Daxing district, in suburban Beijing. According to the *New York Times*, in some markets, florists were told to report anyone displaying interest in the flower, and to record the license plate numbers of anyone inquiring about making purchases.[18]

Here is a simple story of globalization of an idea, and an attempt to fight it. The idea was freedom, and it was represented, because of political circumstances in Tunisia, by the jasmine. Had the Chinese public been completely ignorant of Tunisia's democratic revolution, the jasmine flower would have been devoid of meaning, simply a flower and no more. The minute the jasmine came to symbolize something for people in many places, they had something in common. As basic as that common denominator might have been, it threatened power structures everywhere.

There are aggressive attempts to put the brakes on global consciousness. I once remarked to a Chinese friend that his country's current ambitious president, Xi Jinping, is the most powerful Chinese leader since Mao Zedong. Not true, my friend said. "He's clearly stronger than Mao." I was taken by surprise and asked him how that could be. "Mao was very strong and controlled everything," my friend replied, "but he didn't know what people were thinking inside their heads." He was referring to the way the Chinese Communist Party has been implementing the most ambitious surveillance, supervision, and monitoring policies in history. The Chinese government has the technological capacity to control public discourse, using big data analysis technology. Authoritarian rulers understand that there is no greater threat to political and social power structures than the globalization of consciousness. Ideas are globalization's gunboats.

The critics of today's globalization say that it creates a false consciousness. It actually shores up the oppressive apparatus of the top tenth of a percent, or of the world's single superpower. In fact, they claim, "global" is only code for Americanization, in the form of pervasive Hollywood images and subordination to American consumerism. Most pernicious, they charge, is the malignant expansion of the American concept of happiness.

In 1941, Henry Luce offered a narrative that he called the American Century. In one of the magazines he founded, *Life*, he touted it in an essay that presented the American way of life as a model for the entire world. He advocated for the ideas that were "infinitely precious and especially American—a love of freedom, a feeling for the equality of opportunity, a tradition of self-reliance and independence and also of cooperation."[19] Luce had himself been born in China, the son of Christian missionaries who went there to spread the gospel. His parents' old-style missionary work became, in his hands, a new sort of gospel, a secular one enveloped in the intoxicating spirit of a nation that, as Luce put it, had been "conceived in adventure."

From the first moment, this dazzling proposition menaced local identities, power structures, and traditions around the world. Few doubted the prosperity that globalization had brought, but many rejected the emerging global consciousness, its American influences in particular. Culture has extraordinary economic effects, and vice versa. If, let's say, rice importers sell in Vietnam at attractive prices, local rice growers will most likely suffer a direct hit to their incomes. But if Vietnamese children suddenly decide to eat more French fries, like their Western teenage counterparts do, the threat is graver. If American fast food were to make inroads into Vietnamese culture, the demand for rice would presumably decline. In this scenario, local rice growers are not exposed to competition; they are simply wiped out. A change of taste brought on by cultural integration engenders an event—the elimination of rice cultivation.

International trade can change markets and ways of life, but ideas can invent or destroy them utterly. The emerging global consciousness creates a new world, but at the same time, it is like the god Krishna in the Bhagavad Gita, who declares, "I am all-powerful Time, which destroys all things."[20]

Globalization is a luxury ship that hides its dirty secrets in its interior cabins, the lower deck, and the engine room. In these dark places, the masses are subdued so that the ship can continue to sail. Fittingly, Luce's manifesto for the American Century appeared in *Life*'s inside pages. The cover displayed a Hollywood starlet in an evening gown, along with the headline "HOLLYWOOD PARTY."

CHAPTER 3

The Globalization Wars

Dèyè mòn, gen mòn
("Beyond the mountains, more mountains")

—A HAITIAN PROVERB, USUALLY MEANT TO DESCRIBE HOW AS YOU SOLVE ONE PROBLEM, ANOTHER ONE PRESENTS ITSELF

Globalization improved standards of living, making it possible to eradicate extreme poverty and lay the foundation for global consciousness. In doing so, it threatened tradition and communal power structures everywhere. Simultaneously, trade and capitalism have impelled powerful forces to exploit poorer and weaker societies, and the elites of those latter societies to exploit weaker places and classes within their own countries, often violently. This unsustainable pattern of exploitation manifested itself in a series of wars and conflicts that followed a recurring pattern.

Beijing, 2017

I'm riding a moped in downtown Beijing, my hands gripping the back of its driver in terror. It's not that the traffic lights here are merely recommendations, but that drivers reserve the discretionary power to make a final decision on their own. The driver, a friend, wears no helmet, and studies his smartphone every so often as he steers one-handed at high speed between the cars. Armies of electric-powered scooters crowd the huge Communist roads, their drivers masked in white against the heavy smog.

Beijing immediately surrounds you and drags you under. You feel the great leap forward, the constant hum of development. China may not yet be a superpower, but Beijing is ready to be the capital of one. It was born ready. It lacks the distended clamor of New Delhi, but it's certainly not the earnest freneticism of New York. Beijing is huge but well ordered, teeming but without eruptions of exotic color. Urban planning decisions are made quickly and unsentimentally. The moped slows to a halt across the street from a deserted shopping center. My friend tells me somewhat indifferently that it was built over the wreckage of a small neighborhood. Its residents were evacuated over a period of just a few days, and for all practical purposes simply cast out into the distant suburbs. Business did not go well, the stores closed, and now the shopping center, built over the evacuated neighborhood, was slated for destruction so that a hotel could be built in its place. The developer's sign was already up, trumpeting the new project in large letters. The whole thing, which had changed the lives of thousands of people, took place over the course of just twenty months.

The more fortunate can still take refuge from the city's bustle in the small traditional neighborhoods in the city center. They take the form of *hutongs*, low dwellings that generally surround a common courtyard; sometimes there is even an old well in the middle. In the center, designer shops and boutiques have generally taken over, but here and there you can find a few remaining neighborhoods of this type, green islands in the midst of a gray and brown city. On weekends you can see elderly people strolling in their pajamas, greeting neighbors who are sitting outside watching people go by. They can look skyward, although half the time they won't be able to see the sky, only a dingy haze.

Everyday life in Beijing is a constant battle with air pollution. "It's a good or bad day according to what the egg says," my friend the daredevil moped driver tells me. The "egg" is a domestic device that measures airborne particles; it's a vital appliance for city dwellers. There are days when parents forbid their children to play outside because of the pollution level. Schools that cater to wealthy families advertise that their playgrounds and sports fields are covered by a protective dome that scrubs the soot from the air so that children can breathe normal air when they exercise. An essential appliance in Beijing homes is an air purifier that does its best to filter out the filth. It's a device that signals status and income—there are state-of-the-art ones that cost hundreds of dollars. These high-quality devices, resembling a wall-mounted or portable air-conditioning unit, have filters that are supposed to be good for six months, but they often need to be changed every few weeks because of the huge quantity of soot and other gray dust that clogs them up.

The pollution level was apocalyptic until 2017. For entire weeks, the sky was yellow-gray with such a dense suspension of particles that there were days when it was difficult for drivers to avoid colliding with each other. The newspapers reported that hospitals were flooded with babies and old people desperate for oxygen, pure and simple. A report in the *New York Times* in 2013 said that pollution changed the entire experience of childhood. The headline was "In China, Breathing Becomes a Childhood Risk."[1] The levels on the worst days exceeded that which the World Health Organization defined as an immediate health hazard. Air quality has improved since then, but a face mask remains a necessity at times; on smoggy days, walking without a mask for twenty minutes causes nausea.

Most of the airborne particles come from factories and the coal-powered electricity plants ranged around the huge city, and the industrial cities of Harbin and Hebei as well. The Communist Party, always sensitive to public opinion, forced the factories and electricity plants to adopt a detailed plan to reduce emissions, with strict goals. The plan succeeded, seemingly.

Yet recent studies show that what the government actually did was simply move about half of the electricity production out of the Beijing area. According to estimates provided by one such study, carbon and particulate emissions in China as a whole may have actually *increased* as a result of the program to improve Beijing's air. The government simply relocated the pollution into rural areas, in the already disadvantaged peripheral regions.[2] The pollution levels there get less national and international attention, the people there have less political power, and their plight is less visible. A fascinating documentary, *Under the Dome (2015)*, features a six-year-old girl from the coal mining province of Shanxi, then one of the most polluted places in all of China.

"Have you ever seen stars?" the interviewer asks her. "No," the girl tells her. "Have you ever seen a blue sky?" The girl replies: "I have seen a sky that's a little bit blue."

The interviewer then asks: "But have you ever seen white clouds?" The girl sighs. "No," she says.[3] The film, the independent work of journalist Chai Jung, was viewed 300 million times in less than a week before it was purged from Chinese social media sites.

According to figures from the World Health Organization, 4.2 million people die all over the world each year as a result of air pollution outside their homes, many of them in China.[4] Children are the most at risk—some 1.7 million of them die every year, globally, because of environmental pollution, the majority from poisonous particles and heavy metals that they inhale.[5] Nine out of every ten people who die because of such pollution live in poor countries, mostly in Asia and Africa.[6]

Exploitation Hubs

This is the price of rapid industrial revolution. It is paid today in Beijing just as it was paid in the past in Manchester and London, where the word "smog" was coined. Yet, more than ever before, the factories that operate all over China and sully its sky and rivers produce huge amounts of merchandise, much of which is sold to customers overseas—especially in the West. It's a point that is often missed. The shift of industry to East Asia has relocated the pollution that used to be produced in the global North to developing countries. A study published in *Nature* in 2017 found that in just a single year, 2007, 750,000 people died worldwide as a result of air pollution deriving from the production of goods and services in their home country that were consumed elsewhere.[7] The number has no doubt risen since then. According to the study, another 411,000 people died that same year as a result of particles that reached their home countries from smokestacks and factories elsewhere in the world. Its authors note that "if the cost of imported products is lower because of less stringent air pollution controls in the regions where they are produced, then the consumer savings may come at the expense of lives lost elsewhere."[8]

Smartphones are more affordable because of the cheap wages in Asia, and the pollution from their manufacture remains there and kills people. According to the University of Chicago's Air Quality Life Index, which converts air pollution concentration to its impact on life expectancy, Indians lose 5.2 years of life because of pollution; Chinese lives are shortened by 2.3 years on average.[9] The ultimate globalization is happening in the atmosphere, which knows no passport controls or customs restrictions; it obeys no international authority. The particles move unimpeded across borders, as do greenhouse gases.

Globalization is not just the creation of high-tech hubs but also the fostering of exploitation hubs. High-tech hubs are hives of technology and innovation; exploitation hubs are nexuses of lax local norms, inept or corrupt government oversight, and weakened populations manipulated by outside forces. European and American companies that acquire raw materials from Africa or manufacture goods in Asia exploit not only the gap in the costs of labor and production, which is only natural in capitalism, but also the disparity in norms between developed economies and developing non-Western ones. Hubs of this sort can be based on cheap labor, a monopolistic control of a local consumer market, access to raw materials, low energy prices, or all of the above. Workers in these hubs suffer harsh labor conditions, lack effective representation by unions or workers organizations, are exposed to the consequences of pollution, and are politically disenfranchised. It is exactly these characteristics that make these hubs appealing to foreign and local investors. Weak or poor states in need of foreign currency and employment are caught in a trap. If their institutions grow stronger, or if their governments respond to popular pressure, improve working conditions, and enforce labor statutes and regulation, manufacturers might flee, leaving unemployment, economic ruin, and environmental damage in their wake. Also, factories that export their products to the global North usually offer local workers better conditions than other domestic options. When exploitation hubs emerge, those who benefit from them—local municipalities that earn tax money and businesses that earn profits—have an interest in their preservation. Hubs of this sort thus have a powerful local lobby.

Some analysts, most notably the late Immanuel Wallerstein, have classified the dynamic of similar relations according to a division between center and periphery, or between rural and metropolitan areas.[10] But that way of observation may not be sensitive enough to the essential difference between today's exploitation and that of other eras. Contemporary exploitation hubs are flexible and fluid, just as the movement of capital, production, and labor are. For example, in moving electricity production and other poisonous emissions to politically weak and poor provinces, the Chinese government uses them as exploitation hubs. In fact, what China does to its far-flung provinces is what the industrialized world is doing to China, and to the global South as a whole.

This phenomenon is expanding. In recent years, China has joined Western economies in relocating many of its carbon dioxide–emitting plants to other countries in South Asia, such as Bangladesh and Vietnam.[11] The capital with the most severe air pollution in Asia today is Delhi. These hubs are the Agent Smith of the globalized world. Like the character in the *Matrix* movies, they replicate themselves.

When factories emit lead that makes its way into crops and livestock, poisoning the people who eat them, or spew soot into the air that causes asthma and emphysema in children, they impose costs on society and on the environment. Tainted crops and animals may need to be destroyed. Polluted water needs to be treated. The global market does not factor these costs into the prices advertised on Amazon. Health, environmental, and quality of life costs are what economists call "externalities," and the market is blind to them. The planet and the dead pay the retail price.

■

China is an excellent place to examine globalization's dark side because it has a long history here. Contemporary globalization cannot be understood without reference to the Opium Wars of the nineteenth century.

Economic historians maintain that, during most of the last two millennia, the Indian subcontinent and China together produced almost 60 percent of the world's gross domestic product.[12] Europeans came to appreciate China's economic power after Marco Polo convinced them, around the year 1300, of what the nations of the East had known for centuries about the Middle Kingdom. In later periods, intellectuals and artists sometimes idealized the Middle Kingdom, which they portrayed as an example the West should seek to emulate. The German philosopher Gottfried Leibniz wrote in 1699 that China was the nation with the most advanced legal system and ethics, even if it was behind in mathematics and military capabilities. The latter, he said, was "not so much out of ignorance as by deliberation. For they despise everything which creates or nourishes ferocity in men."[13]

Yet the West's relations with China were by and large commercial. Chinese silk was much in demand in the West as early as Roman times, to the point that in the year 14 CE the Roman Senate had to prohibit men from wearing it, arguing that it is not appropriate for the male sex. More practically, the Roman rulers were worried by the huge sums of money and gold that were being sent eastward to pay for the fashion.

As maritime transport improved, the great mercantile empires sought to buy up large quantities of pretty much everything Chinese—silk, porcelain, some spices, and of course the sensation of the seventeenth century onward, tea. Ships laden with goods set out from Chinese ports to bring goods to enthusiastic Western buyers. But the merchants involved in the trade had a deficit problem, much like the trade deficits of today. There was nothing much the Chinese wanted to buy from the West, nor did they have any interest in equal trade relations, as the Qianlong emperor told King George III of England in 1793. "Our Celestial Empire possesses all things in prolific abundance and lacks no product within its own borders," he wrote. "There was therefore no need to import the manufactures of outside barbarians in exchange for our own produce."[14]

The Chinese emperor instructed the British king, as if he were a bit slow on the uptake, that it was an act of Chinese beneficence to even allow British merchants to buy their superior products. He was prepared to be compassionate because "I do not forget the lonely remoteness of your island, cut off from the world, by intervening wastes of sea."

British merchants desperately sought a commodity they could sell to China that would stanch the flow of silver ingots into the Celestial Empire's coffers. As there was no legal commodity that the Chinese cared to buy, the merchants turned to an illicit one—opium. The Chinese emperors prohibited the cultivation of poppies and trading or using opium, but the prohibition merely drove up demand. The East India Company, one of the first multinational corporations ever formed, was solidifying its control of large swaths of India and the Far East. Chartered by the Crown and employing mercenaries, it established an opium monopoly in Bengal. From there, the drug was exported to, or smuggled into, China by subcontractors. The money the company took in from opium sales in China was then used to buy silk, porcelain, and tea leaves for customers in Britain. The trade deficit was eliminated.

Thus in the early decades of the nineteenth century, the East India Company became the world's largest drug producer, dealer, financier, and provider of military power to support the entire endeavor. Opium was crucial to the economy of the British Empire and was probably the most traded Indian commodity of its time.[15]

With addiction becoming rampant among urban populations, decimating the populations of Chinese cities, the authorities tried to battle the opium trade. They destroyed huge quantities of the drug and opium pipes as well but were unable to quash the demand—just as the US is failing today in its battle against its opioid epidemic.

The First Drug Czar

Walking the streets of the Forbidden City today, amid the tourists and the camera flashes, you can yet feel the sense of insignificance and inferiority that it was designed to instill in ordinary people. The courtiers who frequented these precincts could only tremble as they proceeded, in the footsteps of the emperor himself, through the Gate of Divine Might on the way to the Hall of Supreme Harmony, fearful of crossing a threshold prohibited to them. Here, at the beginning of the nineteenth century, trod the despairing counselors of the Daoguang emperor, a well-meaning monarch who felt his country's anguish. He did his best, from these ancient quarters, to defend his country's sovereignty. The emperor chose Lin Zexu, one of his most assiduous, loyal, and scrupulous officials, to lead China's proud but ultimately misguided and disastrous effort to end the opium trade.

In old paintings, Lin is elegantly attired, his eyes discerning, his beard long, narrow, and white—the archetypical Chinese sage. Lin persuaded the emperor to reject a proposal to legalize the drug. His successful campaign to purge the provinces he governed of opium made him a rising star at court. The emperor appointed him the world's first drug czar. Lin quickly proved that the emperor's confidence in him had not been misplaced. He destroyed huge quantities of opium and was incorruptible. At one point he composed an elegy to the god of the sea to apologize for polluting his dominion with such a foul substance.[16] His now-famous letter to Queen Victoria protested the injustice that Britain was committing against China. "Your foreign ships come hither striving the one with the other for our trade, and for the simple reason of their strong desire to reap a profit," he charged. "By what principle of reason then, should these foreigners send in return a poisonous drug, which involves in destruction those very natives of China? . . . And such being the case, we should like to ask what has become of that conscience which heaven has implanted in the breasts of all men?"[17]

Queen Victoria probably never read the letter. The British Empire mobilized to defend its right to deal in drugs. The First Opium War, the first war of modern globalization, began in 1839. China lost it, and the Second Opium War as well. Vanquished and subjugated, it was forced to make trade and territorial concessions. That was only the beginning of the Chinese catastrophe. China's share of the world's domestic product fell by half.[18] In Beijing today, the ensuing period is referred to as the Century of Humiliation.

Globalization, and the technological progress that energized it, defeated China. There was demand, and supply, and an international commercial enterprise backed by imperial power.[19] The British of course knew that selling opium was wrong. William Jardine, one of the major opium smugglers in China, wrote a letter to a passenger scheduled to board one of his ships. He wanted to explain what the cargo was. "We have no hesitation in stating to you openly that our principal reliance is on opium," he admitted. "By many [it] is considered an immoral traffic, yet such traffic is so absolutely necessary to give any vessel a reasonable chance of defraying her expenses."[20] William Gladstone, who would later become prime minister, opposed the opium trade, warning Parliament that "a war more unjust in its origin, a war more calculated in its progress to cover this country with permanent disgrace, I do not know, and I have not read of."[21]

Gladstone's denunciation did not change much. Capitalism, and China's military weakness, turned the country, over its bitter objections, into an exploitation hub, in the form of a local market in the thrall of a giant drug monopoly.

In 1920, a short, skinny Chinese boy boarded a boat sailing for France as a member of a student exchange delegation. He was sixteen years and four days old, the youngest of the group. His emotional father asked him why he was making the trip. "To learn knowledge and truth from the West in order to save China," the boy replied, quoting his teachers.[22] The sense of dishonor beat so strong in Chinese hearts that for decades after the country's defeat in the Opium Wars Chinese girls and boys grew up with the feeling that their divided and economically backward country needed redemption, and that redemption required modernization. To this day, the Chinese Communist Party views the Qianlong emperor's haughty response to the British and his contempt for their goods as a missed opportunity that left China out of the first stage of economic globalization.[23]

The boy's name was Deng Xiaoping. His reply to his father is the official Chinese version of the story. Deng's studies in Paris led him to Marxism and revolution. Fifty years later he would become China's leader and initiate reforms that integrated it into the world economy.[24] Its share of world gross domestic product rose from under 2 percent in 1979 to 19 percent in 2019. And in 2018, researchers from the Chinese University of Hong Kong estimated that, in 2010, over 1.1 million people died prematurely in China as a result of air pollution.[25]

The Haitian Slave Rebellion

Because of the distances between products and markets, and between workers and consumers, supply and demand are disrupting societal norms and ethics. Exploitation hubs can thus quickly escalate to violence. That was true of the British in China, just as it was true of the horrific regime that King Leopold II of Belgium imposed on Congo. Millions of Congolese were murdered in the reign of terror constructed by his agents in the so-called Congo Free State, which was neither free nor a state. It was Leopold's private domain, not a Belgian government colony, managed for his personal profit, a forced labor camp covering a huge expanse of Central Africa inhabited by people of many cultures and languages. The world wanted rubber, so the populace was forced to collect the coveted sap of the rubber tree. Workers who did not meet their quotas were cruelly punished; sometimes their children were taken as hostages. Soldiers recruited from among the local population, who enforced the slave regime, received their pay and ammunition only when they delivered the severed hands of workers who were not sufficiently productive, as proof that they had used their bullets for the purpose intended. The severed hands actually came to serve as a kind of currency in the territory.

It was an extreme illustration of the dehumanization that foreigners impose on exploitation hubs. If the foreign power has military and economic power, it uses its supremacy to exploit the labor cost differentials and maximize profits with very few, if any, moral compunctions. As the cybernetic civilization-destroying Borg of the Star Trek series say, "We are the Borg . . . We will add your biological and technological distinctiveness to our own. Your culture will adapt to service us. Resistance is futile."

But people do resist. And it is not always futile. The slave rebellion and revolution in Haiti in the eighteenth century was such an event but, unlike the Spartacus revolt in Rome, the uprising on the island of Hispaniola in the Caribbean Sea came in reaction to power structures that are part of globalization to this day. It demonstrates a recurring model of exploitation and confrontation in a globalized world.

Haiti, then called Saint-Domingue, was one of the world's most profitable colonies, and certainly the most profitable for the kingdom of France. It supplied 40 percent of the sugar consumed by Britain and France, and sugar was an expensive item then. Saint-Domingue also produced 60 percent of the world's coffee. At the beginning of the eighteenth century, its exports were equal to those of all the thirteen British colonies in North America.[26] It was a profit paradise for the plantation owners and the investors who backed them, but utter hell for those who produced the wealth—the slaves.

Between 1697 and 1804, some 800,000 slaves were brought to Saint-Domingue from Africa. The importation of slaves was intense because of the high death rate among the people who were brought in and forced to work in harsh conditions. The brutality of the Haitian plantation owners was legendary. Slaves who dared to resist were tortured in a variety of ways—they were immersed in water while tied in sacks, crucified in the middle of swamps, cast alive into the vats in which the cane syrup was boiled down, hung upside down until they died—and these are only some of the testimonies. Over time, some of the slaves were freed from bondage; other free people of color were born from liaisons between whites and African slaves.[27]

Gens de couleur libres, as these people were called by French whites, acquired land, power, and influence, and became an important component of the colony. Most important, they became cognizant of their rights and their role in the island's economy. Haiti developed a caste system—free blacks and mulattoes, wealthy whites, poor whites (called *petits blancs*, "little whites," in French), and under them all an absolute majority of hundreds of thousands of slaves and associations of escaped slaves who lived in the mountains, who were called *marrons* ("maroons").

The rising power of people of color made slave owners anxious. The French administrators of the colony wrote to the French naval ministry in the 1750s to say that "these men are beginning to fill the colony and it is of the greatest perversion to see them, their numbers continually increasing amongst the whites, with fortunes often greater than those of the whites . . . Their strict frugality prompting them to place their profits in the bank every year, they accumulate huge capital sums and become arrogant because they are rich, and their arrogance increases in proportion to their wealth. In this manner, in many districts the best land is owned by the half-castes . . . These coloreds . . . imitate the style of the whites and try to wipe out all memory of their original state."[28]

The whites were understandably frustrated. Those whom they labeled *affranchis* (from the French word for "emancipation"), comprising the freed blacks and their descendants, along with the mulattoes, broke down the whites' conceptualization of slaves and Africans—for example, the idea that the enslavement of blacks was justified because of their origin and their inferior minds. Something similar happened throughout the Caribbean, prompting the French king to establish that nonwhites could never expunge an "indelible stain." As he put it, "they retain forever the imprint of slavery."[29]

The threat that free blacks presented to the power structures of the plantation colony prompted wealthy white merchants and slave estate owners to legislate and enforce what we would today call apartheid legislation, perhaps the first such ever. These restrictions intensified toward the end of the eighteenth century. The *affranchis* were forbidden, among other things, to dress like whites, to eat with whites, to assemble after 9:00 p.m., to gamble, to travel, to belong to the legal or medical professions, and to hold public office.[30] The punishments for violators ranged from fines to amputations.

Then came news of the French Revolution. Rumors that the nobility had been ousted and the monarchy overthrown began to make their way to the colony, brought and spread by the *affranchis*. The news spawned hope for a new order. Some free blacks and mulattoes believed that the liberal principles of the revolutionaries would apply to them as well. One of them was Vincent Ogé, a mulatto who returned from revolutionary Paris with a fierce commitment to ending white supremacy in Saint-Domingue. He led a revolt, was captured, broken on the wheel, his arms and legs chopped off, and then he was beheaded.[31] Later, the revolutionaries in Paris granted some civil rights to the mulattoes and free blacks, but not to slaves. In August 1791, at a secret voodoo ceremony in Bois Caïman (Crocodile Woods), a revolution was declared. It was not an uprising but rather a careful plan by a small and well-established civilization that had managed to grow under the noses of the French colonialists.

In 1938, C. L. R. James published an account of that revolution, which had until then sunk into historical oblivion. *The Black Jacobins* is a critical look into the class-related context of the slave uprising. James describes how the fighting between the plantation owners and their slaves destroyed Haiti. His focus is on the revolution's most important leader, Toussaint L'Ouverture, a freed slave who is almost always portrayed in a French general's uniform. James attributes his success as a revolutionary leader to his realism about European people and the economic system they created. His resistance did not stem only from the tradition and education of liberalism (as with the American founding fathers) but rather from the experience of the radical overthrow of oppression.

"It is Toussaint's supreme merit," James wrote, "that while he saw European civilization as a valuable and necessary thing, and strove to lay its foundations among his people, he never had the illusion that it conferred any moral superiority."[32] L'Ouverture played the region's great powers off against each other, changing alliances frequently from France to Spain to Britain.

It was a brutal revolution of attacks, pitched battles, mass executions, and torture committed by both sides. It was not a simple story of slaves and mulattoes against white Europeans; at first, some of the mulattoes sided with the whites against the rebel slaves, and the whites were divided among themselves.

By the most conservative estimate, more than 200,000 people died. Every agreement was violated, every cease-fire breached. Eventually, L'Ouverture was betrayed and captured. "The rich are only defeated when running for their lives," James wrote.[33]

That's what they did in the end. In 1801, a few years after the establishment of the French Republic, former slaves in Haiti instituted the most radical policy ever attempted up to that time: real equality. They promulgated a constitution based on principles much like those of the United States Constitution. Unlike that document, however, the Haitian constitution rejected invidious racial distinctions: "All men, regardless of color, are eligible to all employment . . . There shall exist no distinction other than those based on virtue and talent, and other superiority afforded by law in the exercise of a public function."[34]

The Haitian revolution ended in 1804 with a horrendous massacre of the white French settlers. The victors established the first modern state in the Caribbean, and the world's first black republic. Slavery was banned, a change the United States would not effect for another half century. The promise of democracy faded in the years following Haitian liberation, as it did in the first decades after the French Revolution. Haitian generals instituted a system of serfdom; the whip itself was banned but much of the colonialist legacy remained.

The territory was devastated; the young country's most pressing problem was survival. It was an independent polity of rebellious black slaves—many of whom were unfit to work after the war—in the midst of a world ruled by white empires and slave merchants. Haiti was shunned by other countries and attacked repeatedly. The American founding fathers, led by Thomas Jefferson, himself a slave owner, instituted a policy of shunning Haiti and banned trade with the pariah state.[35] No other country recognized Haiti, and what had been one of the Western Hemisphere's most prosperous economies became isolated. Cuban slave owners were quick to take advantage of Haitian ruin, and the exploitation hub relocated. Cuba became the Caribbean's largest producer of sugar; its import of slaves quadrupled between 1791 and 1821.

In 1825 the French sent a fleet to exact revenge. Resistance was futile; Haiti's leaders had no choice but to sign a humiliating capitulation to France in return for the latter's recognition of the Haitian state. The treaty forced it to pay reparations to France and to former slave owners, with compounded interest; in the modern era, national debt can replace the lash. Initial payments were made by taking a loan, at an onerous interest rate, from a French bank that had received a monopoly from the French government to collect the debt. Haiti continued to pay this compensation, imposed by military force, until 1947.

In 2003, the Haitian government estimated that the country had forfeited at least $21 billion in funds that could have been used to rebuild it over more than a century. France refuses to this day to discuss restitution. Why should they? Have the British compensated China for the damage done by the Opium Wars? And the Belgians—have they paid reparations to the Congolese for the horrors they inflicted?

Haiti never recovered. It fought for its freedom, but the bonds of servitude were replaced by financial chains.[36] The Haitian revolution was not merely an uprising that secured freedom for slaves; it was the prototype for the devastation of exploitation hubs after they stop serving supply and demand.

Bougainville Rebels

Bougainville, the largest of the Solomon Islands, lies at the northwestern end of the archipelago. Culturally and ethnically, its inhabitants belong to the Solomons. A bargain between colonial powers in 1920 led to its annexation to Papua New Guinea, so it is not part of today's country called the Solomon Islands, a thousand kilometers closer.

At the end of the 1960s, huge deposits of copper and gold were discovered on the island, valued at tens of billions of dollars. The Bougainville Copper Limited (BCL) company, then owned by a large Anglo-Australian corporation, Rio

Tinto, bought the rights to mine the deposits. During the 1970s it established Panguna, one of the world's largest strip mines. The mining involved lopping off the top of a mountain, changing the local landscape entirely. The island's inhabitants claim that the waste products from the mine flowed into the Jaba River and its streams.[37]

It was a strategic economic project for poor Papua, to the extent that at one stage the proceeds it generated constituted almost 45 percent of the country's entire export income. It also, of course, contributed to Bougainville's own economic development, which came along with workers from Papua itself and a major change in the island's social fabric. All this created tension between the local populace and the mother country. The original owners of the land on which the mine was built received negligible compensation. The locals also resented the labor migrants who came to work at the mines, and the pollution of the surrounding area—a by-product of copper extraction. To this day, agricultural communities claim that their rivers are contaminated and their children are being poisoned by the dross the mines left behind.

The island and its inhabitants did not receive any significant share of the value of the metals removed from their land. The issue is all the more explosive because the island's inhabitants from the start felt only a tenuous affinity with the government in distant Papua New Guinea.

By 1988, some islanders had had enough. Several of the original landowners broke into the mine and took the explosives used to extract ore. They used them to blow up the power lines to Panguna. The local leadership, loyal to the government in Port Moresby, the mother country's capital, called in the army. Bougainville soon found itself in the midst of the deadliest conflict in the Pacific since World War II. The Papuan government, assisted by Australia, blockaded the little island for years. Naval vessels encircled its shores, preventing food, medicines, and merchandise from getting in. At the same time, factions on the island began fighting each other, claiming many victims. Estimates are that some 15,000 Bougainville islanders—between 6 and 10 percent of the island's population—lost their lives, among them hundreds of children who succumbed to a malaria epidemic. One fifth of the island's inhabitants were displaced.

The rebels ejected the Papuan army from the island. Displaying considerable resourcefulness, they distilled oil from coconuts to use as motor fuel for boats, automobiles, and trucks. The rebel government built dams in the island's rivers to power hydroelectric plants, and the locals went back to canal irrigation and other traditional farming methods that had been neglected before the rebellion began.

After years of struggle, the islanders won a modest victory that ended the war. They obtained a peace treaty, recognition of their right to self-determination, and recognition of their rights to the island's natural resources. In 2014, Papua New Guinea's prime minister officially apologized to the Bougainville islanders for the war the country had waged against them.[38]

The Bougainville rebellion, like Haiti's, is a tale of resistance and freedom. But it also led to the closing of the mine, devastating the local economy and scarring a fragile and divided community. The islanders continue to wrangle with the Papuan government, and among themselves, over the rights to the mine. Big trucks lie rusting in the open wound on the landscape, in the midst of toxic puddles. Bougainville's autonomous government announced in 2019 that the mine would remain inoperative until further notice, because of concern that reopening it would reignite the conflict.

Rio Tinto, the giant mining corporation, divested itself of its interests in Panguna. The *Sydney Morning Herald* published the multinational's response to Bougainville's demand that it pay compensation for the environmental damage that it caused. "We believe that [the company] was fully compliant with all regulatory requirements and applicable standards at the time," Rio Tinto's management wrote in a letter to John Momis, president of the Bougainville government.[39]

Globalization's adversary is the individual rooted in her local community. She is happy to reap the benefits that globalization brings her—literacy, jobs, smartphones—but holds fast to her individuality, rights, and identity. As Bougainville proves, no island can hope to be overlooked by or remain immune to globalization's forces. Like contemporary Caesars, these forces came, they exploited, and they left.

Yet today the people of Bougainville hold more rights to their natural resources than they ever did before, and enjoy more power over their own lives. In 2019 the island held a referendum and decided on independence from Papua New Guinea.

THE CASES OF CHINA, HAITI, AND BOUGAINVILLE DEMONSTRATE a pattern of trade and globalization. Globalization requires the extraction of raw materials worldwide, the diffusion of labor, and the free flow of capital. It also needs open world markets so that materials and the goods produced from them can be sold to as many people as possible. Supply and demand is the engine of all these stories. In all three cases, the opportunity to extract, grow, or sell a substance for which there was a demand created conditions in which it was profitable, for a time, to crush individual rights, local communities, or national sovereignty. Corporations used state violence and oppression to achieve these aims, enlisting politicians to protect their profits. The foreigners paid little regard to the needs of the people inhabiting the country, colony, or island, in the process inflicting lasting damage to the local ecosystem. As corporations, and often national governments, see it, everything and everyone outside their metropolises are disposable, and even the metropolis itself can become an exploitation hub as manufacturing and capital hold to their quest for efficiency.

But there is also change. In Haiti, centuries went by before the slave rebellion broke out. The oppression of China as a consequence of the Opium Wars lasted a century. The cruelty of Leopold and his agents in Congo caused an international outcry, led by some of the twentieth century's most prominent political and literary figures, causing an abrupt end to the king's monstrous project. Bougainville's resistance commenced after only a few decades.

Contemporary globalization, with its liberal values, empowers the local individual while at the same time using him. It is this empowerment that, in the end, makes exploitation hubs untenable. In a world of unprecedented connectivity and expanding global consciousness, no power or corporation can act with the impunity that Britain and the East India Company did in the Opium Wars.

For instance, until recently, China was the world's most massive recycler of waste materials, accepting delivery of enormous quantities of plastic, paper, and metals from all over the industrialized world. This waste caused major environmental problems, prompting the Chinese government in 2018 to impose a moratorium on imported plastic waste. The corporations that collect the waste then tried to dump what China wouldn't take on Southeast Asian countries. But that same year Malaysia, Vietnam, and Thailand all passed legislation to prohibit the use of their countries as repositories for the discards of the global North. Malaysia's environment minister declared that his country would not serve as the world's trash can. The president of the Philippines was blunter, threatening to cast into Canada's territorial waters 1,500 tons of trash that that country had sent to his.

The life span of exploitation hubs is contracting. Global awareness and local empowerment means that they cannot last long. The result is that these hubs must quickly relocate, again and again. Corporations and state institutions, the initiators of exploitation hubs, are desperately searching for havens where people are unaware of the damage they inflict or too weak to resist it—from the West to Beijing, from Beijing to surrounding cities, and from there to the Chinese countryside or another country entirely. These are intense cycles of boom and bust, hurried attempts to make as much profit as possible before the inevitability of liberal empowerment will destroy another hub of exploitation.

In the long run, it is a positive phenomenon, attesting to the constant improvement in the human condition, in terms of income, life expectancy, and health. But in the short run, it means that exploitation hubs are like the hurricanes of the age of global warming—more intense and violent. Their most severe and long-range effect is on the environment.

CHAPTER 4

The Land of the Last Elephants

In a minute, Sampath Ekanayaka will leap over the muddy channel and I will jump after him. The sun is just about to set, and the shadows of the bush are lengthening under its rays. Once the sun is gone we will be unable to continue our search. Swarms of mosquitoes seem to be soaring out of the huge cracks in the mucky soil, seeking a patch of unsprayed skin. "You can hear them," Ekanayaka says, standing atop a small hummock, gazing into the tangle of the small bit of jungle. His tracker runs before me, proudly bearing a small video camera. I see that the two of them are advancing and retreating, dashing and then standing still. They are restless. "Now we are in the land of the elephants," Ekanayaka suddenly remarks. The smile that he bore throughout our journey is gone. I know that he has issued a warning, and a request: do not continue walking toward the soft sounds of breaking branches.

We are in Galgamuwa, in the northwestern province of Sri Lanka. It is not the country's steamy capital, Colombo, which pulsates with rapid development, nor the beaches packed with sunburned tourists. People in Galgamuwa live off their fields. Close to the rivers, rice paddies stretch out to the horizon, with the narrow leaves of the rice plants waving in the breeze. But where we are the fields are of another kind entirely—jagged plots of modest size, their boundaries determined by the stripping away of woodlands in the dry zone that occupies most of the country's north and east. For families that till this soil, their field is the difference between some level of comfort and rapid decline into perilous poverty. In this season, or perhaps all year, their nemeses are the elephants.

It's a war. Dozens of people are killed by elephants every year in Sri Lanka, and some two hundred elephants die in the battles as well. At the edge of the field in which we are standing, a deep moat prevents the elephants from charging at its sweet crop. Nearby, the farmers have erected a mobile electrified fence. Such fences are in big demand now in Galgamuwa. Scattered around are three or four improvised guard towers that look a lot like the treehouses my friends and I used to build when we were kids. But these are not places to play. They are mounted high up, generally on lone trees. They have low roofs, and their larger opening faces the place of danger, the small and threatening forest. The people scurrying around the fields now hold a variety of instruments to frighten the elephants, from large flashlights to metal pots. Small automobiles chug along the uneven dirt roads, driven by animated young men, on the lookout for elephants crossing into the fields. Rumors fly. Our tracker hears that an elephant has been spotted on the other end of the field. Everyone rushes over there, excited and fearful. As we reach Sampath's small pick-up truck, intending to drive to another location, the tracker runs up to tell us that the elephant is elsewhere, setting off another frantic chase. Here there are large droppings, there tracks; in another place you can hear sounds, and in yet another an elephant appeared the previous night.

I think of all the young men going around in their cars and standing in the guard towers, their jittery enthusiasm, and it brings up the memory of something else, both familiar and foreign. Torches or flashlights. Search and pursuit. Fear and violence. Common defense against a threatening enemy, the Other. Then I suddenly remember. It's like an American film depicting the lead-up to a lynching in a southern town in the 1950s. Except here it is elephants.

I'm immediately ashamed for making the association. The indigent people around me are trying their best to block the elephants, not kill them. The patch of forest is small and vulnerable. It would be very easy for these farmers to attack it with a few tractors and a rifle or two and put an end to their nemeses once and for all. But they do not do that. They do not act like our ancestors in the Northern Hemisphere, the West, or the Middle East would have done. The Hindu and Buddhist principle of *ahimsa*, nonviolence toward all living things, still applies here.

Sri Lanka's elephants are the largest Asian species, *Elephas maximus*. They can reach a height, at the shoulder, of eleven and a half feet and weigh as much as six tons. Less than 10 percent of the males have tusks, which is probably not a matter of chance or normal evolution. The scientific hypothesis is that natural selection against tusked bulls occurred because of widespread hunting for ivory and trophies and the decrease of habitat and export of tuskers to other countries during the British colonial period, when the island was called Ceylon.[1]

The bloodthirst of British hunters was legendary, the most famous Victorian elephant hunter on the island having been Samuel Baker. Baker, a good friend of fellow explorers Henry Morton Stanley and Charles Gordon ("Gordon of Khartoum"). Baker wrote *The Rifle and the Hound in Ceylon*, a tedious book about his smug passion for killing. It is far from George Orwell's deep observations in "Shooting an Elephant." Baker has nothing but contempt for people who pity elephants. "Poor things, indeed!" he writes. "I should like to see the very person who thus expresses his pity, going at his best pace, with a savage elephant after him." He offers repetitive accounts of a subject he seems to have been obsessive about—killing elephant mothers and their calves: "The following evening, we again watched the pool, and once more a mother and her young one came to drink. W. and B. extinguished the young one while I killed the mother." In the most revolting passages, he describes killing a cow elephant whose udder was full of milk, which he then drank straight from her teats, "to the evident disgust of the natives."[2]

Baker was just one example. Imperial legend proudly told of a British officer who killed a thousand elephants. When Sri Lanka liberated itself from colonial rule, the big-game hunting orgy ended, but rapid economic development destroyed enormous parts of the elephants' natural habitat. The horrendous civil war between the Tamil minority and the government was also hugely costly, in terms of the lives of both human beings and animals. According to the World Wildlife Fund, the Sri Lankan elephant population has fallen by nearly 65 percent since the beginning of the twentieth century.

Today, Sri Lankans are concerned about the gradual disappearance of their elephants. Elephants are still kept, in horrible conditions, in many Buddhist temples, where they serve as talismans and local attractions. A unifying factor in the island's complex national identity, elephants are a source of pride to Sri Lankans. They are a regular subject of conversation. Front-page newspaper headlines demand a government response to the plight of the last two surviving elephants in a remote nature reserve. The official punishment for trophy hunting an elephant is death (although the country has not conducted executions in criminal cases since the 1970s). One of the island's biggest tourist attractions is an elephant "orphanage" in Pinnawala.

Ekanayaka is a field worker for a program to resolve conflicts between Sri Lanka's people and its elephants, sponsored by the Center for Conservation and Research in Galgamuwa. In practical terms, it serves as a place where farmers and villagers can lodge complaints against elephants with a patient third party who will try to resolve the problem.

These are places where people still bathe and launder their clothing in the river, keep an eye out for any sign of fever because it may presage a Dengue attack, and walk every week to bring an offering to a shrine. They view Ekanayaka as a representative of the state, development incarnate, and a conversation with him in and of itself allays anxieties, of which there are many.

The essence of the problem is simple. The elephants' roaming grounds have been destroyed to make way for fields and houses; the elephants, in search of food, return to their ancestral haunts. Elephants eat for up to sixteen out of every twenty-four hours, and a Sri Lankan elephant consumes between 300 and 400 pounds of vegetation a day. A natural increase of the human population, along with government programs granting land to relatively poor families, have made conflict between beast and man inevitable. Film clips documenting such clashes, grainy ones taken with mobile phones, are easily available on YouTube. They show a man trying to scare off an elephant crossing his field; the elephant responds by trampling and killing the farmer. Tractors wield their shovels at elephants who are trying to fight them; elephants stand in the middle of roads, forcing cars off. While people are killed every year by elephants in Sri Lanka, in the long run confrontations on habitat always end the same way. Eventually, humans triumph.

The electrified fence is the most popular means of blocking hungry elephants. The fences are everywhere. Fields are encircled by metal wires that give a potent shock to whoever touches them, but not enough to kill. Some entire villages are surrounded by such fences; in others, just a few houses are. Schools and public buildings have their own fences. They are ubiquitous; villages and families live entirely within fences. The only time I see the generally placid Ekanayaka get angry is when he refers to the land the government has granted to new families, without the connection to the electrical grid that such fences require. These poor families erect fake fences that they hope will fool the elephants, but to no avail. Indeed, the elephants are so smart that they have begun to adapt to real fences. Ekanayaka brings the car to a halt at a large house. In broken English, he tells me to take note. There is a lot of food in the garden, including trees bearing bananas and coconuts. That is why the house is surrounded by an electrified fence. "Elephants now learning," he tells me. "If we put up fence, even if there is nothing inside, he tries to break fence because he thinks there is food there. They are adapting." They not only understand that fences mean food, they have developed tactics to deal with them. Young males used to simply knock down a nearby tree so that it fell on the fence, allowing them to cross. In response, Ekanayaka says, the farmers uprooted all the trees near the fence. Now the elephants bring a tree from somewhere else and throw it on the fence. "There are three to five guys who learned it," he explains. "You call them guys?" I ask. He chuckles.

In the well-tended house, surrounded by a large yard, I go with Ekanayaka to visit Somanwathi, a local woman who lost her husband to an elephant in her backyard eight years ago. She serves us cubes of rice and lentils baked in a charcoal oven, along with a spicy hot chutney, and then brings a black-and-white photograph of her late husband. Her current husband gripes, via my interpreter, that it's "gotten much worse." It's not only about food anymore; young males might charge people and cars which they perceive as a threat. If someone is sick and needs to be taken to the doctor, he says, it can't be done in a small car or wagon. Only a big truck deters the elephants. If a truck isn't available, "the whole village" needs to be called out to accompany the sick person the whole way.

"Are you angry?" I ask the widow. "There is no point in getting mad at an animal," she says, with a ringing laugh. She relates that she sometimes stands on the safe side of the fence and watches them for pleasure. I think about how many people in the West would be willing to live behind an electrified fence to protect themselves from an animal weighing many thousands of pounds. It would not take long for them to demand that the authorities move the elephants, or worse.

The elephants did not emerge from the forest that night, although their cries reverberated through the dark. There were muffled calls, and the whimpering of hungry calves, but mostly the unending sound of elephants tearing branches off trees and gobbling them down. At any rate, I was the only person who wanted them to come out.

Ultimately, they would suffer the same fate that had met so many of their brethren. Their population would dwindle, either from starvation or from being killed in conflicts with humans. Some would fall into moats or find themselves alone, far from their herd. If they are lucky, they might be taken to Pinnawala. Hotels, restaurants, and souvenir shops have gone up around this so-called elephant orphanage. One of the most popular shows there for tourists is when the elephants are taken down to the river, twice a day, to bathe. The elephants make their way toward the river, accompanied by spear-bearing guards. The legs of the males are chained, to keep them from stampeding. The procession looks like a parade of prisoners captured from a tribe defeated by the Romans, their chains clinking against the walkway. It sounds like a dirge: "Once we were free and now we are not; once we were in our own land and now we are a spectacle for the masses."

After nightfall we walk alongside a field of feed corn. An elderly woman sits in the guard tower. The beam of her flashlight reveals white hair that frames her face, making her look like a character in an ancient Buddhist legend. I conduct a shouted exchange with her, I below by the electric fence and she above, in the tower. Through my interpreter, she tells me about her poor family, and the long months they spend in the tower driving away the elephants at night. She says she can either babysit her grandchildren, or babysit the cornfield; tonight she is here, staring into the dark patch of woods ahead, armed with pots and pans ready to scare away the elephants.

When your son grows up, I ask Ekanayaka, will he be able to see elephants as you see them today, free ones? He traced out a square on his palm. "Only in pictures," he said. "If we go on this way, only pictures. Maybe in zoo." He offers an uncomfortable smile.

THE GREATEST THREAT CREATED BY GLOBALIZATION IS THE destructive impact of consumerism and industrial production on the earth's ecosystem. The most potent argument against the current world order is simply that it is not sustainable. If left unchanged, animals and humans will not be able to survive it. Human beings and local communities can be exploited and destroyed; civilizations have done this throughout history. But the mortal blows being inflicted by humanity on the planet are something new and might be irreversible.

The decimation of the Sri Lankan elephant is happening, in one way or another, to animals everywhere, all the time. Habitat destruction is the main reason for the unprecedented levels of species extinction we see today. More than 60 percent of all vertebrates have disappeared from their natural habitats since 1970.[3] The number of mammal species has plummeted as their habitats have been eliminated. Some scientists refer to the process as "biological annihilation"; we are witnessing the most precipitous extinction event the earth has seen in 10 million years.[4]

Species loss has obviously been accelerated because of human actions. Ninety percent of the world's cheetahs have disappeared in the past century, a figure a bit worse than those for African elephants: there were 10 million in 1930, and today there are only 415,000. In Mozambique alone, 7,000 elephants were killed by ivory hunters between 2009 and 2011.[5] North America is home to 3 billion fewer birds than in 1970, a population decline of 30 percent.[6] A trailblazing and painstaking study published in Germany in 2017 found that the insect population in the country's nature preserves has collapsed by 75 percent in recent decades. The study was conducted in nature reserves, which are supposed to be relatively protected from environmental damage.[7] We do not know what sort of ecology can be maintained on earth without a flourishing insect population. Insects constitute a critical link in the food chain, pollinating the plants we all depend on. More than 75 percent of the world's agricultural production depends on pollination conducted by animals. Between 2014 and 2018, beekeepers in the United States lost four out of every ten hives to Colony Collapse Disorder, which is devastating the insects most vital to the pollination of flowers.

Extinctions are not limited to the land. The acidity of the oceans has increased by 30 percent in the last two centuries, and fish populations are collapsing. Huge fishing ships and trawlers rake up fish from an area of ocean at least the size of South America. One of the results of industrialized fishing is the indiscriminate destruction of species that have no commercial value, a process known as "by-catch." Bottom-trawling nets destroy breeding grounds, causing the collapse of populations of many fish species. Between 60 and 90 percent of large aquatic carnivores have disappeared since 1950.[8] One study finds that fishing kills 11,000 sharks per hour, about 260,000 each day.[9]

At Gansbaai, in South Africa's Western Cape, I dove in a cage, hoping to spot a white shark. Until not long ago, this was the best place in the world to see them. They have disappeared from the sea near this small town, but no one knows why, or where they have gone. Afterward I watched devoted workers force-feeding African penguins in a facility set up to care for weak ones washed up on the beach. Xolani Lawo, who is in charge of the feeding project, told me that there is no other way to feed them in captivity. Their instincts are only to eat a fish they capture while it is in motion. "They won't of their own volition touch a fish when it isn't moving. Some of them hate to be fed, they hate being on this table." One by one, he holds them by the throat and forces their sharp beaks open, pushing fish into their throats. The central problem the members of this endangered species face, beyond the loss of their natural habitat and nesting areas on the coast, is that they can no longer find sustenance in waters that once teemed with life but which are now overfished.*

I petted one of the young penguins as Xolani held it, running my fingers over its wizened skin. It was starving. "We lose ninety each week," Xolani told me, relating the condition of the entire species. The feeding routine at the facility is so regular that it is easy to forget how crazy the moment is, when the only way to preserve a species of expert swimmers and brilliant hunters is by force-feeding its members. The facility's veterinary nurse, Theanette Staal, spoke to me with tears in her eyes about the desperate struggle to save the African penguin from extinction. "These are not our penguins, or South Africa's," she said. "They are everyone's penguins." At the current rate of decline, the species will die out by 2026.

A report by the Intergovernmental Science-Policy Platform on Biodiversity from 2019 found that a million species of animals and plants face extinction, many of them within just a few decades. The phrase "transformative change" is repeated again and again. Without such a change, the scientists who wrote the report claim, the situation will only get worse. The epidemic of extinction might well so transform the entire world's ecology that humankind will face unprecedented danger. One of the people behind the report, Josef Settele, a German ecologist, writes that "ecosystems, species, wild populations, local varieties and breeds of domesticated plants and animals are shrinking, deteriorating or vanishing. The essential, interconnected web of life on Earth is

---

* African penguins also never recovered from the mining of their guano to be used as fertilizers. The mining operations destroyed their chosen nesting areas, which were safe from predators and sheltered from the weather. Not only that, their eggs have become delicacies. The cafeteria of the South African parliament used to serve penguin eggs once a week.

getting smaller and increasingly frayed. This loss is a direct result of human activity and constitutes a direct threat to human well-being in all regions of the world."[10]

The fundamental tension in this and similar texts is clear. To "sell" the severity of the moment to the public, scientists need to warn that humanity is itself in danger. It's not just frogs, bees, cheetahs, and elephants that will go extinct—*Homo sapiens* will also be vulnerable. The cyclone of destruction and death that humanity has fueled will destroy us as well.

The media and international NGOs stress the danger to humankind in order to gain support. But these marketing strategies are based on the same unstated assumption of Samuel Baker, the animal-murdering colonialist—that nature, including the teats of the she-elephant he shot down, exists to serve humankind. Human civilizations have usually seen the world as an inexhaustible store of resources that they can use for their own needs.

It's easy to imagine what real transformative change would look like. It will happen if voters agree that the mass extinction of animal species as a result of the excesses of human civilization is wrong and unacceptable, even if humans can survive it. It will happen if we become convinced that biological diversity is a supreme social value that needs to be carefully protected. It will happen if we accept that nonhuman animals have rights, real rights.

After all, the corporate system does not provide us with life's most basic needs. It does not create air, or purify it, or maintain water quality, or make it possible to grow food. Those are things that happen because of a diverse, breathing ecological system comprising an unknown number of living creatures, and because of the unfathomable abundance of that system.

It is not a radical position to take, in historical terms. The human species has lived by it in primeval times—but it has been forgotten. Maimonides, the most important philosopher in Judaism, put it succinctly. "The Universe does not exist for man's sake," he wrote in his *Guide for the Perplexed*. Rather, "each being exists for its own sake, and not because of some other thing."[11]

Hoedspruit, South Africa

The electrified fences of Sri Lanka are meant to keep elephants out. In South Africa, they do the opposite—they keep animals from getting out and human beings from getting in. Kruger National Park, more than 7,700 square miles large, is surrounded by such a fence. In Hoedspruit, not far from the park, a number of farmers have established large private safari reserves, augmenting the animals that were already living on their land with others they purchased elsewhere in the country, and have opened luxurious hotels, the kind where you get greeted with a cool, vanilla-scented towel.

These are businesses, of course. You can go to South Africa to see animals, or you can go there to shoot animals. Given that the animals are the private property of the landowner, both experiences are equally legal. Most of the farms around Hoedspruit are not meant for hunting. Their safari areas are crisscrossed by hundreds of dirt roads, covering every possible point and barely enabling the local ecosystem to function in full; the roads are there to allow tourists to get the best angle for their photographs of a leopard, elephant, or rhinoceros. The optimal aim is a selfie, of course.

Rhinoceroses are the real story here, or more precisely, the price of rhinoceros horn on the black market, especially in Asia. Prices range wildly between $50,000 and $100,000 per kilogram in places like Vietnam and China, where the horn of these huge perissodactyls is believed to have extraordinary medicinal properties. (There is no scientific support for this—the horn is comprised almost entirely of keratin, the same type of protein that makes up hair and fingernails.) The high price means that poachers have a significant incentive to trespass on the safaris. In the best case, they brutally saw off horns after partially sedating the beasts. In many cases they simply kill them. The safaris use touch-sensitive electronic fences, helicopters, paramilitary units, networks of cameras and sensors, gliders, intelligence and reconnaissance teams, trackers who live in the bush, and many other means of deterrence. It doesn't work, because demand keeps increasing, and just beyond the fences are shantytowns where people live in oppressive poverty. The average monthly wage in South Africa is $1,400 (as of the fourth quarter of 2018), but the real problem is unemployment—one out of every four members of the eligible workforce can't find a job.

Nakonsiti and Preis are two rangers working for Protrack, a private security service that defends safari animals from poaching. One night I stood with them at an improvised roadblock at the entrance to the reserves, not far from Hoedspruit's small airport. "We can't stop this poaching," Nakonsiti tells me, "because there are no jobs out there." He speaks at length about the social plight, about the feeling that the shantytown dwellers have no part in the safaris' orgy of success. "We can only decrease the poaching, nothing more," he says. The two of them say that they don't talk about their job when they go home to their community. If people knew that they worked in the reserves, a close relative might be kidnapped. To obtain his release, they would be compelled to provide information and to let the poachers into the safaris. And this is just one scenario. "When I go home, I don't wear a uniform," Preis explains. "Only my regular clothes. When they ask me what work I do, I say that I am in maintenance somewhere. I don't say that I work in a wild animal reserve." Their sad need to dissemble shows who is winning this war—and it's not the people who are guarding the animals. Demand is always victorious.

I go to see Karen Trendler, one of South Africa's most prominent conservationists and campaigners for animals, and a senior figure in the National Council of Societies for the Prevention of Cruelty to Animals. We sit in her large backyard. Trendler has for many years served on a rapid response team that saves animals, especially wild animals, from abuse. "The economy in Asia is expanding rapidly," she tells me. "You've got huge population growth, the economy is growing, so disposable income is also growing. They [upper and middle classes in some Asian countries] now have money to buy luxury goods like rhino horns and tiger-bone wine." The latter, a beverage made from the fermented bones of tigers, is in demand throughout the Far East, where it is believed to cure pain and weakness; it is also believed to boost intelligence and male sexual prowess.

As tiger bones are not easily obtained, lion bones have also become a prime commodity on the continent. In the case of lions and other big cats, it is a perfect and utterly efficient commercial circle. Tourists go to South Africa and pay good money to "volunteer" at so-called sanctuaries for lion cubs "abandoned by their mothers," or whose parents were "shot by hunters." They paint a picture that seems inspired by the Disney film *Bambi*, setting an effective sentimental tourist trap. In fact, Trendler says, "the cubs are not abandoned by the mother. Those cubs are intentionally taken from the mother when they are a few days old, because there is commercial value in bringing in paying volunteers and tourists who want to pet and play and take selfies with the cubs. And then the cubs get to a point where they can no longer be used by tourists. They get too big. They get dangerous. They are not as appealing as the cute little cubs." So the best way to profit from them is to move them from sanctuary to safari so they can be hunted by tourists who are less into petting and more into shooting.

Tigers are not native to Africa, but they have been imported, and they are now raised and sold there for hunting. After growing up in these ostensible sanctuaries, the big cats are not frightened by human beings, so it is not a big challenge to shoot them. After the animal is shot, the hunter takes the head to mount as a trophy to impress his house guests. The farmer who runs the safari harvests the bones, which are shipped to the Far East to be made into jewelry, charms, wine, or "tiger cake," also used in traditional Asian medicine. Along with this, females are bred again and again, often with males closely related to them. One former volunteer recounted for me pitiful scenes involving cubs born with defects as a result of this inbreeding. It goes without saying that animals caught up in this system can never be released into nature—they are incapable, temperamentally and genetically, of fending for themselves in the wild.

These stories demonstrate the relentless impact of the international market, as well as the interaction of globalization with local ecology. It is a system of exploitation almost entirely powered by foreigners. Tourists unsuspectingly support the petting nurseries, and volunteers unwittingly are accessories to the raising of animals for the express purpose of harvesting their organs. The great majority of the hunters come from other parts of the world—from the United States, Europe, and Asia. The bones make their way back to Asia, one way or another.

The African lion lives in nature only on this continent, but as an industrial product it lives everywhere. Each stage of its life is exploited in an industrial cycle, until its death, and most profits flow to people outside the community that shares its habitat.

But the big cats are in good shape compared to the rhinoceroses. The world of commerce does not want much from these awesome beasts—no opportunities to pet their babies, or hunt them, or even use their bones. It wants them on exhibit at safaris. And it wants their horns. As I was told by John Hume, a rugged farmer who raises them for just these purposes, "to put it simply, a dead rhino is worth more than a live one." In the global market, rhinoceroses have value primarily in one place, Asia, and to a single end—as a source of powdered horn.

The last northern white male rhinoceros died in 2018. Precisely 769 rhinoceroses of all varieties were poached for their horns in 2018 in South Africa, and more than 1,000 in 2017. These numbers call into question whether any sort of rhinoceros will be able to survive in the wild in Africa. Crime syndicates, some of them from Asia, are behind the poaching operation. They send their own poachers into the reserves, equipped with smartphones so that they will be able to document the source of the horn to potential buyers in Asia.[12]

It's demand that drives the phenomenon, pure demand, alien to the local environment and to the ecological system, and alien to political norms, community needs, and tradition. The world of the revolt is one in which the local is constantly under attack and under challenge by foreign powers that create a sense of arbitrariness. Sometimes, the local is a rhino.

CHAPTER 5

"We Refuse to Die"

The 5 million inhabitants of Colombo, the capital of Sri Lanka, now experience extreme climate events with increasing frequency and unprecedented force. Weather patterns there have changed significantly since the 1950s. Monsoon rains are now shorter but more intense, which leads to flash floods and the inundation of neighborhoods that house the poor. In 2016, Sri Lanka endured its worst drought in forty years. The next year saw record precipitation—11 to 20 inches of rain in just twenty-four hours—causing massive flooding that destroyed a large part of the rice crop. The paddies produced nearly 40 percent less rice than in the previous year, leaving hundreds of thousands of people without food security.[1]

It happens all the time now in poor countries. In the spring of 2019, two cyclones hit the eastern coast of Africa, one after the other, causing heavy damage in Mozambique. Cyclone Idai extensively damaged the country's farms, destroying most of the corn crop on the coastal plain. It was a humanitarian crisis that compelled the country to appeal to United Nations aid agencies for help.

In previous chapters we considered the exploitation hubs that globalization has created, but an exploitation hub is not just a matter of employment, or the extraction of local minerals in a way that harms the people and community near it. The ultimate exploitation is environmental.

Pollution, unbridled consumerism, and the massive carbon emissions spawned by the Industrial Revolution affect the weak more than the strong, and that is not just a figure of speech. Studies have analyzed the glaring inequality of the climate crisis and have produced similar conclusions. The bill is not being paid by the wealthy nations. A recent study by two Stanford University researchers, Marshall Burke and Noah Diffenbaugh, shows that poor nations have been paying for global warming for decades.[2] Using a model that compared how countries performed in warmer and cooler years, they found that between 1961 and 2010, global warming depressed the wealth per person in the poorest nations by up to 40 percent. Furthermore, the GDP per capita of countries that emitted the most greenhouse gases increased by 13 percent.

These figures are astonishing. They show that the global South, which had begun to close the gap with the industrialized world, could have made incredibly rapid progress were it not for the climate crisis. As Burke and Diffenbaugh show, cold-climate countries are getting warmer and are benefiting economically from the greenhouse effect. But countries that were hot to begin with and have been warming even more have been hit hard. Per capita national product in countries such as India, Nigeria, Sudan, Indonesia, and Brazil fell by tens of percentage points because of climate change. In contrast, the GDP per capita of Norway and Canada has benefited substantially on account of warming. The researchers explained how productivity peaks at an annual average temperature of 13°C, yet declines strongly at higher temperatures. "Crops are more productive, people are healthier, and we are more productive at work when temperatures are neither too hot nor too cold," said one of them, Marshal Burke. "This means that in cold countries, a little bit of warming can help."[3] Britain, which was at the vanguard of the Industrial Revolution, saw an additional 9.5 percent increase in GDP per capita with global warming. Another study estimates that by the end of the twenty-first century, the average income of billions of people will be 75 percent less than it would have been without climate change.[4]

This is inequality squared. The countries that contributed the least to the climate crisis have already paid a high price in economic losses, and will continue to do so. They are more vulnerable to extreme climate events because of their weakness and poverty, while, according to climate models, they will experience the strongest effects of extreme events of low and high temperatures.[5]

As economic development in a globalized world is no longer a zero-sum game, individuals and communities can rise from extreme poverty by means of industrialization, trade, and free markets. But a new and dark element has made its way into the equation, paradoxically born of precisely these systems—climate change. Environmental organizations estimate that, in the next few decades, the number of people suffering from hunger will grow by 10 to 20 percent as a result of climate change.[6] One study estimates that by 2050 an additional 1.7 billion people, approximately 20 percent of the world's population, could lack food security due to climate change. The industrial revolution and globalization, the same processes that redeemed people from hunger, might usher their grandchildren back into poverty.

Imagine that an evil politician from, let's say, Canada were to propose spreading a chemical poison that hurts only poor, nonwhite people who live in other countries. The poison would make them poorer or kill them but would also boost the Canadian economy and give it a warmer and sunnier climate, improving the quality of life there. Wouldn't that be despicable? But that is what the wealthy industrialized countries of the global North are doing, knowingly.

The figures bring home how the climate crisis is threatening to reverse the reduction of extreme poverty that has been one of the salutary products of globalization, and to revert the world to its former divided state, or to an even more divided one than ever before. As the Scandinavians take advantage of their balmier climate to plant vineyards, and Greenland, under Danish sovereignty, is looking forward to easier access to mineral deposits, and Britain is saving a lot of money because its inhabitants don't need to spend as much on heating and on trips to warmer climes, the global South is trapped.

Bangladesh is an example. With a population of 168 million, that country has made considerable progress from when eight out of every ten citizens lived in poverty. Life expectancy zoomed up and is now about seventy years. Forty-four percent of its population once lived in extreme poverty; today the number is only 13 percent. Mass famine is almost entirely a thing of the past, and literacy has risen significantly.[7] Much of this is due to technology, industry, and international exports and trade.

But Bangladesh is located by the Bay of Bengal, where the Ganges meets the sea. It's the world's largest river delta. Two-thirds of the country's territory is no more than 4.5 meters (about 15 feet) above sea level, and its population is concentrated on the fertile agricultural land along its rivers. The shallow waters of the Bay of Bengal have warmed significantly. Water expands in volume as it warms, so the warming raises the level of the sea, which has put low-lying levels of the country under water. In recent decades Bangladesh is experiencing more severe tropical storms that drive huge quantities of salt water inland, ruining agricultural land. The Himalayan glaciers are melting, engorging the rivers that flow from the mountains through the delta to the bay. On average, a quarter of the country's land area is flooded at one time or another, and every few years 60 percent of it is flooded.[8] Land has thus become more scarce and expensive for its dense population; there simply is not enough land to live on.

The rural population is migrating from villages in the south and east into the country's capital, Dhaka. Half a million of them migrate into the city each year, settling in poverty-stricken neighborhoods in which 7 million people already live.[9] The World Bank suggests that, by 2050, as many as 13.3 million Bangladeshis could lose their homes due to climate change, and by the end of the century an area home to a third of Bangladesh's population may be permanently submerged below the high tide line.[10] Globalization saved half a billion people from hunger and destitution, but the World Bank calculates that climate change will send 122 million people back into poverty by 2030.[11] A few decades ago, global prosperity began to lift Bangladesh, but immediately, before its inhabitants could enjoy it, they found themselves pounded by monster storms, rising seas, and destruction of farmland.

The Maldive Islands, 2018

The Maldives are on the verge of paying the ultimate price of climate change—the rising sea is inexorably threatening the islands' existence. The tourists who flock to this island nation in the Indian Ocean do not generally bother to visit its crowded capital, Malé. Their planes land there, but visitors head straight for the boats that shuttle them to the exclusive water villas on the islands' heavenly beaches. The islands, their beaches, and their coral reefs are some of the world's priciest and most popular tourist destinations.

On assignment, my cameraman and I land at the busy airport and board a boat that takes us around the island. When you get onto the water, you experience a moment of disorientation, a sort of optical illusion. Malé lies so low that it looks as if its buildings, which are small in any case, rise directly out of the waves. The boat seems to be sailing higher than the land. It hits you just how vulnerable the archipelago is. The islands are an average of a mere 1.2 meters (4 feet) above sea level; the highest point in the country is just twice that.

The motorboat whizzes along, spraying seawater in our faces as we pass island after island. The Maldives are a series of atolls; we are on our way to Thulusdhoo, at the northern end of Malé's atoll. The incredible beauty of the place is evident to all five senses. The beach is golden, the coconut trees soar above us, and there is the fresh salt air. The water is crystal clear, the streets are of white sand, and children run along tiny wharves or play on improvised swings hung from the high trees along the beach. But it is a paradise lost. The country's Coca-Cola factory is located on Thulusdhoo, which is the only place in the world where the beverage is made from desalinated water. It is a symbol of local pride. Coca-Cola signs are everywhere. Globalization is an empire unbounded.

Thulusdhoo is not primarily a tourist island. It has a permanent population of more than 1,500, many of them fishermen who go out every day to spread their nets in the sea. Near the port stand an orderly series of raised wooden lattices, for drying fish. It's evening, and several women are cleaning the lattices, pulling out weeds that have sprouted up under them and sweeping the ground with small brooms. The structures will be ready and clean when their husbands return on their boats, and if they bring in more fish than they can sell, the rest can be preserved in salt.

Early in the morning, we go for a dive to see the coral. The sun rises over the quiet beach; there is a brisk breeze. The proprietor of the little diving store, Aze Ismaiel, puts us on a small boat. The waves are high. We are on our way to a reef a fair distance from the island, one that tourists don't frequent. The reefs are a fundamental part of the Maldives' culture and economy. They are the foundation of the fishing and tourist industries, and they protect the islands from storm swells.

When we reach the reef, Aze drops an anchor, dons a diving mask without a snorkel, and gets in the water. "There is a little current," he tells us. "I stay up on the boat, you jump in, I try to watch you." I realize only when my body makes contact with the water that that was a warning. I feel as if my fins are not functioning, no matter how much force my legs put out. It takes less than twenty seconds for the current to start carrying us swiftly over the reef; Aze keeps repositioning the boat around us.

Not that there's that much to see. Every so often we spot a gorgeous fish, but the corals have become skeletons of what they were. They were bleached a long time ago here; the reef is dying and looks more like a submarine desert. Only at its ends, where the rocky shelf plunges into the ocean, can we see a bit of the variegated and abundant life that flourished in the reef's better days.

When we return to shore after the dive, Aze tells us that about 80 percent of the corals around the island have already died. As a result of global warming, the temperature of the sea has risen. The water has also become more acidic from the absorption of increased atmospheric carbon dioxide. These are the two main causes of coral bleaching and death. These colonial organisms are especially sensitive to such changes, and deteriorate and die quickly. Coral reefs cover a tenth of a percent of the ocean floor but support about a quarter of the globe's marine creatures.[12] Half of Australia's Great Barrier Reef, the world's largest, is already dead; about half of the entire world's reefs have been destroyed since the 1980s.

I tell Aze that global warming might kill all reef-building corals over the next twenty years. When I ask him what that means for his country, he says: "First of all, if there are no coral reefs, there's no Maldives. Our life is very much dependent on these reefs. Everyone comes here to see the underwater world, which is very amazing. The second industry being fisheries. So, if there are no coral reefs then our tourism would end, our fisheries would end, so that is pretty much the end of us, I say."

As Aze and I chat on the gorgeous beach, we can see major construction underway on the next island over. Huge machines are dredging the sea floor to create land on which to build yet another luxury hotel. Our conversation and the construction project seem to exist in different worlds, but in fact they are closely tied. Tourism is thriving in the Maldives, at the same time that global tourism is responsible for about 8 percent of the greenhouse gases that are responsible for the warming that is killing the coral reefs that the tourists come to the Maldives to see.[13] The government of the Maldives has embarked on a project to build artificial islands and reinforce the existing ones. This undertaking includes the construction of 3-meter-high walls to hold back the sea. The next stage will be the organized evacuation of endangered communities and their resettlement on higher islands.

We came to the Maldives, my cameraman and myself, to film a short documentary about the rising seas. We found that together with a sense of national emergency, insecurity, and a fear of displacement, there was also something else. Ironically, Maldivians now feel more connected to the world, less forgotten, part of a larger drama. They hold forth at length on emissions, acidity, and the global economy. My local fixer, while showing me around the island and recommending ideal spots for beauty shots, picks up scraps of plastic on the beach and tells me that they kill sea turtles; he checks to make sure that we use biodegradable straws. I have never encountered such an environmentally aware population, which is hardly surprising given that it is a life-or-death issue for them. The citizens of island nations, backed by legions from the affluent industrialized world, are at the vanguard of the battle to save the planet. The climate crisis impels the inhabitants of the global North to advocate the cause of places like the Maldives. Deep down, we (for I, and many readers of this book, live in the comfortable countries of the Northern Hemisphere) know that the battle will soon engage us, and with a vengeance. Those who lack empathy would see the people of the global South as the canaries in the coal mine. For the rest, there is a sense of common fate, of brotherhood- and sisterhood-in-arms in the campaign to save the globe. It is very much felt in places that engage directly in battle with the ecological crises we are undergoing.

The crisis didn't start yesterday. In Malé, not far from the main mosque, I meet Mohamad Saud. He and his family left their island, Haa Dhaal Maavaidhoo, in the 1990s, because life there became untenable. Saud is one of the world's first climate refugees. Millions of others will join him as the climate crisis worsens. He tells me about terrifying storm surges that washed over the entire island, and the constant erosion of its beaches. "They would surge into the island," he recalled. "The whole island would be flooded. The waves . . . would keep washing into the wetlands in the middle, destroying the plants and wrecking the boats. All the foodstuff would be spoiled."

Mohamad recounted how they could only take boat trips between storms, and how the flooding kept getting worse. After he left, many of the island's other inhabitants also abandoned their homes. In the end, the remaining community met and decided to accept a government evacuation proposal. If the story were made into a film, the screenplay would no doubt reach its climax with a confrontation between those who were determined to stay no matter what and the sober-minded islanders who realized that there was no hope. But Saud said that there was little drama. Everyone agreed that the island had become uninhabitable. When you and your children are slowly drowning, you can't afford to be sentimental. "Since I grew up there, yes, I miss the island," he admits. "But we had no choice."

As he is speaking, I recall a Maldivian legend that I once read. Immense copper walls stand at the limits of the great sea, holding back great quantities of water. Each night, demons lick the walls with their rough tongues. By morning they have managed to lick so much away that the walls are about to cave in. The water they hold back is seconds away from being released to flood the islands and drown all their inhabitants. But the catastrophe is averted by recitation of the morning prayers by the islands' Muslims. At the moment the believers stand and bring their hands to their faces—the *qunut*, the Muslim gesture of supplication—the copper walls regain their former strength and stand strong until nightfall, when the demons will again lick them. The islands are always on the verge of destruction and submersion, the ancient legend warns. But just a moment before the sea claims them, the *shahada*, the Muslim declaration of faith, forestalls the apocalypse.[14] In light of my conversation with Saud, the legend seems like a prophecy.

On a trip to an island not far from Malé, I meet Mohamed Nasheed, a former president of the Maldives. Nasheed, the best-known international figure to emerge from the archipelago, is accompanied by a large entourage. His stride is so quick that I literally need to run to keep up with him. The country's first democratically elected leader, he is a public relations wizard, a master of international public opinion. A large part of credit for the attention lavished on the worsening plight of island nations since 2000 belongs to him. He held a cabinet meeting underwater, gave interviews to journalists including Christiane Amanpour, and signed laws seated at a desk half-submerged in the sea. One of his proposals was that his country be granted an entirely new territory, to be paid for by the industrialized nations. He traveled to international forums, hosted expert visits, produced contingency plans, and made emotional speeches.

But the sea continued to rise.

Nasheed lost his post due to Maldivian politics, not anything to do with the rising sea level, and went into exile. He came home in 2018 and still wields immense influence.

Gulhi, the island where we meet, has a vast and wonderful beach; there is a huge swing in the middle of the water. It seems as if we are on the set of a commercial for a luxury vacation. But just 200 meters away from where the former president and I stand, we can see the signs of the desperate war against the rising sea. Like strata uncovered in an archaeological dig, we can see the layers of the inhabitants' struggle to keep out the waves. There are the artifacts of all the methods that were tried and have failed. There are sacks of concrete petrified on top of boulders. Next to them rise heaps of construction waste that were supposed to avert erosion and keep the waves away. Crumbling concrete and iron barriers look like gaping maws with broken teeth. It's a revealing sight: on one side of the island, tourists with enough money to make the trip to the Maldives are having fun; on the other side, the island's own population is battling the global warming that the tourists have helped bring about.

Nasheed says that he now opposes evacuating the islands. He is focused, he says, on "new technologies" and "groundbreaking" methods to build sea barriers, such as artificial reefs designed to mitigate the severity of storm surges.

"We have a great amount of erosion [of our beaches]," Nasheed says. "We are having a dwindling fish catch because fish are not surfacing as they used to, so we have a food security issue. If water temperatures continue to go up . . . we will cease to exist. The coral reefs will die and our islands will collapse, our livelihood will collapse. Our culture and our people will become destitute. Goodness knows what we will have to do." When I ask him about people who deny climate change, his response is stern. "You can't negotiate with science," he maintains. "You can't do a deal with the facts." Donald Trump hovers in the background of our conversation. "We have a two-thousand-year-old written history," Nasheed notes. "It's difficult for us to go extinct just like that. I do not intend to die from climate change. We refuse to die. We want to live."

Pilgrimage to Fukushima

The conversation with Nasheed reminds me of the two elderly farmers I met near Namie, in the exclusion zone of Japan's Fukushima Prefecture. In 2014 I made my way with a film crew into the quarantined fallout zone closed off by the Japanese government after a nuclear power plant was hit by a powerful tsunami and suffered a severe accident and radiation leaks.

I walked Namie's empty streets, where abandoned laundry hung out to dry had been shredded by the winds but still hung from clotheslines. In the kitchens of the homes, teacups remained on tables, and weeds poked through doorways. At sunset I saw a small family of wild pigs in the center of town. They strode at their leisure, oblivious to the invisible particles released into the air when the power plant's nuclear fuel rods were exposed and the core suffered a meltdown.

At a nearby farm I found Chizuko and Yukio Yamamoto, who wore high work boots. They had permission to spend the daytime hours on their ancestral plot so that they could tend their herd of cows—whose milk and meat could not be used, of course. "Most of the cattle in this area starved to death," they told me. "But for us, the cows are part of the family. We cannot eat and drink sake and enjoy life and simply forget the cows. We want them also to enjoy life until they die." Welcoming me into the home they lived in until the disaster, they showed me heirlooms from the samurai era. He would not abandon his property, Yukio said. Perhaps his grandchildren would be allowed to return, and then they would know that what remained for them was thanks to their grandfather's efforts. The Yamamotos' lush green farm was as beautiful as the beaches of the Maldives. I almost forgot that we were in a radioactive zone, where it's not ideal to inhale the dust kicked up by a tractor.

I placed a Geiger counter on a mossy rock under a plum tree in blossom. The reading was 20 microsieverts per hour, about eighty times greater than normal background radiation. Living permanently in Namie could increase a person's odds of getting cancer, and would be dangerous for children.

At the barrier where we exited the forbidden zone, we were directed to special sheds to be checked for radiation by government agents dressed in protective suits and masks. "You can't leave until you do this," our local guide told us.

MONUMENTS ARE DEDICATED TO THE PAST, BUT FUKUSHIMA and the site of the Chernobyl nuclear disaster are monuments to a possible future. The issue is not nuclear energy—such accidents can happen, but nuclear reactors are actually safer than coal-fired electric plants, which have killed millions by polluting the air. What these two sites symbolize is the capacity of technology to destroy communities, to contaminate the natural environment and force out human beings for generations, or forever. Even in places that are not war zones, human beings can exterminate themselves and their habitats. Visiting Fukushima is a powerful experience; it's an example of how we can drag ourselves down into darkness and collapse.

In the age of responsibility, superpower political leaders and the international community have internalized the memory of the folly of the twentieth century's two world wars. They have acknowledged the need for fundamentally rational thinking, for accepting scientific evidence, and for an international commitment to stability. That includes the prospect of environmental catastrophe. In 1974, scientists realized that chlorofluorocarbons (CFCs) being released into the atmosphere were eating away at the ozone layer that protects life from harmful ultraviolet solar radiation. In 1985, British researchers discovered a huge hole in the ozone over Antarctica, an area of the stratosphere where there is a sizable seasonal reduction in the ozone concentration.[15] Just two years later, over industry objections, the Montreal Protocol was signed, an international treaty that required phasing out the production of substances harmful to the ozone layer. The treaty was ratified by 197 countries. In 2018, NASA announced that, for the first time, its scientists had measured a decline in CFC levels in the stratosphere, and a shrinking of the hole in the ozone layer. In October 2019, the hole was the smallest since it was first discovered.

The international agreement on banning CFCs in 1987 was a simple and necessary story of logic and science, and of the age of responsibility. Fast-forward to May 2016. Candidate Donald Trump spoke to coal miners in West Virginia:

> My hair look ok? . . . Give me a little spray. You know you are not allowed to use hairspray anymore because it affects the ozone, you know that, right? . . . You mean to tell me, because you know hairspray is not like it used to be . . . Give me a mirror. In the old days you put the hairspray on, it was good. Today you put the hairspray on, it is good for twelve minutes, right? You know they say you cannot—I said wait a minute, so if I take hairspray and if I spray it in my apartment which is all sealed, you're telling me that affects the ozone layer? I say, no way, folks. No way.[16]

There's a corporate interest in rejecting the scientific evidence on both the impact of CFCs on the ozone layer and on the human role in global warming. Corporate interests create incentives for casting doubt on the science, as documented at length in *The Merchants of Doubt*, Naomi Oreskes and Erik M. Conway's book on the industrial and oil lobbies. Having learned the lessons of the Montreal treaty, the lobbies do everything in their power to fight similar agreements on fossil fuels. The people who pay the price live in places like Newtok, Alaska, whose inhabitants will soon need to relocate because the permafrost that holds the village up is melting. It's also being paid in Egypt, where the population increases by a million every eight months. The Nile Delta, the country's most fertile farmland, has fed its population for thousands of years. It is in danger of being contaminated by salt water from the rising Mediterranean. Virginia's Tangier Island, in the Chesapeake Bay, is slowly disappearing as the ocean rises. People living in these places represent a larger picture, of a world that is falling apart from the edges inward.

For many, the world has become an increasingly erratic and dangerous place. Greenhouse gas emissions, the demand for red meat, and speculation in commodities all impact individual lives directly and detrimentally, but the modern state's institutions and elected leaders have proved unable to rein them in. The result is a sense of helplessness and mistrust of government, not only on the periphery but also in metropolises, from Beijing to New York. Forces and processes that profoundly affect people's lives are global and interrelated, and may feel completely capricious. It is this sentiment—having no control over much of your life while constantly being told you do—that destroys national and local officials' facade of stability. Climate change has made it worse. People who are no longer alive built industries that released substances that will contribute to deaths in foreign lands of people not yet born. In the face of such complexity, the squabbles, deals, and compromises of everyday politics look irrelevant and trivial, incapable of addressing contemporary challenges. In *The Hitchhiker's Guide to the Galaxy*, Douglas Adams depicts the moment the earth is destroyed:

> "People of Earth, your attention please," a voice said . . .
> "This is Prostetnic Vogon Jeltz of the Galactic Hyperspace Planning Council," the voice continued. "As you will no doubt be aware, the plans for development of the outlying regions of the Galaxy require the building of a hyperspatial express route through your star system, and regrettably your planet is one of those scheduled for demolition. The process will take slightly less than two of your Earth minutes. Thank you."[17]

If national institutions such as governments and parliaments cannot address severe grievances because the communities they represent are subject to global forces beyond their control, whether these forces are financial, cultural, or the effects of climate change, they are seen as either hollow or as collaborators with those forces. Your prime minister can exert only minimal influence on rising food prices, international monetary trends, climate-generated disruptions, or the spread of false information. Yet he usually continues to play the role of the omnipotent leader. But the reality of today's leadership is not potency but paralysis, not control but defeat by globalization. It is amid this failure that people are revolting.

Moral Sensibility

In 2017, Sky Television broadcast an investigation of the mines in the Democratic Republic of the Congo that produce the cobalt used in smartphones. Alex Crawford, a Sky correspondent, revealed that children are employed—virtually enslaved, actually—in the mines from the age of four. They are paid the equivalent of fifteen American cents a day. The footage broadcast on the British network includes the story of an eight-year-old boy identified only by his first name, Dorsen, working in the pouring rain in a muck-filled trench. Standing over him is a bearded man whose hand is raised to strike the boy. Dorsen told Crawford that for the last two days he had not had enough money to buy food, even though he works twelve hours a day.[18]

The cobalt he mined will eventually be transferred through a complex supply chain to the manufacturers, who will use it to make efficient batteries for smartphones. Cobalt is just one example; there is also coltan (columbite-tantalite), a metallic ore also used in smartphones and mined also in Congo. The phones manufactured with these materials are sold worldwide.

Globalization has expanded the geographic, economic, and moral distance between people who buy products and the people responsible for their creation. Manufacturing and marketing have become so decentralized and multi-faceted that it has created a comfortable fog shielding us from seeing the people who toil to make our stuff. Stories such as Dorsen's remind us that our entire physical space is tainted with exploitation and damage to the ecosystem. The unprecedented access to information means that not to know about these wrongs is a moral choice, not merely an innocent ignorance.

Pope Francis once described the world's attitude to the civil war in Syria as the "globalization of indifference." Apathy toward the other, the inclination to turn a blind eye to injustices inflicted on people who are different or foreign, is not new to human experience. But in the past people could have pled ignorance. They cannot do so anymore. In the contemporary world, *the information available to people about the wrongs that they are responsible for is growing as their sense that they can do something about it is diminishing.* The state of humanity is such that we constantly hear a chorus, as if from a Greek tragedy, chanting "exploitation, pollution, shame, shame, shame."

It is a grueling way to live, producing a moral overload. Some will find solace by dissociating themselves from society. Others employ cynical denial, like Trump in his antiregulation tirade before the coal miners ("you're telling me that affects the ozone layer? I say, no way, folks"). Deliberate ignorance is their banner.

Naomi Klein maintains that corporate elites are systematically disseminating the lie that the globalized world in its current form is sustainable and innocuous.[19] But something more profound has shifted, and it relates to moral sensibility itself. When the truth is brutal, people reject it, sincerely, for that precise reason. They may turn instead to imagined imperatives of control, identity, and a glorious mythical past. That can be a demagogue being nostalgic about a better hairspray, or it can be modernism's most ancient rival, radical fundamentalism.

CHAPTER 6

The Rebellion's Harbingers

Fires are burning everywhere. People are dying all over the place. With God's blessing, you've done a brilliant job![1]

—THE HANDLER OF THE TERRORISTS IN MUMBAI, BY PHONE, WHILE THE ATTACK WAS IN PROGRESS
NOVEMBER 26, 2008

As the wave of globalization intensified during the age of responsibility and after the demise of Communism, and interrelationships between far-flung places grew stronger and had a greater impact on people's lives, opposition increased. It had many guises—there were anarchists, radical environmentalists, Marxists, and populists, to name but a few. Among the resisters, the oldest and toughest strain is that of the fundamentalists. Since the Enlightenment, they have braved the storm of change as if standing on a precipice, bracing themselves against falling into an abyss. Contrary to conventional wisdom, fundamentalism is not a natural by-product of ignorance and poverty but rather a radical argument addressing the loss of meaning and sense of alienation brought on by global integration.

Mumbai, 2008

The world economy crashed in the autumn of 2008. It imploded loudly, slowly, and terrifyingly, like a venerable old building undergoing a carefully orchestrated demolition. My wife and I had front-row seats to the scariest show in town, staged in the City of London, usually grand and smug but now suddenly battered and almost hysterical.

In the midst of that crisis, about two months after Lehman Brothers declared bankruptcy, while still in London, I received an urgent call from the editorial offices of the Israeli daily *Ma'ariv*, my employer. Covering the ongoing economic collapse, I barely had time to breathe, but the call had nothing to do with the international economy. "How quickly can you get to India?" the editor asked me. "To Mumbai. Something awful is happening there."

In mid-November, the *Kuber*, a battered fishing trawler, sailed from the port at Porbandar, on India's west coast. Its captain and the four Indian fishermen on board were on their way to a routine destination—Sir Creek, a contested frontier between India and Pakistan. Eels and other fish abound at the creek's mouth, in the Arabian Sea, so its waters are plied intensively by fishermen from both countries, not to mention smugglers. The captain, Solanki Amar Singh, spent most days and nights on the *Kuber*. Singh's employer, the owner of the small boat, paid him about $200 a month. "He was an optimist and worked hard for his children's education," his brother-in-law later told the *Wall Street Journal*.[2]

The journey went without mishap. Singh made sure to speak each day with the owner, who wanted to know that his trawler was in good shape. Then, in the third week of November, Singh disappeared. The next day, several fishermen reported that they had seen three dead bodies floating in the shallows near the Pakistani border. On the evening of November 26, the *Kuber* was abandoned not far from Mumbai's port. Singh's body was found on deck, his throat slit. An investigation would later discover that the trawler had been intercepted by a larger boat coming from Karachi.[3] Most of the fishermen on the trawler were murdered immediately.

That same evening, a small rubber raft made its way to the beach at Mumbai. It ferried ten men: six got off at a place called Budhawar Park, while four headed off to Cuffe Parada, a fashionable business district that stands at the gate of the Indian navy's Naval Command. They were armed to the teeth, and decked out in colorful shirts and cargo pants. They attracted attention immediately; several curious fishermen called out to the new arrivals. The young men ignored the onlookers and strode from the park into the city, bearing large black duffel bags. The hazy light of an Indian sunset began to blur their figures. So began the 2008 terror attack in Mumbai.

The terrorists' first target was the luxurious Taj Mahal Palace hotel. They stormed into the lobby and fired on tourists taking tea by the pool just outside. Other detachments struck at another hotel, a cinema, the Leopold Café, and the city's central train station. They commandeered cabs to take them from one target to another, sometimes murdering the drivers while they were at it. Two attackers burst into Nariman House. It served as the city's Jewish Chabad center—a synagogue and activity center run by the Hasidic movement of that name, where Jewish travelers, especially young backpackers, can celebrate the Sabbath and get a kosher meal.

It was one of the most sophisticated paramilitary operations in the Indian subcontinent's bloodstained history. The terrorists killed 166 people and injured hundreds more; the event is referred to as India's September 11. The operation was not the most lethal that India, or even Mumbai, had ever seen: just two years before, more than 200 people had been killed in an attack on the Mumbai Suburban Railway. But it was a precedent-setter in that its primary victims were not the poor, and it did not take place close to the border. This time the terrorists targeted symbols of power, tourism, and finance in what at first looked strictly like a hostage-taking operation. When the police established contact with the gunmen, some of the attackers said that they wanted to negotiate. But it was a ruse. Their real aim was mass murder; they had been instructed not to return home alive. Another precedent was that the attack was aimed at foreigners no less than at Indians. Hundreds of citizens of other countries found themselves under fire. The international news media covered the events with an intensity almost unheard-of for acts of terror in Asia.

Israelis at Nariman House were taken hostage. My editor asked me to get as close as possible to where they were being held.

It's a strange feeling, trying to get your bearings in a large city when it suddenly falls silent. A relatively brief cab ride from the airport got me into a city devoid of its characteristic roiling traffic. Mumbai, known for its incessant hum of giddy activity, now seemed almost still. I could feel the absence of people who had vanished from the streets. The smell of the congestion hung in the air, bereft of the people who had emitted it. The attack was still in progress; everyone I talked to treated me with suspicion. The hotel clerk who gave me my key advised me to pay careful attention to where the emergency exits were, "in case you need them." It was a symbolic and meaningless gesture, made to assuage the feeling of utter insecurity that had overcome Mumbai, where wild rumors had it that more terrorists were roaming the city, seeking targets. I had previously covered terror attacks, military operations, and a war; I had seen dead bodies, prisoners, and murderers apprehended. But Mumbai was different, and to some extent scarier. There was the feeling of a long-running battle—the attack lasted for days on end. The murderers did not strike and get killed, nor did they kill themselves immediately.

Instead, they continued to fight and to murder—destroying every aspect of normality. When I reached the Taj hotel, I saw it festooned with improvised ladders made from bedding that some of the guests had used to escape. The sheets flapped quietly in the breeze, like banners of death.

Then there was Nariman House. The terrorists had stormed it thirty-six hours before my arrival, at about 9:45 p.m. According to some reports, they immediately shot and murdered Rabbi Gavriel Holtzberg and his wife, Rivka. She was six months pregnant. Before she died, Rivka cried out for help to Sandra Samuel, the couple's Indian nanny, who was hiding in one of the rooms. According to the local press, when Samuel reached the living quarters, she found the bodies of the couple. Alongside them was Moshe, their two-year-old son, whole and unhurt. She took the toddler into her arms and ran toward the exit without looking back, escaping unharmed. In addition to the Holtzbergs, the assailants killed four other people in the house.

Unlike almost every other large terror attack in history, the perception of this incident has been shaped not only by the testimony of the survivors and the investigation conducted by the Indian police. During the course of the attack, the assailants spoke with their handlers in Pakistan via satellite phone, and Fareed Zakaria managed to get hold of the recordings and played some of them on his CNN program.[4] Their communications with colleagues and subordinates evince violent and resolute fundamentalist thinking in action in real time. One of the Pakistani operators, called "Brother Wasi" in the phone calls, provides the terrorists with both religious and tactical guidance throughout the attacks. His instructions are brutal, impatient, and manipulative. The terrorists, young men who grew up in rural Pakistan, are dazzled by the splendor of India's financial capital. "There are computers here with thirty-inch screens!" exclaims one of the terrorists at the Taj. "Computers? Haven't you set fire to them?" Wasi rebukes him. "We're just about to. You'll be able to see the fire any minute," the young militant promises, still stupefied by the opulence of the good life he had been sent to destroy. "It's amazing!" he marvels. "The windows are huge. It's got two kitchens, a bath, and a little shop." Wasi sounds at wit's end. "Start the fire, my brother. Start a proper fire, that's the important thing." He explains to the terrorists that the image is the message here, the sight of the famous Taj Mahal Palace in flames. "My brother," he entreats the terrorist, "yours is the most important target. The media are covering your target, the Taj hotel, more than any other."

The recordings show that a few hours into the attack the terrorists begin losing their drive. They're past the adrenaline rush of the initial slaughter and are clearly starting to get cold feet. Their hostages are starting to look like human beings. At one point, Wasi calls one of the men at Nariman House and tells him to kill all the hostages. "Just shoot them now. Get rid of them. You could come under fire at any time and you risk leaving them behind." The assailant tells him that all is quiet. Wasi will have none of it. "Don't wait any longer. You never know when you might come under attack. Just make sure you don't get hit by a ricochet when you do it." The terrorist replies "*Inshallah*," meaning "God willing." Wasi insists on staying on the line. "Go on. I'm listening. Do it!" he commands. The terrorist plays for time: "What, shoot them?" Wasi issues an order: "Yes, do it. Sit them up and shoot them in the back of the head."

"The thing is," the reluctant killer tells his superior, "Umer is asleep right now. He hasn't been feeling too well." Wasi does not let up and keeps calling over and over again. Finally, gunshots are heard.

Throughout the attack, the operators in Pakistan are focused on how the attack looks, and urge their men in the field to think of the images it produces for the international media. The issues at hand are not only tactical but about image management. Wasi tells one of the assailants that "when people see the flames, they will start to be afraid. And throw some grenades, my brother. There's no harm in throwing a few grenades."

Wasi tries to make use of the Jewish hostages at Nariman House. He even speaks directly to one of them, telling her that if his demands are met, she will be able to "celebrate" the Sabbath with her family. At the same time, he was telling his underlings there that through the killing of Jews, they would he held in highest esteem. "As I told you," he urges them on, "where you are is worth fifty of the ones killed elsewhere." We know Wasi only from his voice in the recordings. He has never been identified.

Bound and Murdered

As a Hebrew-speaking newspaper reporter from Tel Aviv, my central story was the fate of the Israeli hostages at Nariman House. On Friday morning, a day and a half after the terrorists had seized it, Indian commando units circled the building. The operation lasted for many hours. At three minutes to eight in the evening, a final battle with the last of the terrorists commenced. At this hour, on the night of the Sabbath, I was walking along a crowded Mumbai street with a *Ha'aretz* correspondent, Anshel Pfeffer, trying to get as close as we could to the house. It was not the usual frenetic Indian throng but rather a strange sort of vigil by people who knew that the saga's horrifying denouement was approaching. The crowds were like waves heading slowly to shore, breaking every so often into warm applause or slogans like "Long live Mother India!" whenever the crowd saw soldiers approaching or coming out of the building.

We wanted to get there quickly; we didn't know at this point that all the hostages were already dead. But the street was entirely blocked by the crowds, the air thick with anger tightly held in. We could not make any progress. Finally, in desperation, we took out our press cards and started waving them as we called out "Israel, Israel!" South Mumbai's inhabitants gazed at us more or less silently; every so often, one of them took up our cry and shouted "Israel!" with us. A corridor opened up, and a growing group of self-appointed ushers went before us, clearing the way, enabling us to reach the somber Israeli delegation that stood outside the Chabad House. It included the Israeli embassy's security officer, Israel's military attaché in India, and a representative from the office of Israel's prime minister.

It was a grim night. We stood there wordlessly, waiting to enter the ruined house. Smoke and the scent of gunpowder rose from it. We could tell from the faces around us that everyone inside was dead.

A bit after 8:30 p.m. we saw some movement. Having ascertained that all the terrorists had been killed, the commandos handed the building over to the police. "Come, come," an Indian officer said to the Israeli military attaché. The Israeli delegation paced between long lines of Indian warriors wearing black combat vests; having completed the operation, they were eating their dinner. A big pot of dal and rice was placed in front of the building. Some policemen took the Israelis to an ambulance to show them one of the bodies that had been taken out of the house. Then we made our way inside. This is what I wrote then about what we saw:

All the windows shattered as a result of the tremendous explosions heard when the Indian commandoes broke into the house. Several Indian soldiers walk in front of us, one of them with a flashlight. The floor is covered with slivers of windowpanes. The bottom floor is destroyed. The forces blew up the walls and the concrete is exposed. Electrical wiring pokes out from every corner. Grenades were visible in the midst of the chaos; sappers would come in the next day to detonate them. It's best to walk carefully. The main branch of a tropical houseplant of some sort, in a lone planter, remained green, but all its leaves had been severed by the gunfire.

The house is almost totally dark. The delegation climbs a shaky staircase in silence. "Where are the Indians?" one of the Israelis shouts. "They have disappeared on us, we need more flashlights!" The delegation diligently surveys the rooms, but even in the dark you can see that everyone's faces are ashen. It is a moment for which there are no words, only feet dragging through the dust that covers what were once rooms. Everywhere the eye looks, you can see the signs of struggle. There is a strong, acrid scent in the air. The walls are pocked with machine gun fire and the building's skeleton has been exposed; there are bent iron rods. All the small pieces of the life that was lived in the house are flung on the ground in violent disorder. Suitcases. Sheets. Torah scrolls, smeared with blood. A silver plate engraved with a Jewish blessing for a home. On floor after floor the Israelis find bodies. On the next-to-the-top floor we see two terrorists sprawled, their bodies mutilated

> in battle. One of them had been hit by a missile. The soldiers hung a red flag from the window, apparently as a sign of some sort. Two floors below we make out the bodies of three other victims. We see that some of the dead are bound.[5]

As Saturday night approached, I walked to an apartment where Rivka Holtzberg's parents—the grandparents of Moshe, the toddler saved by his Indian nanny—were brought after flying in from Israel. The two-year-old wandered among the people there, shy and quiet. His grandfather took him up in his arms, the boy's golden curls caressing his face. It is an Orthodox Jewish tradition not to give bad news on the Sabbath. The tradition was honored; no formal notification of their daughter and son-in-law's death was offered to Rivka's parents. When the Sabbath ended on Saturday night, they were notified officially. Afterward, everyone in the apartment participated in the evening prayers. Myself included.

The Birth of the Individual Jihad

The horror can be left at that. But to understand the nature of violent fundamentalism and its elemental causes, it's necessary to get down to the small details of the rampage and what motivated it. The Mumbai terrorists were sent by Lashkar-e-Taiba, the Army of the Righteous, which is the military arm of a Pakistani political movement, Jamaat-ud-Dawah (JuD), the Organization for the Calling to Islam.*

---

* Jamaat-ud-Dawah's beliefs derive from Ahl-i-Hadith, a type of Sunni Islamic fundamentalism that emerged in India in the nineteenth century. It rejects all later interpretations of Islam in favor of what it calls the "true source" of the faith.

This Islamist organization funds hospitals, ambulance service, schools, and religious schools, or madrassas, throughout Pakistan. It operates on a model originated by the Muslim Brotherhood that has also been adopted by, for example, Hamas in the Palestinian territories and Hezbollah in Lebanon. Agence France-Press reported in 2015 that the organization's medical clinics offer dental care at the subsidized price of half an American dollar. A senior JuD figure claimed that "laser operations to correct shortsightedness are free."[6] The movement also runs a popular volunteer organization that provides assistance to the victims of natural disasters. Its people often are the first to reach remote places in Pakistan where there has been a mudslide or earthquake.

One of Jamaat-ud-Dawah's founders is Abdullah Yusuf Azzam, also known as the father of global jihad. Azzam was a founder of al-Qaeda and the man who recruited Osama bin Laden. Too little has been said in the West about Azzam, given that there is probably no other radical fundamentalist who has had a greater impact on the international community.

Azzam was a Palestinian cleric who left his village, Silat al-Harithiyah, near Jenin in the northern West Bank, after Israel occupied the area in the war of 1967. He enlisted in militant Palestinian organizations but soon realized that, as a religious man, he did not fit in with the socialist nationalism that characterized the pan-Arab discourse of the 1960s. Azzam completed a PhD at al-Azhar University in Egypt, where he adopted ideas of the Muslim Brotherhood, and in particular of extremist writings that interpreted the Islamic precept of jihad—holy war—as violent struggle against tyrants and their institutions, even if they are themselves Muslim.[7] For a man like Azzam, such rulers and institutions had led to the "disgrace" of Muslims and to their "subjugation by slaves." "Will history repeat itself over us [*sic*]," he cries, "while we swallow degradation, fall into oblivion as those before us did, and lose just as they lost?"[8]

Note the scorn for the recent past, for tradition, for the older generation, perhaps including his own parents in the village of his birth, now occupied by the Israelis. The conservative underpinnings of society must be destroyed so that the fundamentalist can present himself as the only guardian of tradition. Moderate traditionalism refutes fundamentalism's central theme—the claim that it, and no other, represents the true past, as it actually was.

Azzam was the first Islamic authority to rule explicitly that the obligation to engage in jihad to liberate the entire Islamic realm (from Indonesia to Spain, in his view) is an entirely individual one. Perhaps no text had more of an impact on global personal security in the twenty-first century than this passage from his book *Join the Caravan*:

> There is agreement . . . that when the enemy enters an Islamic land or a land that was once part of the Islamic lands, it is obligatory on the inhabitants of that place to go forth to face the enemy. But if they sit back, or are incapable, lazy, or insufficient in number, the individual obligation (*fard ayn)* spreads to those around them. Then if they also fall short or sit back, it goes to those around them; and so on and so on, until the individually obligatory nature of jihad encompasses the whole world . . . to the extent that the son may go out without the permission of his father, the debtor without the permission of the one he owes, the woman without the permission of her

> husband, and the slave without the permission of his master. The individually obligatory nature of jihad remains in effect until the lands are purified from the pollution of the disbelievers.[9]

This text is a cornerstone for groups like ISIS. Notably, radical fundamentalism seeks the destabilization and destruction of the most basic social institutions. It directs wives away from their marital vows; children are released from their duty of obedience to their parents. Conservatives put the family first, but fundamentalists are revolutionaries, not traditionalists. They claim to be saving their societies from alien influences, when they in fact themselves transform it. Azzam's aim is even more ambitious. He takes the duty of jihad against invading infidels and turns it from a local story into a global one, a duty that expands outward like the ripples in a pond. This is a practical reform: The duty to fight the infidels is not regional anymore and does not recognize state or societal boundaries. With this invention of tradition, Azzam proposes an adaptation to a world in which borders and nation-states have become less relevant. He adopts a global fundamentalist approach to a global world.

Azzam's main activity in the 1980s was as part of the Mujahideen in Afghanistan, who fought against the Soviet invasion. They were supported by the United States as part of its Cold War with the USSR. A television report from 1979 shows national security adviser Zbigniew Brzezinski, wearing dark sunglasses and using a translator, speaking to a group of devout warriors training in Pakistan. He urges them into battle against the godless Soviets: "You'll have your homes, your mosques back again. Because your cause is right and God is on your side!"[10] These fields of extremism, fertilized by the Americans, produced a crop that included one of Azzam's allies, Osama bin Laden.

Tactical alliances between liberalism and fundamentalism are doomed. The fundamentalists' goal is not a return to their homes; it is not communitarian or conservative. Azzam seeks not just the liberation of land but rather an all-out war against Enlightenment values. "Jihad and the rifle alone," he declares, "no negotiations, no conferences, and no dialogues."[11] Of course, the vast majority of Muslim religious leaders and Muslim communities reject and condemn these extreme ideas, which are seen as deviating from the teachings of the Prophet. Violent fundamentalists are small cohorts who have not gained effective influence with most Muslims or countries with Muslim majority.

In the pre-internet age, Shaykh Azzam's discourses were distributed on cassette tapes that were played over and over again to jihadists everywhere.[12] It was a harbinger of how radical fundamentalism came to be preached over the media.

JAMAAT-UD-DAWAH IS THE POLITICAL FRONT GROUP FOR Lashkar-e-Taiba. The organization's original goal was to unite Kashmir, a Muslim-majority region in northern India, with Pakistan. Pakistan views the liberation of Kashmir as a supreme national goal. As such, its government's view of Lashkar-e-Taiba is ambiguous at best, but usually positive. It is seen as an arm of the country's popular militia that seeks to promote the idea that Kashmir should be under Pakistani sovereignty, whatever the cost. To this end, the employment of non-state (and sometimes state) violence is seen as legitimate.

For the Islamists, the liberation of Kashmir is a limited tactical goal toward the great mission of establishing a unified Muslim state with Sharia law as its constitution, and with all foreign influence purged from the life of the *umma*, the Muslim community collectively. The Pakistani security forces, for their part, view these religious fanatics as part of their toolbox for achieving their country's national goals in Kashmir.

The Pakistani state itself grapples with the challenge posed by fundamentalism. This was seen perhaps most famously in the attempted assassination of Malala Yousafzai, the teenager who would later receive the Nobel Peace Prize for her campaign against the Pakistani Taliban's prohibition on girls attending school.

The ongoing jihadist struggle is often seen as ideological, but terror is a way of life, a career, a business. After a success—the expulsion of the Soviets from Afghanistan, for example—the terrorists do not repose in the shade of their fig tree. On the contrary, victory feeds the appetite. The Pakistani state is dysfunctional in the socioeconomic realm, leaving a space for "charitable organizations" like JuD to fill. The organization's welfare and military activities are funded by wealthy patrons in the Muslim world, in many cases Saudi Arabia.[13]

Another piece of the mosaic is fundraising through crime. One person who has been cited as a financial supporter of Lashkar-e-Taiba is Dawood Ibrahim, an Indian national born in Mumbai. He leads one of the most powerful crime syndicates in Asia, known as the D Company. Ibrahim is a mafia don, and his cruel reputation precedes him across the subcontinent. He fled to Pakistan when he was suspected of being involved in a series of bombings in Mumbai in 1993 that killed more than 250 people. The syndicate is "an example of the criminal-terrorism 'fusion' model," according to a special report of the US Congressional Research Service, and is one of the largest such operations in the world, with more than 5,000 members.[14] Among other things, Ibrahim shares his smuggling routes, presumably in exchange for a generous payment, with regional and international terror groups. And the US has claimed for many years that he has helped fund Lashkar-e-Taiba. He does so for both business and ideological reasons, but his ideology is not at all ascetic and pietistic.[15] He is India's most wanted man. In 2015, an Indian television station obtained a telephone number said to be that of his home in Pakistan. The woman who answered the phone confirmed that she was Ibrahim's wife; she said he was taking a nap.[16]

The interface between terror and crime is not incidental but a recurring pattern. Another example is al-Murabitoun in Algeria, a jihadist organization that merged with al-Qaeda in the Islamic Maghreb (AQIM) in 2015, creating one of the most dreaded terror groups in Africa.[17] Al-Murabitoun's leader is Mokhtar Belmokhtar, known as the "One-Eyed" following the loss of his left eye in an accident involving explosives. He's also called Mr. Marlboro, because he ran the area's largest cigarette-smuggling ring.[18] As with Hezbollah in Lebanon, which engages in large-scale drug trafficking, the lines are blurred but the contours are clear. The fundamentalist has no compunction about using crime, or wrecking families, because it is the enemy of social conservatism, not its ally.

Video Games and Terror

Salah Abdeslam is the only terrorist to be apprehended after the Bataclan attacks in Paris, which took the lives of some 130 people in 2015. A French citizen, he is the son of Moroccan immigrants who lived in Brussels before moving to France. Until the attacks, he had a police record that included robbery, petty theft, and possession of marijuana. After his arrest for the Bataclan attacks, the French media reported that he had never read the Quran. When his lawyer asked, Abdeslam said that he had "read the interpretation on the internet" without perusing the book itself. His lawyer called him "a little jerk . . . a perfect example of the GTA generation"—he was referring to the online video game *Grand Theft Auto*—"who thinks he lives in a video game."[19]

A year earlier, ISIS had issued a clip in the style of GTA, in which American soldiers were slaughtered, or side charges detonated under trucks bearing soldiers, in sophisticated action sequences. The purpose was to "raise the morale of the Mujahideen and to train children and youth how to battle the West and to strike terror into the hearts of those who oppose the Islamic State," its creators declared. The clip displayed the caption "We do the things you do in games, in real life on the battlefield."[20] At the end of 2015, ISIS issued a horrifying clip showing six children, armed with pistols, entering a ruin that looks like a castle. It is filmed from all angles, suggesting a Big Brother reality show. Each one finds his prisoner, stands him up, and then, after a brief theatrical pause, shoots him dead. The camera angles, the fast action, and the effects all show that the creators are immersed in the gaming culture. Indeed, searching for a prisoner in a castle is a theme in a well-known video game from the early 2000s.

There are people who will deduce from this that there is a causal relationship—that Hollywood violence, gaming, and pop culture produce violence. But that's overly simplistic—studies have failed to prove the existence of a causal relationship between watching violent films or playing violent video games and engaging in violent acts.[21] It would be better to focus on the moral distance such media creates. Video games do not cause violence, but some do reflect how the Other is becoming but a target, a point in the rifle sights. In these games, the world is an imagined, manufactured reality replete with hellish images. The game contains operators and those who are operated, victims and immune players. The player is immersed in the experience of an aesthetic slaughter, yet remains alienated from the images of harm he generates. "Bad taste," Stendhal once wrote, "leads to crime."

This is inspiration far away from the verses of the Quran. Take the young terrorists in Mumbai and their commander, Wasi. He ordered them to shoot people in the back of the head, and demanded to see flames rising from the hotel. Isn't that just like a big video game in which an avid Wasi is the handler with the joystick that controls his players? And is the behavior of the European Islamist terrorist really the product of profound thinking about a Sunni caliphate? Or is it molded by a video fantasy, somewhere on the margins of society?

And, while that fantasy is illegitimate, at the same time, a similar joystick, in the hands of a soldier in a Western army, is used to control a drone that kills suspected terrorists in Pakistan or Afghanistan, often incurring what is euphemistically referred to as "collateral damage." In both cases, that of the terrorist handler and that of the drone handler, dissociation is necessary for the completion of the mission.

Indeed, video games embody globalization. They are globally developed and distributed, and can be played anywhere in the world, online, with anyone. Yet they are just part of the story. While games make images of murder routines exciting, the international media serves as an echo chamber in which acts of real violence reverberate endlessly.

The postmodernist philosopher Jean Baudrillard foresaw the way in which terror attacks would turn into performances, abetted first and foremost by the media, which echoes the terror attack, and the state response to it. "The media make themselves into the vehicle of the moral condemnation of terrorism and of the exploitation of fear for political ends," he wrote, "but simultaneously, in the most complete ambiguity, they propagate the brutal charm of the terrorist act, they are themselves terrorists insofar as they themselves march to the tune of seduction."[22] The fundamentalist project cannot become global if the media does not disseminate the images of terror.

The dark side of globalization has unfurled before us. First, a struggle between empires that have used fundamentalists as proxies; second, a weak state that supplies the substrate on which radicalism and crime can grow; third, the free flow, across borders, of the ideas and capital that support local extremism. The fundamentalist makes extensive use of technology. He exploits the viral potential of ideas and images, and understands that the media creates reality—recall Wasi's gripe that he could not yet see the Taj burning. All these are part of globalization, not nostalgia for the time of the Prophet more than a millennium ago.

Today's fundamentalism is fundamentally modern. The paradox is obvious—radical Islamists use their religion's concept of the *umma* to create a Muslim universe, a global village of strict religiosity united by visible and invisible interconnections. Their call to expunge the boundaries between nations and to establish a single, religion-based community fits in very well with the universal discourse. They do not refrain from using anything they find in globalization's realm, including Hollywood. The blurring of boundaries between states and peoples; the idea that there is a single solution appropriate for all of humanity; their supranational ideology—all these are features of both the current world order and of radical Islam. Tomer Persico, a scholar of contemporary religion, writes that the unfathomable difference between the two visions lies in the fundamentalist rejection of the most basic element of liberal universalism—individualism. The liberal-universalist world dignifies individuals as autonomous, free, and equal. It is a world in which the Other is accepted. Fundamentalist globalization has a comfortable patriarchal hierarchy. It's efficient for those who employ violence. Everything is managed, quite simply, by "a man with a gun over there / Telling me I got to beware," to quote Buffalo Springfield.[23] Ironically, it is not that fundamentalists reject universalism. Rather, they see liberal universal ideas as being in competition with their own universal agenda. Fundamentalism and globalization are not matter and anti-matter. They are two sides of the same coin.

Fundamentalism: Globalization's First Enemy

Fundamentalists bear the bloodiest banner in the war against globalization. Fundamentalism's great advantage is its determination to create a world that accords with its apocalyptic mindset. Abdullah Yusuf Azzam stated that "if the preparation (*idad*)* is considered terrorism—we are terrorists. If defending our dignity is considered extreme—we are extremists. And if fighting the holy war against our enemies is fundamentalism—we are fundamentalists."[24]

---

* Under Azzam's Islamist doctrine, terror attacks are seen as part of the Muslim's duty to prepare for the final battle in which Islam will triumph.

Decades after these words were written, they were used in a video that a suicide bomber recorded before detonating near a Tel Aviv nightclub in 2001; twenty-one people were murdered.[25]

Azzam and his followers rely on a self-fulfilling prophecy—al-Qaeda commits terror attacks, which lead to Western hostility to Islam, and then the organization accuses the West of hating Muslims. All this is aimed at uniting the *umma* against its oppressors. It is a malicious strategy, and a major distortion of the fundamental principles of Islam. Bernard Lewis argued that the Muslim world's attitude toward the West is founded on a sense of humiliation. Islam was left out of the Industrial Revolution and the modernization that transformed the West. According to Lewis, the Islamist suicide bomber is a twentieth-century phenomenon. "It has no antecedents in Islamic history," Lewis claims, "and no justification in terms of Islamic theology, law, or tradition. It is a pity that those who practice this form of terrorism are not better acquainted with their own religion, and with the culture that grew up under the auspices of that religion."[26]

The term "fundamentalism" was coined at the beginning of the twentieth century by American Presbyterians to refer to their proud stand against Darwinism, the critical study of the text of the Bible, and other modern ideas.[27] They issued a list of "fundamentals" that "true" Christians should subscribe to, and began to call themselves "fundamentalists." The concept has broadened since then, and now serves as a general label for those who strictly adhere to religious or political principles that they base on a more or less literal reading of sacred texts, or on dogmatic ideas. They claim to be returning to a primeval pure faith, or of the state of mankind from which individuals—other adherents of their faith or ideology, indeed the entire cosmos—have deviated. Fundamentalists privilege the source text over interpretations of it and refuse to align it with modernism through updating or compromise of any sort. Their values are inevitably binary—there are things that are permitted and other things that are forbidden. Doubts and hesitations are eclipsed by fervent faith in a unitary solution appropriate for everyone everywhere.[28]

Of course, that is in part a pretense. Fundamentalists need to tackle real challenges and dilemmas in the contemporary world. In 1988, the supreme leader of the Islamic Republic of Iran called off the jihad against Iraq. Ayatollah Khomeini had promised to wage a holy war with Iraq until "total victory" was achieved. Yet after eight years of war, he had no choice but to stop despite the lack of such a triumph. He had cancer and was dying, and it was more important to save his larger revolutionary project, the Islamic Republic itself, which suffered huge human and economic losses. "Taking this decision was more deadly than taking poison," he declared. "I submitted myself to God's will and drank this drink for his satisfaction." Yet Iran has never forsaken its ambition to achieve hegemony in the territory of its western neighbor. Indeed, in recent years it has achieved that to a large extent. In order to remain in control of their communities, fundamentalists act pragmatically, while never relinquishing their final aim.

In the mid-nineteenth century, many American Christians sought to reconcile Scripture with science and to explain the supernatural events related in the Bible in rational ways. Fundamentalism was a reaction against this. If the Bible described a miracle, it was a miracle, precisely as recounted. Charles Darwin's *On the Origin of Species* challenged the views of the fundamentalists but also spurred them on. They sortied into ideological and political battle against the teaching of evolution in schools, and against the progressive parts of the political establishment. They largely failed. Even when they won a technical victory, as in the famous Scopes trial of 1925, when William Jennings Bryan persuaded a Tennessee jury to convict a schoolteacher for violating a state law that forbade the teaching of evolution, they lost in the court of public opinion. By the 1930s, American fundamentalists realized that they could not hope to beat the scientists' growing clout. They thus set themselves apart, establishing their own schools, churches, universities, charitable projects, newspapers, radio stations, and missionary operations. The Muslim Brotherhood was doing much the same thing in Egypt at the same time.

Fundamentalism emerged during the nineteenth century in many places, in disparate cultures and religions.[29] For example, in the early part of the century, Rabbi Moshe Sofer (better known as "Hatam Sofer," his work of the same name) famously declared that "the new is forbidden by the Torah." He meant that Jewish observance and law should never allow for any change or revision in light of new circumstances, for any reason. In doing so, he laid the foundation for today's Haredi (ultra-Orthodox) Judaism. The same basic idea was being preached at the same time by Islam's Wahhabist movement.

Paradoxically for a movement that claims to represent the unsullied truth, fundamentalism has a lie at its very heart. It is not distinct from and opposed to modernism, as it presents itself to be. Fundamentalism presents interpretations, stories, and traditions as primeval religious principles that the believer must adhere to or die for. But these principles may well have been formulated by the fundamentalists themselves for the sole purpose of serving its most basic principle—resistance.

FUNDAMENTALISM COMES IN MANY HUES. IN THE MIDDLE East and East Asia it is often a response to state authoritarianism, a sanctuary for those who protest the injustices of dictatorial regimes. Religious devotion might supply a certain degree of protection to the believers because of the importance that traditional societies ascribe to religion. In the Christian world, there are forms of fundamentalism that maintain that the United States, as a Christian nation, has a divine and eternal mission to bring about the final redemption. Most of these fundamentalists belong to the Evangelical movement. In Israel, there are Jewish fundamentalists who preach for a monarchic theocracy on the model described in the Hebrew Bible, with a temple in Jerusalem. Some of them formed a terror group in the 1980s, murdering three Palestinian students of the Islamic college of Hebron in 1983, and injuring dozens in several other attacks. They were arrested as they were planning to blow up the Dome of the Rock, on what Jews call the Temple Mount and Muslims call Haram al-Sharif. Since the beginning of this century, there is a rise of Buddhist violence in Asia, especially in Thailand and Myanmar. The latter country engaged in ethnic cleansing and murder of the Rohingya minority, who are Muslims, since 2016. Buddhism is the religion least associated historically with violence. It stresses compassion and forbids doing harm. One would have thought it would be immune to murderous zealotry. Not so, it turns out.

Most fundamentalists do not engage in political violence, of course—consider the Amish in the United States, for instance. This kind of belief, practiced also by Haredi communities throughout the Jewish world, might be considered a "true" fundamentalism. They may pity the rest of the world, but, following what they believe to be the only true path, they have no interest in subjugating others.

No rebels live the revolt more than fundamentalists do. Jean Baudrillard, quoted above, claimed that the terror attacks of September 11 were aimed at shaking globalization to its foundations. In this attack, fundamentalists acted as representatives of international radical rebellion against capitalist globalization, using globalization's tools. "The unbounded expansion of globalization creates the conditions for its own destruction," Baudrillard declared. He viewed globalization in its entirety as antidemocratic and repressive, a claim hard to support given that it has extricated hundreds of millions of people from poverty and empowered the liberal agenda everywhere. Yet the basic outline of his argument has stood the test of time. Radical Islamists see capitalism as a menace and globalization as their enemy. So do their mirror images—ethnic and racist nationalists.

CHAPTER 7

Talking with Nationalists

Globalization is when immigrants enter and jobs leave.

—FORMER SLOGAN OF THE NATIONAL FRONT,
FRANCE'S ULTRANATIONALIST PARTY

Depicting business at the City of London stock exchange in the eighteenth century, Voltaire foresees today's globalization and the diversity that is an inseparable part of it.

> Take a view of the Royal Exchange in London, a place more venerable than many courts of justice, where the representatives of all nations meet for the benefit of mankind. There the Jew, the Mahometan, and the Christian transact together, as though they all professed the same religion, and give the name of infidel to none but bankrupts. There the Presbyterian confides in the Anabaptist, and the Churchman depends on the Quaker's word. At the breaking up of this pacific and free assembly, some withdraw to the synagogue, and others to take a glass. This man goes and is baptized in a great tub, in the name of the Father, Son, and Holy Ghost: that man has his son's foreskin cut off, whilst a set of Hebrew words (quite unintelligible to him) are mumbled over his child. Others retire to their churches, and there wait for the inspiration of heaven with their hats on, and

> all are satisfied. If one religion only were allowed in England, the Government would very possibly become arbitrary; if there were but two, the people would cut one another's throats; but as there are such a multitude, they all live happy and in peace.[1]

This passage declares that people are no longer defined according to some primal identity or theological stance, or the directive of some political entity. Everyone can unite around the practical and materialistic aspects of life, except for the bankrupt, of course. Voltaire brings home globalization's threat to identity, and to those who do not benefit from its bounty. Who is threatened in Voltaire's universe? Those who do not feel that the religious or national distinctions among them are obviated by commerce, and who have maintained a deep resentment for the Other. The poor do not have a place at the groaning board of the stock exchange's capitalists, and can thus be expected to reject the universal order that international trade gives rise to.

Voltaire concluded that the multiplication of religious sects in eighteenth-century Britain facilitated coexistence and prosperity. It was a provocative claim, typical of the Age of Enlightenment in that it challenged all sorts of isolationism, whether religious, ethnic, or national. Were he alive today, Voltaire would no doubt inveigh against Brexit. He would argue that diversity and economic cooperation among many different kinds of people lead to prosperity and thus to greater happiness. Since World War II, that has been the position taken by decision makers in Britain and the rest of Europe, both those of the mainstream right and those of the left.

But that began to change in a fundamental way soon after the advent of the new millennium. The people Voltaire ignored—those for whom identity is crucial and those not reached by the ripples of prosperity—began to fight back. Nationalism's faithful agents returned to center stage, decades after being shunted off to the wings.

Brexit remains a conundrum if the resurgence of British nationalism is regarded as a fringe phenomenon. The heroes of this dark resurrection were the leaders of the extreme right. Their project was surprisingly successful, given that they had been marginalized by mainstream politicians for decades. In the years since the Brexit referendum, the far rightists have returned to obscurity, but their ideas live on. The historic parties of the center-right have appropriated their radical right-wing ideas.

Welshpool is a small town in Wales, just a few miles from the English border. It is ringed by green hills and proud to be home to one of Europe's largest sheep markets. It's also the community of Nick Griffin, a former leader of the British National Party (BNP). Before his expulsion from the party, Griffin did all he could to sow the seeds of xenophobia. He failed as leader of the BNP and his personal political fortunes foundered, but other politicians have enjoyed the fruits of his intolerance.

When the world's financial system slipped into crisis in 2007 and 2008, Griffin immediately grasped its potential for the extreme right. Seizing the moment, he aggressively promoted an anti-immigrant agenda, aiming his arrows in particular at the European Union, which is loathed in the United Kingdom not only by extremists but also by many on the left and the mainstream right.

We meet in 2008 in a small, quaint pub. He's a chubby man of average height and middle age. I found it hard not to think of his haircut as Hitlerian. One of his eyes is glass, purportedly the consequence of a shotgun cartridge that exploded in a campfire. His voice is abrasive and admonitory, but otherwise his demeanor is bland. He does not exactly radiate charisma.

Griffin belonged to the new generation of the extreme right. He's a graduate of Cambridge, a family man with a full head of hair that sets him apart from the skinhead look of the 1970s and 1980s. A spark of malice flashes occasionally during our conversation—for example, when he chuckles in response to a question of mine about Holocaust revisionism, saying that he "is not well versed in Jewish terminology." In 1998 he was convicted of incitement of racial hatred in a far-right newspaper he edited. At his trial he declared, "I'm well aware that the orthodox opinion is that six million Jews were gassed and cremated and turned into lampshades. Orthodox opinion also once held that the world is flat."[2] He also penned an anti-Semitic manifesto on Jewish influence, primarily in the media, entitled "Who Are the Mind Benders?"

Griffin told me that he was delighted with the financial crisis. "Every political change in European history, every significant change, came out of nowhere, almost from nothing. For example, the communist and fascist revolutions of the past century. It happened very quickly, and I think you can now see in the West that the entire liberal consensus—which wasn't based on the correctness of liberal ideas and certainly not on public support for those ideas—has zero support." He grew excited. "The consensus was based on the goods the liberal system provided: food in the refrigerator, holidays abroad. There's nothing wrong with all this, as long as they continue." He took a long sip of his beer. "But now it's over. Everything the liberal elite tried to do, which aimed to change our society in Britain, in Europe, and in the United States too, was based on the ordinary citizen telling himself,

'I don't accept the ideas, the liberal values, but I'm doing okay, I have food to eat.' In other words: materialism."

As he spoke, he seemed to become confident that a racial revolution is around the corner. "The wheels of the capitalist system have fallen off," he proclaimed. "It's finished. And it's a crisis that will last for generations. What remains of the system is destructive chaos." He looked around at the pub's patrons. "Therefore, everything is about to change. When you say that we won't succeed in gaining power—that might have been a correct assessment when we were living in stable economic times. But today, everything is possible."

I listened to his monologue while I was still under the spell of Barack Obama's election less than a month previously. I had attended Obama's victory speech in Chicago's Grant Park, where I managed to get through the security perimeter and join the small crowd surrounding the stage. Obama made his entrance and stood directly above me. I looked around and saw people in tears, trembling with emotion and pulsating with excitement. That this man could be president of the United States was compelling evidence, I believed, of America's stunning flexibility and the openness that has long been its lifeblood.

In light of Obama's election, Griffin appeared delusional. In many people's eyes, America had ascended to a new height of embracing diversity and rejecting bigotry. Now the man with the glass eye and rasping voice wanted to celebrate the return of racial nationalism.

Griffin's fundamentalism calls for a return to an imagined primeval and pure Britain. His doomsday prophecy is of civil war that can only be avoided by reestablishing a world in which religious or ethnic community is the supreme value. Griffin advocates separating schoolchildren in this way. British Jewish children shouldn't, he argues, study side by side with British Christian children or with children of Indian descent. "Children suffer from this more than anyone else," he maintains. "We oppose multiculturalism and multicultural integration in principle. We believe that human diversity is a good thing. Your identity is created not because you're a citizen of the world but because you come from a particular place, from a certain culture, from your people." Note the glib use of the word "diversity" to justify separation, or segregation.

For people like Griffin, biological metaphors and threats of "extinction" are the secret sauce of every conversation. "There's an ongoing erasure of human diversity, and if this was happening to another species, ants for example, this country would move highways to save a rare ant. The whole inclination of the global capitalistic system is to erase diversity, and it's more effective than any action by a dictator in the past—Stalin, for example, who deported all of the Tatars. Capitalism does it better. Capitalism is a campaign of annihilation."

Griffin asserts that a school shared by children from different religions or communities is "multiculturalism." It's not. It's simply the implementation of the principle of equality within a national framework. The concept of citizenship requires equality between people of different religions and various ethnic minorities who live in the same nation-state. Griffin doesn't recognize modern citizenship because it means the acceptance, adoption, and internalization of liberal and national values rather than a primal religious, ethnic, or racial identity. He also contends that putting children from different backgrounds in the same educational setting "erases human diversity," because for him, diversity is a matter of primary colors, categories taken out of an imagined glorious past.

One of the advantages radicals enjoy (radicals of any sort, not necessarily fundamentalists) is the ability to rip off the mask of liberal hypocrisy. Such hypocrisy proves the disingenuousness of mainstream discourse. This lack of honesty is extremely empowering for radicals. Hypocrisy is a primal sin if authenticity is the ultimate virtue.

Griffin accuses the British elite of deciding to turn the country into a migration state without a mandate from the public to do so. There's some truth in that. For many years, the British government refrained from providing detailed migration data to the public, and politicians rarely discussed the policy in the open. Griffin is also correct when he notes that some of the best schools in Britain have a religious affiliation, are private rather than run by the state, and serve the wealthy. "What you have here, in fact, is segregation that the elite and their children enjoy, but this doesn't exist in schools for ordinary people," he insists. "So, either integration should be applied to everyone, including the rich, or, if segregation is good for the rich, it's good for everyone! I don't understand why integration is imposed on the poor in this country while the wealthy are exempt from it."

Griffin was on the far fringes when we met, and his movement later crashed and he was ousted. But much of the agenda he touted more than ten years ago has been adopted by the British and European right. When he called for Britain to bolt from the European Union, he was denounced as delusionary, but that's exactly what the British people have decided to do. Protectionist economic policy, aimed at shielding local production, is back in fashion. Talk of sending Muslims "back to where they came from" is rising. And those who advocate this, like Griffin before them, don't care if the Muslims in question were born in Britain and are citizens. When I asked him how he would expel millions of people,

Griffin had a simple answer: "We'll pay them." It's either that, he said, or civil war.

I think back on our conversation from time to time. In retrospect, this ultranationalist apparently had a better reading of the future than did the journalist who had just witnessed Obama's election and believed in his soaring rhetoric during the campaign about America's "better angels."[3] President Obama will be remembered always as an inspiration to progressive leadership, but he failed to reshape the spirit of our era. Griffin was more astute in grasping the consequences of the financial crisis. Did Obama really believe in his optimistic vision? Or did he think that if he told a story that was sufficiently credible and persuasive it would come true? In his reelection campaign, in 2012, Obama said that his biggest mistake during his first years as president was focusing exclusively on "getting the policy right," while forgetting "to tell a story to the American public that gives them a sense of unity and purpose and optimism, especially during tough times."[4] Griffin, unlike the American president, understood that liberalism is more vulnerable than it seemed. Perhaps the failure is not the way the story is told, but the story itself.

Over a decade ago, I started meeting with neo-Nazis, Holocaust deniers, members of antigovernment militias in the US, and leaders of Europe's extreme right, from Marine Le Pen to the spiritual leader of the extreme right in Greece. Nick Griffin was just one of them. It was a decade of ascendance for them, more so than at any other time since World War II.

The extreme right spans a spectrum from institutional politics to violent gangs. There was an impression that the racist extreme right does little more than spout slogans, in contrast with Islamic fundamentalists who commit terror. The past decade has proven this wrong. According to a US government study presented to Congress in 2017, a full 85 fatal terror attacks were carried out in the country after those of September 11, 2001.[5] Right-wing extremist individuals and groups were responsible for 73 percent of these incidents, while Islamists conducted 27 percent. The two groups were almost equally deadly. Right-wing extremists murdered 106 people, while Muslim extremists murdered 119. Notably, 41 percent of the victims of radical Islam were murdered in a single incident, the nightclub shooting in Orlando, Florida.

Of course, the single most murderous terror attack in US history was 9/11. Yet the deadliest attack before al-Qaeda's mass murder was the bombing of the federal building in Oklahoma City in 1995 by Timothy McVeigh, who killed 168 people. McVeigh's political affiliations are a matter of heated debate. He definitely saw the federal government as an imminent threat,[6] holding to a sort of deviate form of libertarianism. He attended several meetings of a Michigan militia, that is still active today. In 2016, I joined some of its members on an armed patrol; they'd rather not discuss McVeigh but still talk obsessively about the tyranny of the federal government. Unlike radical Islam, extreme-right politics usually distances itself publicly from terror while building a comprehensive ecosystem that supports violence.

Much of the extreme right-wing ideology in the West is a type of contemporary fundamentalism, displaying some of its classic traits: the longing for a unifying, constant, and unchanging ideal; the invention of a magnificent and imagined past; an abhorrence of impurity, and an attempt to create a pure world; a binary and rigid view of the world; an exceptionalist distinction between believers and nonbelievers; and opposition to modernism and the values associated with it.

Paris, 2010

I met Marine Le Pen for the first time in 2010, at the entrance to a TV studio, where she was about to appear on a prestigious live interview program. She arrived in a swing coat with a small retinue, walking briskly. Her father, Jean-Marie Le Pen, never even dreamed that he could become a regular guest in such a slot. He was and remains a political leper, convicted for belittling the Holocaust. His obnoxious personality is legendary.[7] For the French left, he was a wannabe Hitler.

When his daughter entered the public arena, there were public calls to ostracize her as well. But Marine Le Pen is more sophisticated and less direct in her rhetoric. She was never caught making an anti-Semitic statement or violating anti-incitement laws. At first, the political relations between father and daughter were those of the founder of a movement and his political heir; then they soured.

When I first made her acquaintance, Le Pen was a leading figure in the National Front, the party her father had founded. She was competing to succeed him as party leader, and had his support. The elderly Jean-Marie was still unaware that she planned to remake the party's image—to recast his nationalist ideology and, ultimately, show him the door. Marine Le Pen was not just an ideologue; she actually wanted to get elected. France's presidential election was two years away, but she was already laying the groundwork for her campaign. She was eager to be seen as part of the mainstream, to detoxify the legacy she inherited from her controversial father and gain legitimacy. As a Jewish Israeli journalist, I was a useful tool for her. My interview with her was scheduled, postponed, and rescheduled. It started out pleasantly and ended abruptly, without an *au revoir*.

At first we converse in the makeup room. A makeup artist is fluttering around. President Nicolas Sarkozy had adopted some of Le Pen's rhetoric about "law and order" and "integrating into France," and Le Pen is worried. She's concerned that the mainstream right is growing more populist and stealing her platform. Just outside the studio, she and the interviewer rehearse their entrance. They joke about the French voters who are now rejecting Sarkozy. "It's a trilogy," she says. "The French voter fawns, abandons, and in the end lynches." In French it channels Julius Caesar's conquest of Gaul: "Lèche, lâche, lynche." "Now, with Sarkozy, we're at the lynching phase!" The interviewer giggles and thanks his guest for having taught him something new. Le Pen is pleased.

A decade later, she has good reason to be satisfied, and frustrated. In 2017, she scored one of the greatest successes achieved by the European extreme right since World War II. She won 34 percent of the votes in the second round of the presidential elections, in which she lost to Emmanuel Macron. It was a respectable result, but not much more than that. She ran into the glass ceiling that other European nationalists have encountered—memories of their descent from the fascist political forces that devastated Europe in the mid-twentieth century. In her case, her family name, which paved her way into politics, now holds her back. It keeps reminding voters of the party's origins, its caudillo, and the backing he received from the most reviled corners of French society, including supporters of the Vichy regime and the diehards who maintain that the French government sold out the army and the French nation when it decided to pull out of Algeria. On the other hand, Macron, her liberal opponent, encountered serious difficulties as soon as he became president, and his attempts at rational policy-making have exacerbated his troubles. By sophisticated planning, he made these troubles even worse. The yellow vest protests of 2018–19, triggered by Macron's tax initiatives and other reforms, seem to show that a large portion of the public may well be sympathetic to Le Pen's rhetoric about the threats of globalization. So far, she has managed to tread the fine line between the crude hatred spewed by her father and her positive messages about the importance of French culture and civilization and her paeans to "our beautiful France." Yet Le Pen is not a French Donald Trump. She is much more sophisticated, refined, and ideologically disciplined. My conversation with her in 2010 augured the future, in particular the nationalist revolt against globalization.

We're less than five minutes into our conversation and she's already quoting Vladimir Putin, the de facto godfather of European white nationalism. "You know," she says, as if the Russian president were a reliable source of sociological analysis, "Putin said that in another twenty years France will be a colony of its former colonies, and there are other leaders of states who believe that Europe will become Muslim."

Le Pen is important. Years before Breitbart News and Trump adviser Steve Bannon began to formulate their doctrine of economic nationalism, she grasped its two pillars: antiglobalization and hatred of Muslims. "There are new totalitarianisms in the twenty-first century," she explains. "The first is Islamization, which says that everything is religious. The second is globalization, for which trade is everything. And if we do nothing, if we don't insist on the values and laws of the French Republic, our civilization will face grave danger."

Le Pen says it's a choice between survival and extinction. Of course the National Front will protect the Jews while this existential drama unfolds! "French Jews do not have to consider the National Front as their enemy . . . They won't find a better protector than the National Front . . . We'll defend the French against extinction." French Jewry greets her promise with skepticism, to say the least. For a brief moment in our conversation, she flashes a threat. "I think they'll assume responsibility in the future," she cautions, talking about France's main Jewish organization, the CRIF. She accuses it of sabotaging the effort to stop "Islamization" by supporting immigration. "They were part of creating the danger facing our civilization, because of their positions."

It leaves open the question of what France under Le Pen might do to those who will menace French civilization. Earlier, when she announced that she intended to visit Israel, another step in her campaign to win legitimacy, she was informed that she was persona non grata and would be denied entry. She gave me a sharp look when I asked her about that. "I came to the conclusion that Israel apparently does not have enough enemies," she said.

I ask her if she understands why Jews are frightened by nationalists. "Excuse me, not in Israel," she declares. "Israelis are patriots, they defend their nation. They want to be sovereign in their territory, to ensure the security of their people. We're not asking for anything less." I note that I'm speaking about Jews and that she had spoken about Israelis. As Jews are a minority almost everywhere, I suggest, they are uncomfortable with nationalism. "Pardon me," she says, laughing sarcastically. "We'll also become a minority!" My questions start to irritate her. The interview is cut short.

At the end of the same day, I hear her speak at a National Front meeting in a basement in a Paris suburb. People are lined up outside waiting to get in. There's demand for what Le Pen stands for. Party activists hand out cheap red wine in paper cups to the crowd. (France may be the only place in the world where wine is served at political rallies.) She reminds the crowd of a famous National Front poster bearing the slogan "Globalization is when immigrants enter and jobs leave." She surveys supporters. "You know this deep down. We're not a political party but a resistance movement . . . We must take action for the sake of our children. A possible troubling future of horrific violence awaits them. For the sake of our beloved country, which is heading toward decline and depletion, for the sake of our faltering civilization, there's no alternative but victory!" Le Pen said to me more than once she is not an anti-Semite, but many of those I speak with in the audience afterward refer to "Jewish power" and say something about the Zionists having too much influence.

It's no coincidence that the most popular nationalist extremist in Europe since Adolf Hitler chooses globalization itself as her target. Le Pen recognizes what earlier ultranationalists, of her father's type, failed to understand. Economic globalization poses a significant threat to identity. It inevitably injects universal values into the local discourse because of its need for supranational relations. Prosperity cannot be achieved alone, and the need for the economy to interact globally does not coexist easily with exclusive national power structures and community values. Marine Le Pen was the first ultranationalist politician in the twenty-first century to create a coherent narrative for our era: Our good and righteous ethnoreligious nation is threatened by outsiders, migrants, and others. Our corrupt elites (she says the EU bureaucrats "govern" France and describes them as "dummies, incompetents . . . morons") are bound by an all-encompassing economic and political globalized system that we have not empowered. Their talk of norms and morality is thinly veiled hypocrisy, and the lying media serves them.

The main issue is not government and its powers. Rather, the central question is: Will the nation be able to resist and repel an invasion of foreign ideas, foreign interests and merchandise, and foreign people?

If you think about it, it's exactly what Donald Trump always stood for.

As always with nationalism, the focus is external; politics is not about an in-depth inquiry into the correct use of government and its powers. It's about an enemy lurking in the dark. What Le Pen does not say, but which can be inferred, is that for the nation to endure, the world order must be upended. This is a common misperception when dealing with nationalists—that they are focused on their nation-states. But actually, they need to revise or destroy fundamental international norms, from the treaties that require countries to give asylum to refugees to agreements governing international financial institutions and free trade. Le Pen calls these norms "the silent dictatorship of globalism." The aims of nationalists are local, but their project is global.

IF LIBERALS WANT TO COUNTER THE NATIONALIST PROJECT, they need to take heed of what nationalist populists are saying and doing. While nationalists seek to destroy globalization in its current form, liberals treat globalization as an unwanted stepdaughter of the Industrial Revolution, or dirty linen, or an expletive that refers solely to material exploitation. Nationalist extremists attack globalization because they see the rise of universal identity and realize that, if they are to survive, they must destroy the material foundation on which that identity stands—globalization itself. They do not sense from the center of the political spectrum a resolve to fight for globalization. Instead, they get one excuse after another, vague promises, and vacillation. The reason that Marine

Le Pen, Griffin, and others constantly stress the dichotomy between the global and the local is that they estimate that, when push comes to shove, people will naturally gravitate toward a particular identity, family, and community. As Albert Camus is reported to have said, "Between justice and my mother, I choose my mother." Donald Trump, who calls himself a nationalist, has his own definition of what a globalist is: "A person that wants the globe to do well . . . not caring about our country so much."[8]

This false dichotomy, which implies that people cannot be both citizens of the world and patriots in their own homelands, is the cornerstone of contemporary extreme nationalist rhetoric.

Liberals are not making an effort to understand the extremists' tactics for ending globalization, but in any case, they will not defend globalization with passion. As the progressive left sees it, globalization has oppressed workers and catalyzed environmentally dangerous industries that are contributing to mass extinctions. For the right-wing mainstream, the current global order might be a threat to national sovereignty and community structures. These approaches created a vacuum the extremists are only too happy to fill in order to achieve their overriding goal—to destroy, not to repair, our era's globalization and thus pave their way to power.

The left has also neglected its core ideas about the importance of material well-being. A crucial assumption of the post–World War II order is, to put it simply, that people need to have a good life in order to *be* good. Yet since the fall of Soviet Communism, liberals have convinced themselves that the values of an open society have become the uncontested fabric of industrialized countries. While the values are indeed of intrinsic worth, they are not uncontested, and cannot necessarily survive an era of economic hardship. The great architects of the postwar order—among them Roosevelt, Churchill, de Gaulle, and Adenauer—were well aware of that simple truth. "It goes without saying," Sigmund Freud wrote in the 1920s, "that a civilization which leaves so large a number of its participants unsatisfied and drives them into revolt neither has nor deserves the prospect of a lasting existence."[9] Mainstream politics should always assume that, when their materialistic condition deteriorates, people will turn to their very worst. If democracies do not provide just institutions, education, and social safety nets, they cannot withstand grave military or economic crises. However, even safety nets can only buy time, and nationalists or other authoritarians will prevail if the crisis endures and deepens.

The ultranationalists I spoke with had little interest in economics, or specifically trade policies. For them it was just a charade. While nationalists eagerly exploit economic grievances, they are only passionate about the nostalgic sanctification of the national community and the rejection of those who are not part of their imagined pure and original community.

The liberal mainstream woke up late. It continues to engage in a discourse that has long lost its relevance—it tries to understand nationalism exclusively through an economic prism, pushing income redistribution. Yet a growing body of research shows that the link between economic crisis and rising nationalist-populism is unsubstantiated. A fascinating long-term study of political life in Holland, for example, conducted by Noam Gidron and Jonathan Mijs, has shown that people whose personal economic position worsened during the last economic crisis did not swing toward the extreme right. In fact, there was a slight movement toward the radical left. Neither did they display a significant rise in nativist attitudes.[10] While there is a correlation between adverse economic conditions and support for radical parties, those who support those parties are not the members of society *directly affected* by the actual material hardship. The crisis opens the door to destabilization, to the revolt itself. Those who are attracted by the extreme right are not necessarily those who were hit economically by the crisis.

What, then, is going on?

Like an autoimmune disease, nationalism breaks out in a democracy when the body politic is weakened by socioeconomic ills. But when the body recovers, it does not necessarily eliminate the disease. In the past decade it has worked like this: First, nationalist rhetoric enters the political discourse at a time of severe economic or security crisis. In the second stage, the ultranationalist malady continues to progress, no longer needing the underlying crisis to do so. In Hungary, for example, the influx of immigrants and refugees has slowed considerably since 2016, but the country's nationalist prime minister, Viktor Orbán, has continued to rail against them. This foreign threat continues to boost him and his party electorally.

Another study shows that there was indeed a correlation between worsening economies and the rise of extremist nationalist parties in Europe, but only until 2013. Even though the economy began improving, support for nationalist parties continued to strengthen.[11] Obviously, the improvement in the US economy between 2016 and until the outbreak of the coronavirus pandemic in 2020 did not significantly weaken President Trump or the agenda he promotes, in particular his measures against immigrants.

Steve Bannon used to talk about what he called economic nationalism. But he has changed his tune since then and in 2019 announced the establishment of an "academy" that, as he puts it, will be a "gladiator school for cultural warriors."

That doesn't sound like a person whose main concern is trade barriers or the decline of the middle class; it is about an imagined threat posed by Others.

Populist nationalism's attraction is not its economic agenda but rather its message about identity, against immigration, and for a chauvinist approach claiming to restore personal security.[12] The twentieth-century French novelist (and war hero) Romain Gary put it simply: "Patriotism is the love of your own people; nationalism is the hatred of others." Glorifying the national community and underlining identity can remain in the realm of mainstream politics, but increasingly these ideas empower sentiments of nativism, xenophobia, and racism everywhere.

CHAPTER 8

A Nazi Revival

It is said about us that Jews cannot speak with us, they leave here only as a pile of ash. But look, you're standing here, and everything is normal.

—CONVERSATION WITH A GERMAN NEO-NAZI, 2014[1]

Constantine Plevris is one of Europe's most prolific racist polemicists. I interviewed him in 2014, when the streets of Athens were full of angry protesters, as they often were during Greece's great depression. I was working on a documentary on the rise of racial hatred in Europe. Arriving for my appointment, I am greeted at the entrance to the building where Plevris works by a suspicious guard who checks my passport and accompanies me in the packed elevator up to the office of the ideologue of Greece's extreme right. Plevris, an unrepentant Holocaust denier, is generous with his time and courteous in the answers he gives to my questions. At first it is hard for him to believe that he is actually talking to an Israeli. "I've never visited Israel," he says. "I can't imagine that you have," I cautiously reply.

Plevris proudly shows me his heavy tome *The Jews: The Whole Truth*. The cover photograph is of a pious Jew dressed in black, with flashing and threatening eyes. Plevris then begins to show me pictures that, from his perspective, prove that Israel is a brutal country. I ask him why he's showing them to me. "Because the Jews pretend they are innocent," he says. "They say, 'We never kill anybody.' Even your God sent an angel to kill the children of the Egyptians. The Jews are the first people in the universe who say that God commands them to kill other people."

He clearly has no problem with people branding him an anti-Semite. When I mention that Jesus himself was a Jew, Plevris shoots back: "He was a Galilean, not a Jew." I make another try: "You write about Jewish control of the banks."

"It's true. Nobody can deny it."

"I deny it."

"So name one bank that is not controlled by a Jew. One!"

"Okay. Barclays Bank."

"What?"

"Barclays."

"Listen," he says abruptly.

"Citibank. JPMorgan. Alpha Bank. I can go on."

Plevris does a double take when I mention Alpha, a Greek bank. He's clearly terrified that I'll quote him saying that this local financial institution is owned by Jews. "No, no, of course not Alpha," he says. "About the others I don't know, but not that one!"

The conversation is hanging somewhere in the abyss between the grotesque and the amusing. To my chagrin, Plevris suddenly says that he admires Israel, and that "I believe they are an inferior race." I ask him, "Who are 'they'?" I'm trying to stay on course in his hatred's stream of consciousness.

He is utterly astounded by my question. "The Arabs, of course." Then he asks me a question. "Do you believe we are all equal?"

"I actually do believe we are all equal," I answer.

"So you believe that the mind is the same?"

"Yes."

"You know, in nature, it is like dogs and horses. There are dogs for shepherding. Dogs for hunting, luxury. In nature there are no two equal things. There are none." He hates gays, too.

When I show a video of my meeting with Plevris in Europe, North America, and Israel, audiences generally go into gales of laughter. The Greek extremist, whose political career has long since withered, seems ridiculous. Spouting lies that are simply too numerous and too outrageous to take seriously, he comes off as a clown. But he is the author of dozens of books that present a comprehensive ultranationalist philosophy. His vapid generalizations and oversimplifications are used to great advantage by his disciples. Out-and-out lies, boldly asserted, are particularly potent in a world molded by social media. And liberal arrogance, which does not want to honor lies by engaging them, serves his supporters well. On Greece's extreme right, they are all his sons. The most important of these followers is Nikolaos Michaloliakos, the founder of Golden Dawn, the Greek neo-Nazi political party, who started out his political career in Plevris's 4th of August Party during the 1970s. Of all Europe's neo-Nazi movements, Golden Dawn was, until 2019, the most popular in its home country.

Racist fundamentalism and ultranationalism are the ingredients of the Nazi and white supremacist resurgence. Their contribution to mass violence is today greater than it has been since the end of World War II. The mass murder committed by Anders Behring Breivik in Norway in 2011—he killed seventy-seven people—has become a prototype that has been imitated by other terrorists of the extreme right. They accumulate weapons and gather intelligence, train, and photograph their attack or broadcast it live on the internet. They leave electronic testaments meant to mobilize further attackers, just as Islamist terrorists do. The model was followed by, for example, Brenton Tarrant in his attack on mosques in New Zealand in March 2019, in which he killed fifty-one people, and by John T. Earnest, who attempted a mass shooting in a San Diego synagogue in April of that year, killing one person. Robert Bowers, who in 2018 committed the deadliest hate crime against Jews in American history when he massacred worshippers at the Tree of Life synagogue in Pittsburgh, also followed the model in part.

These racist fundamentalists are more dangerous than any other kind. They come from the white Christian majority, so they naturally have larger constituencies to recruit from and echo chambers to justify and amplify their message. Institutional racism within law enforcement agencies means that if they engage in illegal activities, they are less likely to be under suspicion, in comparison with militants that are not white. Public demonstrations by such groups have become more outspoken than in the 1980s and 1990s; they make no effort to conceal their agenda. They have also turned violent, as at the notorious Unite the Right rally of the summer of 2017 in Charlottesville, Virginia, where white supremacists walked with torches and chanted, "Jews will not replace us!" One of the neo-Nazis hit a thirty-two-year-old counter-demonstrator, Heather Heyer, with a car, killing her.

The ideology uniting those groups and their online platforms, with its toxic mix of conspiracy theories and racist ultranationalism, is supranational. In every country, these movements reflect anxiety that whites are liable to lose their position as a majority because minorities are "taking over" and "distorting the natural order." The claim has been voiced by a series of killers. Dylann Roof killed eight worshippers at a church in Charleston, South Carolina, in 2015, because, he claimed, blacks are taking over the United States. Patrick Crusius, who murdered twenty-two people at a Walmart in El Paso, Texas, spoke of a "Hispanic invasion." These fears are profoundly linked to the way the world has become more integrated since the beginning of the twentieth century, and to burgeoning connections between communities and individuals of different racial, ethnic, and religious affiliations—in other words, to globalization.

Such fears were born along with modernity and the multinational utopia envisioned by people like Voltaire in his paean to the London stock exchange. They were shaped on cinema screens in movies like *The Birth of a Nation*, which contributed to the rebirth of the modern Ku Klux Klan, and presented as coherent ideologies. A prime example is Lothrop Stoddard, an infamous white supremacist who published his book *The Rising Tide of Color Against White World-Supremacy* in 1920. Today's extremist talk of the "great replacement" and a "white genocide," two of the most popular conspiracy theories of today's online world, are simply the current versions of much older fictions claiming to document a threat to the "white race." Jews hold a special place in all of them. They draw from classic anti-Semitism and charge that a Jewish cabal is using its financial and intellectual power to annihilate white Christians and replace them with foreigners and heathens.

At the end of 2016, while in the US, I tried to get an interview with a Ku Klux Klan leader. I wasn't too surprised by the answer I received, from a man who identified himself as an imperial wizard: "To have you at a Klan rally would be tempting. To nail you to a cross like your people did [to] Christ. Then light your ass on fire and watch the light of Christ shine down upon our faces." I decided to pass.

Former East Germany, 2014

It's pouring rain when I arrive, with my producer, Antonia Yamin, in Schleusingen, a small town in what was once East

Germany. We drive to the address we had been given, but no one is there. Our news organization is nervous about this meeting with neo-Nazis. We have told our editor that, if we do not call in at the time of the meeting to tell him that all is well, he is to alert the German police.

We wait for a few anxious minutes and a black car appears. Out steps Patrick Schroeder, tall and blond, with a square jaw and narrow-set eyes. His companion is Tommy Frenck, a neo-Nazi who ran for a seat on the local council. Frenck was once convicted for involvement in incitement.

The two young men look uneasy. The police won't allow them to hold the meeting here in town, they report. "Follow us," they say.

We drive behind their car over an unpaved road. We wonder whether we're being led into a Nazi ambush. But then, to our relief, we're stopped by a German police car. Never were we happier to see a German in uniform. The authorities clearly know about the meeting.

The policeman asks for identification. "This is Tommy Frenck's place," he informs us in passable English. "Do you know what they do there? Führer? Nazi? We in Germany have a problem with the extreme right." He explains that we'll be required to testify in court if there are displays of support for Nazism at the meeting. "It's Germany's secret service that told them we're here," Schroeder charges, standing next to the policeman, clearly delighted by the attention.

Allegedly, Schroeder is a new breed of neo-Nazi. The German media has even coined a word for people like him—they refer to him as a "nipster," a Nazi hipster. *Rolling Stone* ran an article about Schroeder's online program and campaign to persuade Nazi groups that the hypernationalist scene must also accommodate young people who prefer a hip-hop or hipster lifestyle.[2] Schroeder seeks to expand the circle of Nazis in Germany and gain legitimacy. That's why the neo-Nazis of Schleusingen agreed to meet with me, a Jew.

German law prohibits all expressions of support for racism and Nazism, including display of a tattoo with a Nazi motif. Schroeder knows this very well. Like other German radical nationalists, he speaks in coded language to circumvent the law. A former Nazi once explained to me how it's done. "For example, I can print 'anti-Zionist' on shirts in a Nazi Gothic font," he said, "and there's no problem with this because it ostensibly expresses opposition to Zionism, which is a political movement. But my audience knows the law. They understand that it actually means 'against Jews.'"

We enter a small yard where the neo-Nazis meet. A barbeque is heating up and there is a large bucket of pork marinating, waiting to be grilled. Young local extremists stand around, mostly dressed in black. They look at us with curiosity and whisper among themselves. I once swam with sharks, protected inside a metal cage. Being surrounded by neo-Nazis is just like being surrounded by sharks, except that there's no cage. We mingle cautiously, sticking close to each other. Ostensibly, this is a meeting to prepare for local elections, but the date, May 8, is a significant one and possibly gives away the real reason for the barbeque. It's actually V-E (Victory in Europe) Day, the anniversary of Nazi Germany's surrender to the Allies in 1945. For the Nazis, it's a day of mourning.

Standing rigidly, with his hands clasped behind his back, Schroeder shoots the bullet points of his credo at us in fluent English. First, "the Muslims, to put it simply, are taking over this country." Second, "In Berlin, there are schools where it's difficult to find a single German child." Third, "In a few decades, this will happen everywhere in this country." I ask him how the "takeover" is being pursued. He offers an example:

"They prohibit German children from eating pork in school. You don't do such things if you want to be part of Germany. If there's someone who wants that type of Germany in another fifty years, like the Neukölln borough in Berlin, then okay. But I'm one of those who say we won't let this happen." Neukölln has a large Muslim population. The so-called Muslim threat is a recurrent theme in conversations with extreme rightists in both Europe and North America. Muslims have perpetrated most of the deadly terror attacks in Europe in recent years, and the extreme right has effectively exploited that to drum up votes. Muslims account for only 5 to 6 percent of Europe's population. In France, the European country with the largest Muslim community, the percentage is 7.5–10 percent.[3] Even if all these Muslims wanted to change Europe's character—a debatable statement at best—they hardly have the political power to do so. No Muslim party is represented in any of the parliaments of the large European countries. There is little tolerance for religious fanaticism in secular France, and certainly not in Germany, which has long rejected multiculturalism. The Muslim minority in Europe has grown; projections for 2050 put the population at about 10 percent. If the continent were to permit intensive immigration (which it does not at present), Muslims would reach 14 percent.[4] In short, Islamic fundamentalists have no avenue to gaining substantial political power in Europe in the coming decades. But the prospects of the extreme right in Europe are another story. Not only is the far right confident about its future, it also knows that it has *already been* in power before. It's much better placed to be voted into power than Muslims are.

Here, in the small towns of eastern Germany, neo-Nazism is less an ideology than a scene, a lifestyle. The young people around us, chowing down on their barbequed pork, sport t-shirts promoting extreme right-wing heavy rock bands and silver pendants in the shape of five-pointed stars. Neo-Nazism is a way to belong to something. Schroeder and Tommy Frenck are the group's ideologues.

"The German era ended on May eighth, on the day of Germany's surrender in 1945," Schroeder says. His comrade in arms Tommy Frenck adds: "Two thousand years of German history came to an end on that day." Schroeder expounds: "We've never been as enslaved as we are today. So pinned to the floor. It began with the Nazis' defeat and continues today. Sovereignty died that day. *Volksgemeinschaft** was at its peak, and since then has only gone downhill . . . my grandparents told me, if you don't look at the treatment of Jews, it was completely perfect . . . The Third Reich society was great . . . for the normal guy, who just lived in that state."

"Who was not a dissident, a homosexual, a Jew, a Roma," I remark.

"Yes," he confirms.

Schroeder's Thousand-Year Reich nostalgia is a species of fundamentalism. True, the term "fundamentalism" generally refers to a religion, and Nazism is not a religion in the conventional sense. The sociologist of religion James D. Hunter explains that "all fundamentalist sects share the deep and worrisome sense that history has gone awry. What 'went wrong' with history is modernity in its various guises. The calling of the fundamentalist, therefore, is to make history right again."[5] Alon Confino, a historian of modern Europe, writes about the way the Nazis imagined a world without Jews, and how this idea of racial purity drove them to craft a

---

* *Volksgemeinschaft* was the term Hitler used in his speeches to refer to the spirit of the people as a racial community, organized hierarchically under the Führer.

"new genesis," an utterly new tradition of the origins of a new world order. In other words, they sought not just to conquer territory but also to conquer memory and history while expunging the Jews from them. Nothing could be more fundamentalist than this fantasy of primeval purity. The fantasy mandates that the modern unclean world must be purged by destruction, murder, and genocide.[6]

I ask Schroeder if he accepts the historical fact of the Holocaust. "In Germany, you can address the question of how many witches were burned at the stake but you can't discuss other questions," he dissembles.

Like many fundamentalists, Schroeder is unable, or unwilling, to grapple with arguments that distinguish between identity and political opinion; people are merely pawns of their ancestry and always play the role written for them by their race or religion. Nazis are being denied their democratic right to free speech, he maintains: "I don't think there's a difference between us as a group, when they say all neo-Nazis are extremists and they should be in jail or if you say, in my opinion, that Jews have a lot of power in the banking sector." I respond that Nazis are people who have decided to support a political idea. Jews are born as Jews; it's not an identity they choose for themselves. "Yeah," he says, uncomfortably. But that doesn't stop him. "If Hitler had won the war," Schroeder proclaims, "he would've become a great hero in the world. The one who loses is always denounced and becomes the bad guy. What would be the historical fact if Hitler had won the war? You wouldn't be reading then about six million Jews or something. You'd read in history books that Hitler was great, the biggest German hero ever."

But six million Jews would still be dead, I respond.

"But you wouldn't hear about it," he replies, triumphant, "It doesn't matter what really happened."

I am reminded of Heinrich Himmler's infamous Posen speeches. His SS troops were, he claimed, maintaining their decency while they carried out the secret Final Solution. "Most of *you* must know what it means when 100 corpses are lying side by side, or 500 or 1000. To have stuck it out and at the same time—apart from exceptions caused by human weakness—to have remained decent fellows, that is what has made us hard. This is a page of glory in our history which has never been written and is never to be written."[7] In Himmler's world, and in that of the neo-Nazi hipster, truth has no inherent meaning and facts die an agonizing death. Truth is simply a statement that accords with Nazi ideology. What matters is how history is written, not what actually happened. Power alone accords meaning.

"There are a lot of stigmas about us," Tommy Frenck complains. "It is said about us that Jews cannot speak with us, they leave here only as a pile of ash. But look, you're standing here, and everything is normal."

I don't feel like I'm in a normal situation. I tell Schroeder and his friends that Israelis are flocking to live in Berlin. "If they start to take over Berlin," the neo-Nazi says with a motionless expression, "it's a problem. It's not what I want." At that point, we decide it's time to go.

The Fundamentalist as Sisyphus

The fundamentalists who appear in this and the previous chapters are prophets of the revolt against globalization. The concept is much broader than its conventional usage, and applies not only to religious extremists. Nazis, both the old and new versions, benefit from one of fundamentalism's great strengths—consistency. In a world of rapid change, where globalization threatens jobs and traditional values, the fundamentalist finds meaning through identity. Islam is the solution, the Muslim Brotherhood preaches. Happiness is racial purity, say the Nazis. England for the English, France for the French, America for Americans, ultranationalists shout. Fundamentalism takes pride in its dissonance with the world around it. It insists on going back to an original state, real or imagined. That requires a source of original truth and being—pure blood, a sublime soul, the direct word of God, always taken literally. It seeks a revolt against multiculturalism, against heterogeneity, a restoration of patriarchy, and the reassertion of what it claims are traditional values.[8]

Elites are reluctant to recognize that religion gives meaning to people's lives, even in a globalized world—and that serves the fundamentalist cause. "The learned have their superstitions, prominent among them a belief that superstition is evaporating," notes the American writer on religion Garry Wills. "Since science has explained the world in secular terms, there is no more need for religion, which will wither away . . . Every time religiosity catches the attention of intellectuals, it is as if a shooting star has appeared in the sky. One could hardly guess, from this, that nothing has been more stable in our history, nothing less budgeable, than religious belief and practice."[9]

The fundamentalist benefits from this superstition because it allows him to position himself as the ultimate guardian of religion or community in general, not only of its extremist interpretations. In the face of what they call "secular coercion" and "liberal oppression," the fundamentalists' power grows in communities that already feel threatened by globalization and liberal values. Trying to use science to disparage faith and tradition is a weapon in the hands of the very forces liberals most fear.

Framing fundamentalism as a product of distress is another typical mistake. The fundamentalist ideologue is often not a product of ignorance; rather, his ideas are a deliberate, even thoughtful, response to modernism. It's his familiarity with the global world that makes him want to stay away from, rectify, or completely destroy it.

Globalization is a reality that all players in the modern political arena contend with. Mainstream conservatives devoutly advocate free trade and flow of capital but decry globalization's threat to local identity. The mainstream left denounces the global economy's erosion of workers' rights, even as it praises one-world universalism and the progressive values that come with it. Greens bewail carbon emissions and destruction of habitat—by-products of international production, consumption, and trade. At the same time, they tout international cooperation as the only way to avert environmental catastrophe. Marxists believe that today's global corporations are the epitome of unfettered capitalism, but their utopia is one in which the workers of the world are united. "This is the final struggle / Let us group together, and tomorrow / The Internationale / will be the human race," promises "The Internationale."

The fundamentalist, whether the radical Islamist or his nemesis, the white supremacist, is the exception. In his worldview, globalization inherently threatens to marginalize him. He's right. Since the 1990s, the number of Americans declaring themselves to be of "no religion" has more than doubled. Today it stands at slightly over 23 percent—more than any other faith.[10] This is an epic change. In the face of such attrition, some traditionalists swing toward extremism. Globalization signals to the fundamentalist that his fate will be like that of the few and diminishing Amazon tribes who shoot arrows at the helicopters that are documenting their extinction from above. The projects of Le Pen and Griffin all want to invert that. They seek to change the world so that the helpless tribe will be the liberals, relegated to shrinking enclaves of democracy; and the past decade has shown that they are not at all delusional. The liberal assumption that ultranationalism has no future grows out of our powerful recollection of the 1940s and the triumph of good over evil—a central narrative of the post–World War II era. But that victory offers little guidance about how to win today's battles.

In the language of technology, fundamentalism is a closed operating system in an open world full of software, files, and ports. A closed system, as iPhone users know, has advantages: it's simple, predictable and internally stable. The fundamentalist looks for purity in a diverse world, and exclusive truth in a reality full of truths.

Albert Camus suggests in "The Myth of Sisyphus" that we imagine a happy Sisyphus. The gods condemned him to an eternity of pushing a large rock up a hill in Hades, only to see it roll down the slope again each time he almost reached the top. But Camus argued that it is the struggle, not the achievement of a goal, that might give meaning to life. When Sisyphus accepts the total absurdity and futility of the human condition in the modern world, he achieves happiness, because the struggle itself satisfies him, even though he knows that the rock will inexorably roll down the hill. Such is the paradox of the fundamentalist. He is part of, and a response to, the very globalization he seeks to slay. Like Sisyphus, the fundamentalist pushes his bloody boulder to the top of the hill, but it can't stay there. His imagined world—glorious from a religious or racial perspective—will never exist. It will always be polluted by the impurities of a complex global reality and collapse into itself. But it is the journey itself that changes the world, even if it fails. The fundamentalist is happy not because he is coping with the futility of the modern world but because he is determined that the modern world be destroyed. Violent fundamentalism threatens the most basic demand citizens have of their governments—personal security. In a global world, the rock of fundamentalism will always roll down again, killing people along the way. At the bottom, Sisyphus will begin again, determined to push the rock back up the hill in his Hades.

CHAPTER 9

The Middle-Class Mutinies

Q: It is an internal Goldman document . . . "Boy, that Timberwolf was one shitty deal." How much of that shitty deal did you sell to your clients after June 22, 2007?

A: Mr. Chairman, I don't know the answer to that, but the price would have reflected levels that they wanted to invest at that time.

Q: Oh, but they don't know—you didn't tell them you thought it was a shitty deal!

A: Well, I didn't say that.

—FROM THE TESTIMONY OF THE HEAD OF GOLDMAN SACHS'S MORTGAGE DEPARTMENT BEFORE A SENATE SUBCOMMITTEE, 2010[1]

I remember hearing, at the age of three, the sound of my father's army boots descending the spiral staircase in our home, thumping on the worn wood of the steps. He kissed us goodbye, and my mother, brother, and I remained at home with the painful knowledge that he had gone to war. It was

1982, and the government of Menachem Begin had decided to invade Lebanon. The goal was to defeat the Palestinian militants who were using that country as a base from which to launch rockets against Israel's northern communities. My father spent many long months in that war, the most controversial one in Israel's history. As he fought on the outskirts of Beirut, his friends demonstrated against the incursion in Tel Aviv's central square. He agreed with them.

I was eleven years old when I first heard the rise and fall of an air-raid siren. My parents took me and my brother to the bomb shelter in our house's basement. The Gulf War of 1991 was underway, and Iraq's dictator, Saddam Hussein, was firing ballistic missiles at Israel. During the initial days of the war, there was fear that he would try to hit Israel with chemical and biological warheads. Every Israeli adult and child had been issued a civil defense kit and was told to carry it at all times. It contained a gas mask and an atropine injection to be used in case of a nerve gas attack. When that first air-raid siren went off, some Israeli civilians panicked and administered themselves the atropine injection on the spot.

My daughter was two years old when I first carried her into our apartment's safe room (Israeli building codes now require each apartment to have one, with walls reinforced with steel) during a rocket attack from the Gaza Strip. Two years later, I did the same thing, but with two children. Then again with three.

This sense of insecurity is only my side of the story. The other side suffered more profoundly. In 1982, thousands of Lebanese civilians found themselves on the front lines. A Lebanese Christian militia allied with Israel carried out a horrific massacre in Sabra and Shatila, two Palestinian refugee camps in Beirut. Saddam Hussein killed only a few Israelis with his missiles, but his oppressive regime killed tens of thousands of Iraqis. And the American occupation of Iraq set off a chain of violence and misery that continued for decades thereafter and which still has not come to an end. About 2 million Palestinians live in abominable conditions in Gaza, most of them as virtual hostages of the fundamentalist Islamist organizations that rule there. They have no bomb-proof rooms to run to with their children when the Israeli army fires on areas where Hamas fighters are hiding.

The Middle East is one of the world's most unstable regions, rife with conflicts and tensions. Actually, however, the lives of many of the world's people are afflicted by violent conflict. Among them are Syrians, Israelis, Palestinians, Iraqis, Mexicans, Colombians, Indians, Pakistanis, Sri Lankans, the people of more than half of Africa's countries and all the countries that used to be part of Yugoslavia, and many of those that used to be part of the Soviet Union. Wars that have no defined front lines and no real rears have dictated the lives of many of the inhabitants of these countries in recent decades.

Western Europe, in contrast, enjoys relative tranquility. But it's a post-traumatic tranquility. While Western Europe may have not experienced destructive violent conflict since 1945, it remains scarred, and the concrete marks of the destructiveness of war can still be seen from London to Berlin, via Auschwitz.

The United States stands apart. It's hard not to envy the sense of security its citizens exude. The US is an anomaly of history, a country protected by two oceans from the storms of the current era. Wars? Of course Americans have seen them. Economic crises? Those too, and crime as well. But they have not been victims of the modern breaking down of the lines between the civilian and the military, and between the rear and the front lines, that has characterized the world since the French Revolution.

American mothers could send their children to school without fear in the middle of a war—even a world war. Their husbands, the fathers of their children, might be in danger on the battlefield, but the home front was safe. American children didn't see classmates killed by a German, Japanese, or Soviet bomb.* Americans engaged in nuclear attack drills, but they never had to run to their bomb shelters for real. America's civilians, as opposed to its soldiers, never learned to recognize the wail of a falling rocket, or the dull whoosh of an enemy bullet. No neighborhood in Philadelphia, San Francisco, or New York was rebuilt after being flattened in a bombing, as was the case with European, Asian, and African cities.

It would be easy to argue that Americans suffer from a different form of insecurity, caused by the violence that has for years roiled beneath the surface of their lives. Is there a difference, people might ask, between a child killed by a stray bullet fired by a criminal or a policeman on the streets of Chicago and a Somalian child killed in an attack by a local militia in Mogadishu?

There is. The motivation for the carnage is important. The feeling that you and your family are menaced by an external enemy makes a difference. And people in other countries suffered from crime along with all the other dangers from which Americans were exempt. Pax Americana was not felt in Vietnam, the Middle East, or Central America, but it prevailed in North America. Until.

On September 11, 2001, Americans witnessed the murder of 2,977 people in the largest terror attack in history. It was

* The only time civilians were killed in an enemy attack on US continental territory during World War II was when a Japanese incendiary balloon reached Oregon in 1945.

a moment of tremendous success for violent fundamentalism, which assiduously seeks to undermine the security of contemporary life. Suddenly America was jerked out of the alternative universe it had occupied in the modern era. The tools of globalization—airplanes, immigration, technology, the media—were suddenly turned by the assailants against the personal security of Americans. It marked the beginning of a new era—the day on which the seeds of the revolt were sown.

A human life in an industrialized nation measures about eighty years, but profound social changes take much longer. Perhaps because the world now operates at a faster clip, we have trouble taking in the larger landscape of our lives. It is as if we are hacking a path through a tall cornfield without being able to see the pattern that a tractor has already made there, visible only from above. This is what the landscape looks like: 9/11 is not over. We are still living at the end of the moment that began when the first plane hit the north tower that awful morning. The reverberations are still palpable—a crisis in the markets, two wars, the rise of fundamentalism, interest-rate cuts and burgeoning consumption, a real estate bubble and its aftermath, the Great Recession.

YEARS HAVE GONE BY, CHILDREN HAVE BEEN BORN AND HAVE matured, but that dreadful moment has still not passed. The 9/11 attacks cracked the foundational guarantee of personal security that Americans had always taken for granted. They happened close on the heels of the economic slowdown brought on by the collapse of the dot-com bubble at the end of the 1990s. Huge accounting frauds came to light. Enron, a giant American energy firm, went bankrupt, as did two other firms of mythic proportions, WorldCom and Tyco. The blow that bin Laden dealt to America exacerbated the economic crisis. Two and a half million Americans lost their jobs in the second half of 2001. The markets tanked, and for a long time. In fact, the American stock market recovered more quickly from the assassination of President Kennedy, and even from Pearl Harbor and the American entry into World War II, than it did from the September 11 attacks. In the meantime, the George W. Bush administration declared its war on terror, launched a war in Afghanistan and then one in Iraq, and began spending trillions of dollars (while refusing to raise taxes to pay for it) in a desperate attempt to restore the citizenry's lost sense of security.

Alan Greenspan, chair of the Federal Reserve, who was seen then by the markets as a genius, prescribed a potent antibiotic against the plague of mistrust—he advised the Federal Reserve to drastically cut interest rates, and it did. In 2000, the Fed's interest rate was almost always more than 6 percent; at the beginning of 2002 it had plummeted to less than 2 percent. It was not just artificial respiration, it was electric-shock resuscitation. The message sent by the political and economic elite to the public could not have been clearer: take risks. Jump off the cliff, you've got a parachute, everything will be fine. Coming on top of President George W. Bush's massive tax cuts beginning in 2001, the result was an economy on steroids.

In other words, suppress those images of planes that blew up in Manhattan and start thinking about buying a new TV on your credit card's installment plan, or refinancing your mortgage and buying a bigger house. Spend, spend, spend—we'll give you all the credit you need. It was a national mission in the United States, and it set the tone for the rest of the world. In October 2001, less than a month after Manhattan was covered with a cloud of dust made up of powdered concrete from the towers and human remains, *USA Today* reported that pins were being handed out in New York City with the slogan "Fight Back, NY! Go Shop!"[2] This is not just an odd fact. It gets to the fundamentals of what happened. The terror attack sought to disrupt the American way of life. As the American way of life had come to center on consumption, America had a profound need to save and protect this salient part of its national identity.

Americans returned to the shopping malls, despite high unemployment, which persisted until 2003. But the biggest spender of all was the American security apparatus. The Pentagon's budget doubled between 2001 and 2008. During the decade following al-Qaeda's attacks, American military and defense expenditures grew by about 50 percent as a percentage of GDP.[3] At the same time, the American economic leadership reduced the regulation and oversight of Wall Street firms. The central lessons learned from the outbreak of the Great Depression in 1929 were forgotten. Nearly all markets went into high gear. The restrictions imposed by the Glass-Steagall Act of 1932, which mandated a separation between banks and financial institutions that issue, underwrite, and trade in securities, were largely revoked at the end of the Clinton administration. Deregulation continued under President George W. Bush, who dismissed the chairman of the Securities and Exchange Commission, eventually replacing him with a man who agreed to rescind other rules that had been designed to maintain the financial system's stability.

From that point on, the script largely wrote itself: unfettered expansion of credit, an explosion of leverage in the residential real estate market, the invention of complex financial instruments aimed at enlarging profits while at the same time hedging and distributing risk throughout financial markets. A study from 2018 shows the position of the American middle class during the four decades preceding the economic crisis. It shows that the increase in net worth that the middle class experienced over this period was almost entirely the product of the rise in value of the real estate it owned. There had been no significant increase in wages. The middle class's capital is held mostly in the form of real estate, while the upper 1 percent's capital is predominantly stocks and other capital investments.[4]

It took less than seven years after al-Qaeda's terror offensive against the US for the next great economic crisis to explode, in 2008. At the very least, the 9/11 attacks set off a chain reaction that exacerbated the middle class's severe grievances. When the crisis broke out, homeowners' illusion of wealth was punctured, and they found themselves facing a truth long suppressed: capitalism had not worked well for them. It was a humiliating moment for the middle class. The magnitude of the physical and economic insecurity engendered by the attacks made them the most successful terror operation in history.

London, Autumn 2008

The master's degree programs in London that my wife and I enrolled in in 2007 were an expensive adventure for us, and would have been impossible had we not received scholarships. We first lived in the dormitories and saved every pound we could. London was elegant and well-off, but not benevolent toward us. On one of our first evenings in the city we were invited to dinner at the home of friends who had been living there for several years. They had found work in the flourishing financial industry. When we entered their beautiful penthouse in Marylebone after the experience of our room and a half with its tattered rug, we felt like Julia Roberts discovering the bubble bath in *Pretty Woman*. All the other guests worked in banks, investment funds, or the high-tech companies that had also started springing up in the City. They were all older than us, and much posher. I sat across the table from a young investment banker—she looked to be thirtysomething—who had just received her British citizenship after several years of work in London. Back then it was easy to be naturalized. She griped about her endless hours and the pressure of working in such a competitive financial firm. It was the first time in my life that I had heard a young person speak seriously about her impending retirement, in the sense of the end of the working part of her life. At first I thought I hadn't heard her well, or that she was speaking only about leaving her profession, not actually retiring and living off her wealth. When I grasped what she meant, I was so astonished that I had no idea how to participate in the conversation. I did know, however, that the bonuses awarded in such workplaces were legendary. "Fine," I remarked. "Work a few more years and you'll be able to afford a house in London." She raised her eyebrows. "I already have a house. Two, in fact." There was a brief silence. I heard the swirling of the Chardonnay in the wine glasses around me.

We were smitten with London, but we also felt how swell-headed it was. Those who earn little are the first to sense the hubris of the bubble. In downtown restaurants we were watched critically by hostesses drumming the floor with their high heels. I don't know if it is really true that they can pinpoint a diner's net worth by a quick look at his shoes, but it certainly felt as if they could. We were generally seated at a table in the back, close to the wall and the lavatory. A leading real estate chain in the city tried to rent flats to several of our friends without allowing them to see the places from the inside. The debonair agents apologetically explained that there are people who are prepared to close a rental on the telephone, sight unseen. Even though we insisted on seeing our flat first, we neglected to check the toilets and discovered only after moving in that one of them did not work. We protested to the agent, but he made it clear that if we didn't want the place, someone else would jump at the opportunity. Use a bucket or fix it yourself, she suggested.

I WAS BORN IN LONDON TO ISRAELI PARENTS AFTER THE Winter of Discontent of 1979, the name given to the months of strikes and economic disruption during which Britain's Labor government faced off with the trade unions. This was Britain as it was prior to the great privatizations led by Margaret Thatcher. My parents' London was modest, sometimes barely functional, smoggy, insufficiently heated, and there were huge piles of uncollected garbage left by sanitation workers who were striking against the James Callaghan government. My mother and father, who lived on kibbutzim, had come to London from the Middle East for a period of study and life overseas. They stayed there for several years because they joined a spiritual group, a not uncommon occurrence in the 1970s.

Nearly thirty years later, the meager, scarred, and perhaps idealistic city of their experience had been replaced by something else entirely. Despite all its flaws, Britain was once a country that manufactured automobiles, dishwashers, cake mixers—and mined coal to provide energy to its industries. After the Thatcher and New Labor years, it became a playground for investment bankers. Up until the 1970s, the balance sheets of British banks showed value equal to about half of the country's GDP. In the 2000s the financial sector grew until the balance sheets of the banks had inflated to five times the GDP of the entire British economy, a figure in the trillions of pounds sterling.[5] In the 1970s the top centile of the British population earned 5 percent of the national income; on the eve of the economic crisis it was taking in 15 percent.

The City had come to be defined by avarice. The winds of the crisis were already blowing in 2007, but Ferrari reported that year that it had a three-year waiting list for the supply of its cars in Britain. A spokesman for Aston Martin told *The Guardian* that its waiting list was seven months long. But financiers had trouble waiting. "People who get bonuses in the City are generally looking for instant gratification," he said. There were more modest luxuries as well, such as a collector's edition of a book on New York weighing thirty-five pounds, with hand-stitched silk binding. The publisher said that, despite the developing crisis, he had no doubts that people would pay the book's price of $7,500, given that "your private equity guy or equity derivatives guy at Morgan Stanley or Goldman Sachs is not really affected."[6]

The fact is that, despite the bad name earned by American extravagance, consumer credit in Britain was, before the crisis, on the average higher than in the United States. The British were addicted to an orgy of debt-funded consumption more than any other developed nation.

Then the skies of prosperity darkened. As students and then journalists, we enjoyed the privilege of standing somewhat apart from the catastrophe and watch it slowly unfold. Northern Rock, a British bank, was the first to hit the headlines, in September 2007. A severe liquidity crunch impelled it to seek government assistance, setting off a three-day run on the bank as panicked savers withdrew £2 billion.[7] The government was forced to nationalize the bank in order to save the public's deposits. It was the beginning of a turbulent year that climaxed with the Lehman Brothers bankruptcy. The City of London, on no less wild a spree than Wall Street, had been infected by the subprime virus. The chief symptom was a lack of mutual trust among banking institutions, which led to a credit crunch.

The British government had to make available £200 billion in loans from the Bank of England and another £50 billion in government loans, half of the latter in exchange for preference shares. It was a partial and precedent-setting nationalization of the British financial sector. The crisis percolated down to the real economy—restaurants, hair salons, and wine and cigar shops. The effect was profound and palpable. Within three years, the number of the unemployed soared by 50 percent. Fine restaurants put signs in their windows offering a Credit Crunch Lunch. A local Birmingham newspaper reported that six unemployed men, some with paunches, were rehearsing a strip show, a live version of the film *The Full Monty*.[8] It was a fun story, an anecdote that dispelled some of the long shadows that darkened the pages of the press at the time. The media reported, for example, the suicide of a German billionaire whose capital was wiped out in the stock market crash. At his high point, Adolf Merckle had been worth $12 billion, but his investment company collapsed. He threw himself under a train in January 2009.[9]

Globalization turned an American crisis into an international one. In each location it revealed different iniquities. In Spain and Ireland the real estate markets were in an outrageously dangerous position; France had a rigid job market, in Italy and Portugal the problem was national debt; the East Asian economies had become utterly dependent on exports. The citizens of industrialized nations were confronted with the same fact—commercial financial institutions were evading public accountability, because if they were allowed to go bankrupt they would drag the entire economy down with them. They were, as the mantra went, too big to fail. Saving those institutions was not a demonstration of wild capitalism but a betrayal of its celebrated principles—a nationalization of the personal failures of bankers who continued to spend astronomical sums in the most fashionable bars in Soho.[10]

A perfect storm ensued—a crisis in the real economy, rising unemployment, and, as a result of both, state debt crises. It was a globalization of entanglement, born in the US.

Americans generally did not understand then, and most likely still fail to grasp, just how profoundly the crisis affected how the world sees their country. For Americans, the subprime crisis was ultimately a local problem rooted in their own financial system. For the world, it was an American problem transfused into the international bloodstream by a made-in-America global order. As such, people outside the United States felt much more powerfully than Americans did that their communities were under the thumb of arbitrary external forces that were toying with their lives.

Americans explained to themselves, in their own terms, how they had ended up in crisis. It was simple. After all, they owned it. The rest of the world found itself ensnared in a story told by foreigners. Europeans and Asians of course blamed the US, but it ran deeper than mere recrimination and blame games. As soon as economies were joined and interdependent, decision makers could no longer quarantine their countries off from the crisis. The attempts to do so were pitiful, even when the local banking sector acted with cautious conservatism and did not invest in toxic securities. Had sovereign governments resolved to dramatically increase their public spending with the aim of maintaining demand and economic activity in their countries, they would have been punished immediately in the markets. The central banks of other countries had very limited monetary freedom, because the international economy was utterly dependent on the dollar. The Fed presented a monetary policy for America, but for central banks everywhere that policy was, in practical terms, a strict edict.

In a moment of truth, it all turned out to be a puppet show. Legitimate institutions and elected representatives proved to have little control over the fate of their communities. The affected public felt that the political discourse in their own countries was a mere facade. As they saw it, real power resided, in the best case, in Washington or, worse, in the hands of plutocrats—or, in the absolute worst case, in a market that was in the throes of arbitrary forces that no one controlled at all. In 2009 stories began appearing in international media about ATMs where you could receive gold ingots instead of cash,[11] a glistening indication of just how worried the middle class was, from China to the US.

The financial crisis was so poisonous because it was so diffused. The removal of barriers amplified interrelations. At a time of global crisis, these connections could turn into a stranglehold. Actually, globalization created a situation in which America lost control of the dollar market. The 1950s saw the beginning of a Euro-dollar market, in which non-American banks, in particular European ones, began offering loans and financial instruments denominated in US currency. This market grew quickly, to the point that when the crisis developed, European dollar investors began to flee in panic. The Federal Reserve had to provide guarantees worth at least $1 trillion in foreign countries and to foreign banks. In other words, as the US itself stood on the verge of financial collapse, it had to step in to save foreign markets as well.*

European Steroids

Tragically and tellingly, the American economy was not the only one on steroids. On the other side of the Atlantic another crisis brewed, similar in many ways. About three months after the September 11 attacks, in one of the most ambitious projects in modern history, twelve member states of the European Union adopted a single unified currency, the euro. In January 2002, euro banknotes and coins went into circulation. The currency had been launched on paper in 1999 but served only as a unit of accounting until the changeover occurred.

Something strange happened during this transition period—countries such as Greece, Italy, Spain, Ireland, and Portugal suddenly benefited from the aura of being strong, wealthy economies, as if they were the equal of Germany, Holland, and France. This aura made it easier for the EU's poorer countries to raise money. The bond yields of the eurozone countries aligned almost completely. It was as if a bank manager were willing to give a loan to a family in financial straits at the same low interest rates it offered to the rich family next door, which had solid assets and an excellent history of repaying its loans, simply because the two were neighbors. By any reasonable economic standard, this should not have been allowed, but the European project sought to bend the conventions of economics to its political fantasy.

---

* Since the 2008 crisis, the total dollar credit outside the US has nearly doubled, to about $13 trillion as of January 2019.

Back in April 1977, Donald MacDougall, then chief economic adviser to the chancellor of the exchequer, submitted a report arguing that an economic union would require a consolidation of budgetary policy. In his view, a single currency could not survive unless there were massive capital transfers from the European north to its poor south.[12] To this end, some kind of European government would need to be put in place, overseeing expenditures of between 7.5 and 10 percent of Europe's GDP. That was a more modest level than the prevailing one in federal or confederal systems around the world. MacDougall was not alone. At the beginning of the 1990s, Chancellor Helmut Kohl of Germany maintained that a monetary union would be impossible without a political union. Milton Friedman, the leading voice of the monetarist school of American economists, wrote in *The Times* of London in 1997: "Europe exemplifies a situation unfavourable to a common currency. It is composed of separate nations, speaking different languages, with different customs, and having citizens feeling far greater loyalty and attachment to their own country than to a common market or to the idea of Europe."[13]

This cautious approach was rejected by the bureaucrats of the European Commission, who thought that fiscal discipline would be sufficient. They chose the more dangerous option because they knew very well that European taxpayers would never consent to grant them the authority to make outlays on the level of a real federal political entity. On the eve of the euro's launch, the Greek government needed to pay interest rates 50 percent higher than those of Germany in order to borrow money on the bond market—in other words, the market considered the Greeks more risky. Two years later, Greek bonds were being traded at the same yield as Germany's. Recall that when the Federal Reserve lowered interest rates, it spurred its citizens to take cheap loans. The convergence of European interest rates at a low rate injected money into countries like Spain, Ireland, Portugal, and Greece—until they burst under the weight of their loans when crisis hit. Governments borrowed money to fund welfare programs, roads, and inflated bureaucracies. And of course, the cheap money made its way into private markets and to consumers. As in the case of the US mortgage market, everyone thought that the music would never stop.

Fortune Cookie Globalization

Americans had perhaps the ultimate excuses for their pumped-up economy—a massive terror attack, two wars, and a doubling of defense expenditures. European leaders were both determined and deluded when they constructed a shared political and economic future. In both cases, the leaders tried to treat fundamental economic and societal problems with a globalization elixir. EU leaders believed that the disparities between southern and northern Europe could be erased by a common currency. It was a kind of fake-it-till-we-make-it. "Making it" meant creating a strong supranational European authority that also left the member states with considerable autonomy.

The result was a monetary union with no joint fiscal responsibility, European solidarity while at the same time reinforcing national identity. If it sounds like a paradox, looks like a paradox, and acts like one, then it is, apparently, the European Union. Europe wanted to be united, so it hid its head in the sand in hope that the risks it was taking would wane and disappear.

On the other side of the Atlantic, American leaders were convinced that the fundamental problems of the US economy in the wake of 9/11 could be ameliorated by injecting cheap money directly into a deregulated financial system, and by a focus on consumer confidence. The great optimism that followed the fall of the Berlin Wall and the sense that the end of history, as Francis Fukuyama famously put it, had arrived helped drive the euphoria. Political slogans became policy.

Had the US had a strong and prosperous middle class prior to the al-Qaeda attacks, it might have worked. But many of its blue-collar workers were already suffering from the trauma brought on by free trade, the decline in interest rates since the 1980s, the growing dominance of the financial sectors, the stagnation of wages, and the fracturing of the American dream itself. Once the supposed invulnerability of the home front was smashed when the World Trade Center crumbled, the bleeding could not be stanched with economic bandages. It was no longer a matter of it's "the economy, stupid," as the Bill Clinton campaign put it on the way to the White House. Rather than addressing the profound insecurities of the American middle class, the country's decision makers plied them with loans.

It was a kind of fortune cookie globalization, offering banal platitudes wrapped in sweet pastry. The cautious and measured style of the age of responsibility was replaced by wild leaps of faith and vast overconfidence in policies based on positive thinking. The European middle class was enticed into a political union that, it was assured, would have only benefits. The American middle class was promised prosperity, low interest rates, low taxes, and an expanding empire. In both cases, communities were driven to ever-growing risk-taking while taking bites of a fortune cookie. But it turned out that the cookie was poisoned and the fortune inside utterly specious; it was only natural for the middle classes to reconsider their long-held adherence to mainstream politics.

Determined to regain control, many searched for new answers. It was a spiral of reaction following reaction, a turn to radicalism that was met with counter-radicalism. All that was needed was a spark to set off the revolt against the current globalized order. No place exemplified this better than tormented Greece.

CHAPTER 10

Anarchists with Ferraris

I hope for nothing. I fear nothing. I am free.

—NIKOS KAZANTZAKIS'S EPITAPH

ATHENS, MARCH 2009

Four months after the fundamentalists struck in India, I set out for Athens to cover the crash of the Greek economy. The crash was the inevitable result of the global economic crisis. Suddenly the revolt was brewing everywhere, all at once.

A DILAPIDATED HOUSE STANDS IN THE CITY'S CENTER, NOT far from the National Archaeological Museum, with its treasures from Greek antiquity. The weather is balmy and spring is in the air. The walls of this place, whose grand days are long past, have been stripped of everything of value. Its wooden door, heavy and fractured, is generally chained from the inside. Three guards in jeans and t-shirts sprawl on the steps. If you knock very hard, the door opens a crack and the house's squatters peek out and ask to know who you are. "We received the email," they said. "Come on in." Inside, the house reeks of beer, cigarettes, and something else that would later become clear. "My name is Yiannis," says one of our hosts, a courteous young man in a black windbreaker. "Welcome to the commune."

The West was born in Greece. That is not just a historical statement but also a geopolitical fact. Ancient Greece was the wellspring of Western culture. Greece is the source of a huge river of ideas: Plato and Aristotle, whose opposing schools of thought, along with Judeo-Christian traditions, would make up the mosaic of Western civilization. Modern Greece is the eastern boundary of the European Union.

In 1974, the military junta that had ruled Greece since 1967 collapsed. Constantine Karamanlis, the new prime minister, enthusiastically promoted Greek integration into Europe. He told one of his fellow European leaders that "Europe is a Greek word." The founders of the European Community, the forerunner of the EU, felt on their part a sense of historical responsibility and reconciliation. France's then-president, Valéry Giscard d'Estaing, put it simply: "It was impossible to exclude Greece, the mother of all democracies, from Europe."[1] Apart from its symbolic import, that wasn't a real justification for welcoming Greece into the EU. Greece was, after all, a relatively poor country that was far behind the other members in terms of industrialization. The historical pretext concealed geostrategic considerations that were spoken of less openly—for example, the importance of creating a model of relative economic stability in southern Europe with the aim of blocking potential Communist influence. At first it looked as if Greece's membership in the community was a big success. Portugal followed, creating a sense of balance between southern and northern Europe.

It quickly turned out to be a mirage. Decades later, in a double interview, Giscard d'Estaing and the man who had been his German counterpart at the time, former chancellor Helmut Schmidt, told *Der Spiegel* that it had been a mistake. "Greece is basically an Oriental country," Giscard d'Estaing told the magazine, using "Oriental" pejoratively. "Helmut, I recall that you expressed skepticism before Greece was accepted into the European Community in 1981. You were wiser than me."[2] Such arrogant language, the Greeks say, was always the EU's problem in Greece.

When I made my visit, Athens was suffering from a deep economic crisis; demonstrators were throwing Molotov cocktails at the police and vowed to destroy "the oppressive system." Inside the derelict house were electrical cables, empty beer crates, bags of garbage, and boxes of food. We ascend a staircase to the building's large and impressive central space. Two young women, students, sit on a threadbare couch. They are drafting a manifesto. Their previous manifesto is available on the internet. "We have once again demanded this derelict house," they wrote, "so as to turn it into a free social zone in which, by means of self and collective organization, we will all participate in the plot to *destroy this world* [emphasis in the original] . . . The imminent uprising is everywhere!"

The two revolutionary women do not look very subversive. They are sipping café frappé, a Greek form of very strong iced instant coffee. They drink it *glyko me gala*, sweetened, with milk. Yiannis tells me: "We received your email and we know that you are a journalist, but we did not have a chance to discuss it. You are invited to stay for coffee." He makes one thing clear: "We do not speak with the mainstream media. Only dissident. Do you work for a dissident outlet?" I respond cautiously: "Not exactly." Yiannis says, "We'll need to discuss it at a general meeting of the comrades."

A pudgy boy with dreadlocks sits on a bench not far from me, smiling. "*Shalom*," he says in Hebrew. "I'm Dimitris." It turns out that he has some Israeli anarchist friends and they've taught him some Hebrew. "Listen, we're all very busy," he explains. "The girls are drafting a resolution. I have to be someplace in ten more minutes. Come back afterward." As we speak, I notice that Dimitris is doing something with several bottles that are standing on the bench next to him, but my brain doesn't entirely process it. He is filling them with yellowish fluid to the halfway point, and then inserting wads of cloth or paper. There's an odd smell in the air. Then I realize that he's preparing Molotov cocktails.

In the weeks after my visit, the Greek media reported a series of acts of arson committed by anarchists. One of the fires was set in a store in a suburb of the Greek capital that sold gear for luxury yachts. An anarchist website issued a statement accepting responsibility:

> Many have been indulging in displaying their wealth, wasting themselves in a race for more wealth, the only way to fill up emptiness inside. Luxury and expensive stuff are there to be worshipped by those who can have them—and make people who cannot afford them dream about them. This is building the foundation of a hollow society, a society of illusions and spectacle. This is why we attacked the yacht store in Argyroupoli, and we are determined to leave behind, each time, only ash and debris to agitate the nervous system of all those submissive defenders of wealth.[3]

Perhaps you are smiling. The labored and rhetorical extremism of the political fringe, with its rigid and unbridled fervor, can verge on the ridiculous. But radicals play a key role in the societies they denounce. They mark the boundaries of public discourse, both to the right and to the left. In recent years, radicals have crossed from the margins into the mainstream, either by being elected to positions of power, or by having their ideas adopted by central political parties and turned into policy. The texts employed by Trump and the Brexiteers are prime examples. This is understandable: the political center failed to avert the great recession and the ocean of insecurities that followed, so the crisis has brought on a renaissance of extremism.

The anarchists in the ruined Greek house belong to the longstanding core of opposition to globalization, and in particular to the capitalism with which the global system created by the United States is imbued. They have not changed. It's the sons and daughters of the middle class who, faced by threats to their security, identity, and employment, began a passionate affair with radicalism. This happened in many countries, taking on local characteristics as it connected with old beliefs and new fissures in society. In Greece, one of the weakest countries in the EU, the fractures ran wide and deep.

The Greek Fantasy

Greece has been part of my experience since my boyhood. My family had a small business there, and my father lived in the country for many years. We traveled there on summer vacations but also experienced the local stormy winters. We rambled along the empty beaches of the Peloponnese, drove along the shaded roads of the Pelion peninsula, wandered the old cities of the green north, from Ioannina to the beaches—where you can see the island of Corfu—but we also went to university towns, like Larissa. I spent my summer vacations with Greek and Cypriot teenagers, committing pranks that none of us will ever forget. Once, a rowdy truth or dare game went wrong and I found myself being chased by a Greek man in his underwear, waving a sandal that had slipped off my foot as I fled. He didn't catch me and I never got my sandal back.

When I got older and no longer disturbed the sleep of good, decent Greeks, I learned from them the art of relaxing in cafés or *ouzeri* and having idle conversation while nibbling *meze*. Henry Miller's paean to Greece, *The Colossus of Maroussi*, may be guilty of Orientalist arrogance, but it also touches on a truth:

> Everybody goes the wrong way, everything is confused, chaotic, disorderly. But nobody is ever lost or hurt, nothing is stolen, no blows are exchanged. It is a kind of ferment which is created by reason of the fact that for a Greek every event, no matter how stale, is always unique. He is always doing the same thing for the first time: he is curious, avidly curious, and experimental.[4]

For many people, Miller's depiction remains on the mark. They sense it on their August vacations on the Aegean islands, where life is unhurried. The feeling is that, despite Greeks' avid patriotism, the state itself is tenuous and haphazard, and thus allows people's ancient freedoms to survive undisturbed.

When I went to Athens to cover the protests brought on by the economic crisis, I found myself driving my tiny rental car through the middle of mass disturbances in the city center. Tear gas grenades thudded around me. I needed to transmit a report to my television news program, but I couldn't find a parking space anywhere. My deadline was uncomfortably close. So I made an appalling decision—I circumvented Syntagma, the city's central square, and parked my car (devoid of a press sticker) directly in front of the Parliament building. I returned an hour later and the car was still there. It had not been towed, there was no parking ticket on the windshield, and no irate police officer stood next to it waiting to arrest me. This Greece, unorganized but safe and easygoing, began to fade over the last decade. Its deterioration was a direct result of the country's contention with globalization, and its chief representative in the country—the European Union.

From 1980 until the economic crisis began, Greece received some €200 billion ($212 billion) from the EU in aid and grants.[5] Blue signs proclaiming projects funded by the EU could be seen all over Greece, especially close to main transport arteries. The Greeks have claimed for years that most of the money, such as infrastructure grants, found their way into the pockets of European companies that bid on the tenders and sucked out the money that the EU sent in. In that sense, the funding was actually a European fiscal expansion, and the rich northern countries greatly benefited from it. By any account, the local economy grew, but the replacement of the modest drachma induced a frenzy of consumption funded mostly by credit.

Anyone who knew Athens could see it. The price of a cup of coffee jumped from a few hundred drachmas (about $1.70) to several times that. EU membership gave Greece a radiant aura of development, but the Greek economy remained saddled with severe structural problems. For example, its labor market was ossified, granting workers a month's vacation, along with one and sometimes two bonus salaries a year; it was very difficult for businesses to fire workers. In some cases, women with children were eligible for pensions at the age of fifty.[6] The public sector was bloated and tax collections low.[7] The *New York Times* reported in 2010 that in all Greece, a country with a population of about 11 million, only a few thousand citizens reported an income of more than $132,000 a year. The real number was at least in the hundreds of thousands. Only 324 homeowners in Greater Athens reported having a swimming pool, but aerial photographs commissioned by the Greek tax authorities after the crisis showed that the real number was more than fifty times higher, 16,974.[8]

The EU set as a basic rule that its member countries could not run a deficit higher than 3 percent of their GDP. The purpose was to ensure stability, but there was no enforcement. It was the epitome of fortune cookie globalization—the EU created a monetary union that included rules about deficits, but there was no effective monitoring of those deficits. The Greek government used creative accounting to present a false picture to European institutions. On the basis of that picture, the Greeks were able to borrow more money on the bond markets. Large investment banks helped by selling to countries like Greece "sophisticated financial products" that served mostly to hide huge deficits and postpone debt repayment. The banks of course made huge profits off these "products," leaving the Greeks with the debt.[9]

When the great crisis of 2008 struck, the Ponzi scheme collapsed. Greece's deficit was revealed to be almost 15 percent of its GDP. The Greek state could no longer raise money from the markets to finance its debt and pay its rocketing costs; the state faced bankruptcy. Private-sector businesses soon found themselves in the same situation. A classic remedy would have been for Greece to devalue its currency and tighten its budgetary belt. Devaluation lowers real wages without lowering nominal salaries, reduces the costs of production, and encourages exports. That's what Iceland, not a member of the eurozone, did in 2008 and again in 2011, contributing to its rapid economic recovery. But Greece was stuck with the euro, and thus found itself without the economic tool it needed to respond to the crisis.

The country's creditors—the International Monetary Fund (IMF), the European Central Bank (ECB), and the European Commission, collectively known as the troika—demanded that the Greek government impose severe austerity measures. The EU transferred to Greece enormous sums as part of a bailout program. But most of the money went to pay the country's debts and interest on those debts rather than being invested in the devastated economy. The austerity regime instituted to cope with the crisis was cruel. It included deep cuts in pensions, welfare payments, education, and the health system. It tore Greek society apart. Under the constraints of the program, the government imposed an austerity tax that had to be paid along with electric bills. To avoid paying it, thousands of homes disconnected themselves from the electrical grid. Many others simply left their homes. For years, citizens had a daily limit on the amount of money they could withdraw from ATM's and bank accounts. In 2017, half a million Greek homes, most of them in Athens, stood empty.[10] As of the end of 2018, 39 percent of Greece's young people were unemployed,[11] and almost 17 percent of the population lived in a state of severe deprivation.[12]

The EU did Greece no favor with its rescue program. By one estimate, the cost to French and German taxpayers of a Greek default would have been €60 billion, and the leaders of the EU were determined to save their own heads. Had Greece gone into bankruptcy, it would have opened a Pandora's box, revealing dirty secrets, rigged bids, and secret partnerships.

Shitting on Europe

Oddly, that morning, in the revolutionary squatters' house in Athens, was full of hope. For the young anarchists I chatted with as they manufactured Molotov cocktails, the economic crisis was simply an excuse for completing the democratic revolution that their parents had begun when they took over Athens Polytechnic as part of their uprising against the junta in 1974. As they saw it, the entire global system was a disease. Greece's failure was proof of its repressive nature. The existing political system, with its established parties, were part of the regime that had duped the Greek public and the world, taking their cut from fat deals before sending the country careening toward disaster. Their analysis of the facts was accurate enough; the solution they proposed was no less extreme—revolutionary solidary and a deconstruction of the entire political system. Some factions turned to violence.

Other homes in Athens hosted meetings of Golden Dawn, whose logo resembles a swastika. Golden Dawn spoke of Greek superiority and committed harsh acts of violence, mainly against migrants. For the extreme right, the country's crisis proved that its dependence on foreigners was the source of all evil. The solutions the Nazis offered were old and morally defective, but their time had come—again. In the elections of 1996, the party had received only a tenth of a percent of the vote. In 2015, it became the third-largest party in Greece, winning 7 percent and hundreds of thousands of votes. In the elections to the European Parliament in 2014, Golden Dawn received the support of nearly 10 percent of the Greek population. An elderly Greek man with a cigarette in his mouth told me, during a demonstration in front of the Parliament building, that he voted for them so that they could go to Brussels and "shit on Europe." The party collapsed in 2019, as a result of a relative improvement in the economic situation, a police investigation against it, and because some parts of the mainstream right adopted planks from Golden Dawn's platform. The era of the revolt against globalization is one of extreme swings and distortions in an unstable political system.

Maria Axioglou is an official in a commercial bank. When the crisis broke out, we sat in a fashionable café, one of several in the center of Glyfada, a tony Athens suburb. The road there, Vouliagmenis Avenue, is lined with car companies—Porsche, Lexus, BMW, Mercedes, Alfa Romeo, to name a few. Glyfada is green, elegant, and obsessed with brand names—its main street is often described as Athens's fashion mile. Not far from the cafés lies a large marina where white yachts are arranged one after another. Axioglou, born at the end of the 1970s, is a Glyfada native—her family was one of the first to move into the suburb. Her well-off parents built an entire building, giving her and each of her siblings a floor. After a decade working for her bank, she earns €1,300 a month ($1,800 in 2009 dollars) before taxes. She has a college degree, of course. It's Maria's friends who took to the streets in December 2008. They coined a name for their cohort—the 700 euro generation.[13]

"The 700 euro generation," she says, "are people who completed degrees but don't find work in their professions. They have no choice but to work as, say, waiters. If they do find work in their professions, they don't earn more than 700 euros. Understand, that's a wage you can't live on, and it certainly won't buy you an apartment in Athens. In December 2008 there were riots and demonstrations. I didn't go, but I had friends who took part. People took to the streets because of a feeling that they have nowhere to advance to. They feel that everything is stuck, frozen, and that the economic crisis is just making it worse. Our feeling, all of us, is that we will get less than our parents did."

There is a palpable contradiction between the wealth and tranquility that Glyfada radiates and what Maria says. "Our parents' Greece was a different Greece. We want to squander. To consume. To have a good time. On the one hand, because Greece is part of the European Union, we have more opportunities. You see the display windows in this neighborhood. On the other hand, we don't have money. We also know that we never will. Everyone here lives on their credit cards. I know, I see them at the bank. People drive Porsches, but they don't have a cent in their bank account. Young people lived from loan to loan. And the political system is corrupt, so corrupt. I can understand why people went to demonstrate. But there's a difference between protesting against the government and the system and destroying everything. Don't expect a revolution here. We're too wimpy. Maybe we'll believe in it, but we don't do it."

Her monologue voices just how distressed—but also just how conventional—this entire European, even global, generation is. At least it is the "voice of *a* generation," to quote Lena Dunham.[14]

In every country, the global crisis loosened the seams and allowed tightly reined-in sentiments to break free. Anarchy and the rejection of hierarchy have been fundamental to all of Greece's liberation struggles—against the Turks, the Nazis, and the dictatorship. "The Greeks," Maria tells me, "do not like it when someone rules them. But more than that, they want to be anarchists with Ferraris." Between my conversations with her in the café in 2009 and 2018, when Greek salaries plummeted by 20 percent in real terms, the Greeks have lost much more than their Ferraris.

A Financial Virus in a Global World

At the beginning of 2015, the radical left alliance Syriza won the Greek elections. The two parties that had ruled Greece for generations, the social democratic PASOK and the conservative New Democrats, were sent into the wilderness of the opposition. Later that year came a referendum in which the Greeks said no to the European Union's bailout program and the extreme austerity measures it demanded.

What had they discovered in the years between the onset of the crisis and 2015?

They realized that, as far as they were concerned, the EU was a con for the benefit of well-fed elites. As they saw it, it was a con combined with a Ponzi scheme in which Berlin and Paris handed out money to their tiny subject nations, making sure that it would flow straight back to Berlin and Paris. It was an outlay that was entered on the books as a pan-European gesture, but in fact it was simply a budgetary expansion aimed at entrenching the dominance of Germany and France. A study from 2016 showed that only 5 percent of the money from the bailout programs actually reached Greece. All the rest made its way to financial institutions, private and public.[15]

Something unusual happened in the referendum. I was in Athens when the Greeks issued their resounding *ohi*, Greek for "no!" It wasn't the first time they had done so. Each year on October 28, Greece observes Ohi Day, the anniversary of the people's rejection of Mussolini's ultimatum that Greece allow Axis troops to enter the country in 1940. Greece valiantly resisted the subsequent invasion, and underground forces continued the resistance after the Axis occupation of the country.

It was an impressive thing to watch. For Greeks battered by the economic crisis, the referendum was a chance to stand tall and declare their patriotism. It didn't matter whether their *ohi* was radical, ultranationalist, or communist. What was important was that they spit in the face of the European Commission, the IMF, and the ECB.

Souvlaki stands sprang up not far from Syntagma Square, and Greeks of all incomes, faiths, and political persuasions drank ouzo and raki and chanted slogans together, some of them from the Bakunin school of radical anarchism, others from the Greek Orthodox community, and still others from the Greek nationalist movement of which Lord Byron was so enamored. Luxury hotels locked down their reinforced shutters and waited for the wave of celebration and rage to pass. But instead of shattering windows, Greece shattered united Europe and the mantle of arrogance it had draped itself in. It was a mess, yes, but a democratic mess. The Greeks said: "You wanted globalization, you wanted monetary union? And you wanted firewalls that would keep our troubles separate from yours? You can't have everything."

Then, less than a week after this mutiny against the European Union, which activists saw as a version of the Prague Spring of 1968, Greeks capitulated unconditionally. The ostensibly radical Greek government accepted a European austerity plan even more draconian than the one that voters had overwhelmingly rejected in the referendum. The Greeks discovered that, as in the epitaph chosen by the modern Greek literary giant Nikos Kazantzakis, they hoped for nothing. But unlike him, they fear everything. And they are not free.

RIGHT NOW IT LOOKS AS IF IT'S ALL IN THE PAST. INVESTORS have been pleased with Greece since 2017, and it has even been able to borrow money by selling bonds. The prices of its private-sector assets have been attracting investors, who see an opportunity to make money off of Greece's plight. Greece has met the strict conditions of the bailout program, avoiding deficits. Unemployment has declined significantly, and the economy had been growing moderately up until the Coronavirus crisis in 2020. The government is collecting taxes effectively, in part thanks to electronic monitoring methods imposed by its European creditors. The conservative New Democrats won the elections in 2019, returning to the halls of government they had been ejected from by the protest movement. They won thanks to a resurgence of enthusiasm, among young people as well, for a neoliberal market economy.

But the revolt is still underway, because the ground from which it grows is still fertile. After all the talk of the 700 euro generation, the Greek government in 2012 set a special minimum wage for young people up to the age of twenty-four—€511 per month.[16] It was revoked only in 2019. What about their parents' wages? They declined, or at best remained stagnant. Hundreds of thousands of Greeks were reliant on soup kitchens and food banks. One food bank in Athens had 26,000 registered clients in 2017.[17]

The fertility rate has declined to 1.35 children per woman. The low birthrate and negative net migration is driving a decline in the population.[18] Between 2008 and 2016, four out of every one hundred Greeks left the country. Most of those who left were young.[19] It's easy to understand why. A third of minors live in poverty or are at risk of falling into it, and some 40 percent of young people between the ages of fifteen and twenty-four were unemployed in 2018. Greece traditionally had a low suicide rate, but between 2010 and 2015 it rose by 40 percent; during the same period, the budget for the country's mental health services fell to half of what it had been before.[20] At one point, 12 percent of the population reported being depressed for more than a month.[21] A Council of Europe report explained that many of the people who had reported depression were forty or older and had no history of mental issues. They were "unemployed persons, bankrupt businessmen, or parents who have no means of taking care of or feeding their children."[22]

Greece served at the front lines of the crisis in that its citizens experienced its most dire economic effects more acutely than pretty much any other citizenry. As they saw it, the global economic forces affecting their lives were capricious, and they were powerless in the face of them. The country had become a province taking orders from Berlin, Paris, and Brussels on everything to do with its material well-being. The (former) mainstream political parties understand that, and, following the failed revolt of the 2015 referendum, they no longer pretend that they can engage in substantive economic debate over issues that are beyond their control. All they can do is fight about identity, culture, and perceived external threats, such as whether the use of the name "Macedonia" by a tiny country to the north is a threat to Greece's sovereignty and national pride. (Greece has a province of the same name.)

The right has a clear advantage over the left—identity is basic to its code. The more extreme elements on the right lash out against immigrants, or the "Muslim invasion," as they call it, and trumpet conspiracy theories about Jews taking over the country. In 2018, for example, they accused Greece's foreign minister, Nikos Kotzias, of being an agent of the Jewish Hungarian-American investor and philanthropist George Soros. New Democracy, the mainstream conservative party, has incorporated some extreme right-wing elements. The heirs of Constantine Plevris, the Greek racist I wrote of above, are now respectable leaders of New Democracy. Adonis Georgiadis, the party's deputy leader and one of its cabinet ministers, was in the past associated with racist remarks and enthusiastically promoted one of Plevris's books, *The Truth About the Jews*. Plevris's son now serves as a New Democracy parliamentarian. Another extreme rightist, Makis Voridis, who according to his own admission has "coexisted politically" with Holocaust deniers has been appointed to the cabinet.[23]

Greece is merely a showcase for what is happening all around. The mainstream French right has taken up Le Pen's positions on immigration; Britain's Conservatives surrendered to Brexit; and in the United States the Republican Party's conservatism has been replaced by Trumpism.

EXTREMISM ALWAYS BENEFITS FROM THE EXPOSURE OF HYpocrisy. That happened not only in Greece but throughout the West. For example, there was the hypocrisy of the Wall Street bankers who peddled to their customers ostensibly reliable and secure investments that were in fact anything but, while at the same time investing their own money in instruments that would pay out if their customers lost money.[24] There was the hypocrisy of world leaders who backed a new order in which the two sides were clearly drawn—there were the weakened classes, including the middle class; and there were the big winners. The latter could be politicians, contractors, high-tech giants, or financiers—anyone who benefited from the age of exceptional inequality. This inequality was of course not a matter of chance but the result of considered tax and interest policy. Beginning in the 1980s, interest rates were slashed again and again all over the world, encouraging consumption and discouraging savings. Some argue that low interest rates reflected reduced risk and widening economic confidence. "The spice must flow!" the Emperor in the film *Dune* is warned.[25] In the real world, instead of the spice, it was cheap money. The West was addicted, so the drug had to be supplied at any price. For that to happen, the rules had to change, everywhere.

This is not the wisdom of hindsight. In 2004, a brilliant professor at Harvard, Elizabeth Warren, told PBS that, since 2000, there had been an increase of 55 percent in the number of people who were defaulting on their credit card debt. There had also been a 45 percent jump in home foreclosures because people could not make their mortgage payments. She warned against "creative" financial products that were robbing the middle class of its future. "Alan Greenspan, our national economic leader," she warned, "has stood up for the last four years and told Americans, 'Borrow against your house. If you can't close the gap at the end of the month, just borrow against your house.' Now, he never called it borrow against your house. He said fancy things like 'tap your home equity' which sounds like some kind of dance . . . That's really scary financial advice for someone to be giving American families. And what frightens me most is millions of American families have taken that advice."[26]

Warren addressed one angle of the financial takeover of the real economy. To put it simply, large parts of the industrialized north, Germany being the main exception, turned from a place where things like cars and television sets were manufactured to places that "manufactured" bonds, stocks, derivatives, and of course debt.

In 1940, financial sector income was equal to only 3 percent of American GDP; by the 2000s, it had almost quadrupled.[27] The level was the highest since just before the stock market crash of 1929, which marked the beginning of the Great Depression. That's not all. In the mid-2000s, 40 percent of all American private-sector profit came from banks, investment firms, private funds, and other parts of the financial sector.[28] At the same time, between 1999 and 2008, American household debt doubled nominally, growing from 67 percent of GDP to almost 100 percent, largely due to mortgages.[29] As Elizabeth Warren predicted, it ended in a crash. But there are a lot of ticking bombs to set off the next crisis—for example, student loan debt.

At the end of 2019, American household debt stood at $14.15 trillion, an average of about $110,000 per household, higher than it had been at the outset of the Great Recession of 2007–2009.[30] When debt grows, so do the profits of the financial industry. But those profits are not always good news for the rest of the economy. Martin Wolf, the noted columnist for the *Financial Times*, wrote, "The huge expansion in finance since 1980 has not brought commensurate economic gains."[31]

So what did we have? A policy of reducing interest rates, cheap money, financial industry expansion, households enthusiastically saddling themselves with debt, asset bubbles, and unsustainable growth. The Day of Judgment in 2008 laid the foundation for the revolt, and the revival of growth after the crisis did not strengthen the middle class. On the contrary, the position of the middle class continued to erode. The US unemployment rate plunged beginning in 2010 and nine years later had reached a historic low. Yet at the same time, four out of every ten Americans were unable to cover an urgent onetime expense of $400 except by taking a loan, or selling something—or not at all.[32] In other words, large parts of American society are frighteningly, even catastrophically, fragile; during a time of economic prosperity, they cannot regain the financial security they lost. The most avid supporters of the populist and extreme left or right are not the poor who suffered the most from the crisis but in fact members of the middle class who have not yet fallen but are gazing in trepidation over the precipice.

In these precarious circumstances, extremists amassed and set their gazes on a new world. The Jeremiahs can all pat themselves on the back, transform mainstream political parties into hives of populism, set up Nazi movements, fill bottles with gasoline, promise a state of utter equality, or solemnly engage in revolutionary neo-Marxism dialectics.

They can post bogus stories on Facebook and spout conspiracy theories on Twitter, all with the certainty that extremism has been welcomed again to public discourse. In most cases, these people will not themselves enter the halls of power. But they will intimidate those at the helm and pull the mainstream political camps to the extremes. In such challenging times, existential elements of social identity and society's sense of meaning are potentially explosive. The most important of these is the family.

CHAPTER 11

Disappearing Children

I'd like there to be 120 children in my class.

—KAGAMI REIKA, JAPANESE FIRST GRADER

About sixty miles out of Tokyo, the mountains glitter in the light of the rising sun. The steep route runs along a large, softly flowing river, and the morning is crisp. The trees are displaying their first flowers. Hanami, the cherry blossom holiday, is approaching, and at the small shrine near Nanmoku, a small band of villagers is praying and supplicating. They are all elderly. That makes sense. According to figures supplied by the prefecture in which the village is located, Nanmoku has a population of 1,666.[1] In 1997, just a little more than twenty years ago, 4,000 people lived there. That's a decline of 58 percent and explains why the village has the dubious distinction of being the most elderly village in the oldest country in the world. The median age of its inhabitants is seventy.[2] The few stores on the main street have few customers. The place looks as if it has been swept almost clean of its inhabitants. Many buildings are unoccupied, although they are well kept up. This is Japan as it ages—still clean and upright, like a cherry tree.

The local elementary school is a structure of impressive size, built for many hundreds of students. In 1959 Nanmoku had three schools attended by 1,600 children. Today there is just this one. It has a total student body of twenty-four.

There is a huge disparity between the stories our minds fashion and the complex contours of the real world. It's so simple to let stock images dictate the story we tell, and to set aside some of the evidence that our senses provide. Doing so means falling into the trap of depicting the school as dismal and depressing, lonely and craving more children. While the schoolchildren are indeed few in number, the school itself is warm and human. As is customary throughout Japan, you take your shoes off at the entrance and replace them with slippers. Pictures and assignments that teachers and pupils have worked on together are posted on the walls. Because there are so few children, the older ones play with the younger before the bell sounds for class, and the teachers play with them as well. The classrooms are well stocked. The arts and crafts workshops are equipped with cutting-edge engraving tools; there are computer rooms, and there is a kitchen for cooking classes. It's just that most of it, most of the time, is not in use.

In the first-grade room we met the entire class—two girls. Kagami Reika and Hisoki Miaka are Nanmoku's tiny future. Their faces round and serious, they are deep into a workbook. Theirs are the only two desks; perhaps no unoccupied ones have been left there so that they will not feel that other children are missing. The room is large, so there is a lot of empty space. The walls are decorated with drawings in which the two girls depict their experiences in their first year of school. Kagami wears a paper face mask, covering the mouth and nose, that is so common in Asia—and with the coronavirus—now everywhere. She has a bit of a cold.

It's an awkward moment. How can I ask them about this school with hardly any children? How do I do it without embarrassing them?

"Can you tell me what you like about school?" I venture, and wait for our interpreter to translate the question. Kagami's face lights up immediately: "I like lunch!" Hisoki says that she likes to plant flowers.

"And what's good and less good in a school with not so many children?"

"I'd like to learn together with second grade," Kagami says; her teacher asks her why. "Because it's sad to be just the two of us." She says that if it were up to her, she'd bring 120 more children into the class. Hisoki objects that that is too many. "Then ninety," Kagami proposes. "Still too many," Hisoki says solemnly, but her classmate continues to bargain. She wants to be with a lot more children.

But there aren't more children. First grade is better off than third grade, which has no pupils at all. The school, like the rest of the village's public institutions, receives special subsidies. Although the children could be bused to a school with more students and more activity in another town, the prefecture and national governments want this one to stay open at all costs.

Nanmoku found itself facing a series of misfortunes characteristic of Japan's non-urban areas. The severest blow was the loss of local livelihoods, the most important of which was the cultivation of konjac, a tuber better known as devil's tongue. Asians use all parts of this plant, from the flowers to the bulb, from which they make gelatinous noodles and the famous and beloved Japanese dish *konnyaku*. The area around the village is considered one of the best places to grow it in all the Japanese islands, so its produce was in great demand and fetched high prices. But 90 percent of the Nanmoku municipality's land is mountainous, so cultivating it is difficult. With the industrialization of agriculture, huge corporate farms were able to produce more konjac at lower prices, undercutting the village's farmers. Many of the young people left for the big cities. The crisis was exacerbated by the plummeting birthrate, which has its most disastrous effect on areas that from the start were sparsely populated.

The children, however, receive a thorough and well-funded education. On the morning of my visit the students made pancakes in their cooking lesson. It's not just hands-on learning. From a young age Japanese children are taught the principles of balancing food ingredients and colors in their foods; they prepare meals according to these traditions. When the lesson ended, the cooks, all six of them, at their teacher's suggestion, scampered down the hall to offer their pancakes to the principal and assistant principal, who exclaimed "Oh!" and smacked their lips, as Japanese people do when presented with a gift. They and the teachers radiated exceptional warmth.

The principal said that there were advantages to having only twenty-four students in his school, since each one receives personal attention. He explained to me that the children would not be able to spend much time in his office. They had to go back to clean up the classroom. In Japanese public schools there are no janitors. The students—all of them, with no exceptions—clean the school, including the bathrooms, themselves. At one end of each classroom there is a stand for drying out wet rags that the children use to clean their school during the course of the day. According to the principal, that is the biggest challenge of running a school with such a small number of children. "It's a heavy burden for so few students to have to clean such a large building. But the teachers also take part," he explains. He stepped out into the corridor with me to show me a poster explaining the benefits of the method: cleaning develops the students' physical fitness, makes them feel responsible for their school, allows them quiet time to think, and provides the satisfaction of doing a job well.

It is easy to write pessimistically about a civilization from a distance, but close up you see the fruits of its accomplishments. In the Japanese case it is the sense of collective partnership and responsibility imbued in children from birth. Japan consistently has among the highest scores on international exams used to measure educational achievement. The extreme contradiction between the quality of Japanese education and the emptying of Japan's schools illustrates the unique crossroads Japan stands at. The country that was an economic miracle after World War II is shrinking. In 2019 alone the number of inhabitants fell by about 512,000,[3] approximately the population of Tucson, Arizona, in the US.

The Virgin Nation

The birthrate crisis is a national issue of paramount importance, so the Japanese government is battling to keep Nanmoku alive. That means providing not only for the needs of its few children, but also for its growing number of elderly. After the school, I visited the village's senior home, located not far from the school. Unlike the latter, it is crowded and lively. The residents seem to be mostly around the age of ninety. I spoke with some of them, including many who helped rebuild Japan after the war. They receive good but frugal care. The steep decline in births has come along with an impressive increase in longevity. As a Japanese friend bitterly joked with me, "Our problem is not just that people have forgotten to make children, but that they have also forgotten to die."

The decrease in the number of children means that fewer and fewer public playgrounds are being built in aging cities. Far more drugstore shelves are taken up by diapers for adults than by those for babies; since 2011, more adult diapers have been purchased each year in Japan than baby diapers. According to Japanese government figures, 18.5 million dogs and cats are kept as pets in a country in which there are only 15.5 million children. It's apparent on Japanese streets, which are full of strollers that look like they are meant for babies but actually hold small dogs. Dogs, and sometimes cats, are considered not only integral parts of the family but are at times its only scions. Some couples decide to raise a pet rather than a child. "The dog has its own room, with an air conditioner specially adapted to the body temperature of small mammals," a Japanese acquaintance told me. "And of course we leave the television on for it when we go out."

The facts are extreme and staggering. Japan has more than 8 million empty homes, and the number continues to rise. More than a quarter of the population is over sixty-five, the highest level in the world—the comparable figure for the US is 16 percent. Thirty years from now, they are projected to make up four out of every ten Japanese citizens. Japan's population is forecast to crash to 88 million by 2065, down from 125 million in 2019.[4] If current trends continue, in less than a hundred years, Japan will be home to only 50 million people.

WHAT HAS HAPPENED TO FAMILIES AND COUPLES IN JAPAN is the foundation of the crisis. I specify couples because one salient part of the Japanese problem is that single-parent families, whether led by women or by men, are rare in this country. In 2015, only about 12 percent of the country's children lived in one-parent families (in Britain a quarter do, and in the US the figure is 27 percent).[5] While 85 percent of single-parent mothers participate in the workforce, more than half of them live under the poverty line. That's the worst figure among all the countries of the OECD.[6] Only 2 percent of Japan's children are born out of wedlock; in contrast, more than half of France's children are born to unmarried parents. Japan's divorce rate has jumped by tens of percentage points since the 1980s, but in Japan alternative family models are less accepted and assisted, even as the classic model disintegrates. According to the country's National Institute of Population and Social Security Research, in 2015 a quarter of Japanese men under the age of fifty were unmarried, and one out of every seven women. These numbers do not include the divorced and the widowed. In comparison, in 1970 only 1.7 percent of men were not married at the age of fifty.[7]

The whole world of romance in Japan, whether heterosexual or LGBT, is in great distress. "In Sexless Japan, Almost Half of Single Young Men and Women are Virgins," shouted a headline in the *Japan Times* in 2016, with reference to a study done by a government-linked research institute.[8] Seventy percent of single men and 60 percent of single women between the ages of eighteen and thirty-four were not in relationships. More than 40 percent of single men and women in this age cohort had never had sex.[9] A study published in 2019 showed that, since 1992, the proportion of virgins among both heterosexual men and women up to the age of thirty-nine had grown. For example, by the study's conservative estimates, the number of women between the ages of thirty-five and thirty-nine who had never had sex doubled between 1992 and 2015.[10]

Western media loves to report on "explanations" for the incredible shrinking Japan—for example, young people having relationships with characters from the comics or with virtual online characters. In the buzzing Akihabara district in Tokyo I met Sanpei Mihira, who introduced me to his wife. We sat at an anime-themed pub; the man's own chosen name is of a famous Japanese cartoon character. He removed his spouse from a small paper sack. She was a Tsukasa Hiiragi doll, a familiar character from these Japanese comics. Her personality is "forgetful, positive, and sweet, and never angry," to quote a comics site. This young man is perhaps an extreme example of an *otaku*, an obsessive fan, generally of anime and manga comics.

He showed me a snapshot of them together at what he referred to as their wedding anniversary, the chocolate he bought her on a holiday, and the trip they took to Japan's Disneyland. He was fully aware that this relationship existed only inside his head; when I mentioned that, he replied, "Doesn't the entire world exist only in our minds?" When I pointed out the obvious, that it's easy to carry on a relationship with a man or woman who doesn't talk back, he told me about the conversations they have. Mihira said that, as a forty-year-old man, he is mature enough to have "a partner who exists only in my head." It was an interesting answer, dialogue with a Zen Buddhist quality. The Western media presents such phenomena as a curiosity show, a sensation. By this account, Japan is a technological society with a highly developed digital culture characterized by industrial alienation. Having thus lost its libido, Japanese society is dying, the Western media claims.

But the couple seemed to me to be very much of their time. The man was rational, perhaps overly so. Living in a time of addiction to smartphones and social networks, in an era of cutting-edge artificial intelligence, he deliberately chose an imaginary romance. It is a quite natural extension of a world in which large chunks of a person's life are virtual and disconnected from concrete reality. An Israeli friend of mine whose marriage was in crisis, in part because of pressure at work, suggested that Mihira had chosen a bogus relationship, but one that *felt* real, over a real marriage that would have felt bogus.

Even if it is true that Japanese society is grappling with dissociation and alienation,* [11] such presumptuous anthropological diagnoses tend toward hollow sensationalism. Any approach to birthrate decline needs to take into account that the phenomena is not confined to Japan but is worldwide. With regard to Japan specifically, here are factors to take into account if one is to understand the crisis: work hours, day care centers, the treatment of women in the workplace, wage differentials, the social safety net, economic stability, and economic security. Japan is not merely an outlier island nation. In many ways—in its handling of industrialization, computerization, entertainment, transport—the country is innovative, and its habits and solutions point to a possible future for the entire world. And this is no less true with regard to its birthrate.

The worsening demographic crisis is creating political and budgetary problems for Japanese society as a whole, in everything from pensions to the housing market. Fewer children were born in Japan in 2019 than in any year since the country began counting, in 1899. It broke the previous record, set the year before. In the same year, the total fertility rate in Japan was 1.36 births per woman. In developed countries, replacement-level fertility—the total fertility rate at which

---

* A good example of this phenomenon is the *johatsu*, or "evaporated people," who (often suddenly) change their identities and sever all their family and professional ties. A 2016 book claims that some 100,000 people disappear this way each year in Japan.

the population level is preserved (without migration)—is 2.1 births per woman. The demography of Japanese society is approaching a state in which the workforce will be the same size as the portion of the population dependent on it—that is, pensioners and children. Its demographic time bomb has already exploded, and the shrapnel is flying everywhere. The country has been in a severe economic slump since the 1990s, which has, in a feedback loop, further lowered the birthrate, which in turn depresses the economy even more.

In industrial societies, economic downturns push the birthrate down. In the United States, the birthrate began declining in 2008, and it continued to do so even after the recession. In Japan's case, the economic trauma, in the form of deflation and stagnation, began earlier, in the 1990s. Stock markets in the US and the rest of the world made a full recovery from the crashes of 2000 and 2008 and have reached new highs, but the Tokyo stock market's Nikkei index stood in 2019 at just 60 percent of its highest level in the 1980s (in nominal terms, in 2016 dollars). Between 1980 and 1995, GDP per capita in terms of buying power increased in real terms by 60 percent; between 1995 and 2010, it rose by only 10 percent. The result has been an intensifying burden on Japanese households.

Up until the 1980s, the fundamental compact on which Japanese society was based expected workers to put in long hours at their corporate jobs. In exchange, they had job stability, received regular raises, and enjoyed good working conditions. In other words, a Japanese man who found a job after completing his studies could assume that he would remain in it until retirement, after which he would receive an excellent pension that would provide for all his needs. Anything that deviated from this path, like being laid off, was exceptional. The corollary was that the labor market was inflexible: workers would not switch jobs to improve their salaries, nor could employers fire inefficient workers. Few people worked at temporary jobs or for subcontractors, and the typical salary was enough to support a family. Women's participation in the labor force was low; men were supposed to earn enough not to "force" their wives to take even part-time jobs.[12] Indeed, after 1945, Japan rejected the idea that women were a normal part of the working world, perpetuating patriarchal attitudes in the modern labor market. The percentage of women in the workforce declined from the 1950s onward.

When the Japanese economy went into shock in the 1980s and 1990s, that trend was reversed. Women began entering the workforce, while also retaining their traditional roles of having and raising children, as society demanded. In 2007, for example, in a speech before a convention of the ruling party, Hakuo Yanagisawa, Japan's health minister, bemoaned the fact that women were not having enough children. "The number of women aged between fifteen and fifty is fixed," he declared. "Because the number of birth-giving machines and devices is fixed, all we can do is ask them to do their best per head . . . although it may not be so appropriate to call them machines."[13] Yanagisawa was simply stating the traditional established view in his country—women should stay home, and if the birthrate was low, it was their fault.

Globalization, and especially the rise of Chinese manufacturing and technology, forced Japanese businesses to become more efficient, and that required them to revise the social contract. Large employers lobbied the government to loosen labor norms. The government responded, allowing more flexibility to hire and dismiss workers, and especially to make it easier to use subcontractors and labor agencies. This created a new class of what are referred to as "nonregular" workers, who enjoy few protections even if they remain in the same job for many years.[14] Nonregular workers were a way for companies and corporations to pay workers much less than "regular" workers received, even if they performed exactly the same tasks. A nonregular worker might have precisely the same job as a regular colleague, but earns 60 percent less.[15] Nonregular workers do not receive many social benefits, such as contributions to continuing education and pension funds. By now, four out of every ten Japanese workers are nonregular, double the level of twenty-five years ago. The number of regular workers has declined by 43 million during this same quarter century, while the number of nonregular workers has grown by 12 million.[16]

Nonregular workers have not only poorer work conditions but poorer love lives. The study analyzing trends in heterosexual inexperience shows that, among men aged twenty-five to thirty-nine, those at the lowest income level were ten to twenty times more likely to be virgins than those at the highest level. Unemployed men were eight times more likely to be virgins than those with regular employment; part-time or temporarily employed men were four times more likely. In other words, according to the study, "money and social status matter for men."

When young Japanese are asked why they do not have more than one child, or decide not to have any, they frequently say that they have nonregular jobs and thus are not confident in their ability to provide for a family. Parents often discourage children with nonregular work from marrying, the social assumption being that raising a family requires a regular job.

Women cannot compensate for the employment security that men have lost, because of their insecure position in a society with entrenched discrimination. About 50 percent of mothers do not rejoin the labor market after having their first child.[17] If they do, they do not receive fair pay; the differential between them and men doing the same work is one of the highest in the world—Japan is worse on this count than, for example, Azerbaijan.[18] Their families apply immense pressure on them to stay home. "A Japanese woman does not want another woman to care for her children—it's not accepted," I was told by an Israeli woman who has lived in Japan for many years, working in childcare at a private center. "There is private and public day care. But the idea of hiring a nanny is rare. That means that the minute children are born, [being a mother] is a full-time job."

At any given moment, tens of thousands of children are waiting for a place to open up in a public day care center; while they wait, their mothers have limited work options.[19] Women who become pregnant or take maternity leave suffer discrimination at work, especially if they hold a nonregular position. In Japan they call it *matahara*, maternity harassment. It may include salary cuts, demotion, and verbal abuse. A study published in 2015 found that half the women in nonregular positions reported abuse after returning to work in the wake of maternity leave, or after becoming pregnant.[20] In the past, Japanese health insurance did not even cover births, on the grounds that it was not a disease but a choice. Just giving birth used to cost families thousands of dollars out of pocket.

After a birth, many men are not involved in child-rearing or household chores. In Germany, the US, and Sweden, men devote about three hours a day to such tasks; Japanese men contribute only an hour and a half. On average, Japanese men spend only one hour a day with their children.[21] About a third of married American men head home from work at 5:00 p.m., more than at any other hour.[22] Mary Brinton, a sociologist and Japan expert, relates that when a Japanese colleague of hers at Harvard looked at these figures about American men, she refused to believe them. According to Brinton, "When she calculated this she thought she made an error in the data analysis, she said—married American men can't be leaving at five o'clock!" The reason is that more Japanese men leave work at 8:30 p.m. than at any other hour, and many leave even later.[23] The total commitment to work customary in Japan requires men to work endlessly; after work, instead of heading home, they are expected to go out with their colleagues—for example, to karaoke evenings, seen as an inseparable part of their work. Addiction—or subjugation—to work is a national epidemic; indeed, hundreds of deaths a year are attributed to overwork (*karoshi*). A fifth of Japanese work fifty hours or more a week.[24] Workers often do not take the paid vacation days they earn, as a result of implied or overt pressure from their employers.[25]

The question is not so much why the birthrate crashed but why it did not reach an even lower level. In the face of the situation, the Japanese government has begun to institute reforms. In October 2019 it adopted a new policy of free education for ages three to five and free day care for low-income households from the age of two (day cares were charging between 100 to 800 dollars per month, dependent on whether they were classified as private or public). Japan has also lowered medical costs for births, and it is trying to reduce pressure in the workplace. It is even sponsoring "singles seminars" and funding singles parties in the hope of boosting the number of babies.[26]

MANY IN JAPAN MAINTAIN THAT, UNDER CURRENT CONDITIONS, the government's birthrate goals demonstrate the narrow-mindedness of the country's patriarchal establishment. Ueno Chizuko, a sociologist whom the local press terms

Japan's "best-known feminist," gives me her position on the subject right at the start of our meeting in her Tokyo office. "I don't understand what the problem is and why it's important to you," she says. As she sees it, just raising the issue of a "need" for children is motivated by patriarchal impulses. She once published an article showing that the birthrate is plunging in all the former Axis states—Germany, Italy, and Japan. In it, she argued that it was a rebellion against the belligerent masochism entrenched in these societies.[27] "The first question you need to ask yourself," she tells me, "is for whom the birthrate is important. Because having children is a personal decision when it comes down to it. The answer is the government, employers, and other elements who want Japanese economic growth. I belong to the group of people who are asking what is wrong with a shrinking population."

"But you won't be able to maintain your older population and the social structure will collapse," I point out. "Population size moves around," she says. "Even if the population shrinks, why is that important? In the worst case, other people will come. The size of the population can grow naturally, by births or by immigration. Japan's population has shrunk because the Japanese government is focusing only on encouraging women to have babies." Ueno is referring to Japan's immigration policy, which has traditionally nearly banned foreigners from settling in the country. The country could have benefited from millions of immigrants from all over Asia, but it insists on maintaining a strict national and ethnic identity and refuses to naturalize foreigners.

How does that connect to the birthrate? Young immigrants, even if they are not educated and lack advanced skills, contribute to economies by paying taxes, buying goods, and having more children than the general population. But this is Japan, where many still refer to black people with the local version of the N-word, *kurombo*, and where the government is willing to allow foreign care workers for the elderly to enter the country but tries to keep it quiet.

Even without immigration, Ueno says, the conservative male government's interest is in maintaining all current power structures, at women's expense. "Most women who work are in nonregular positions," she adds, "and in Japan the central factor in having children is economic security. Since they are nonregular, most of them will not go back to work after giving birth. What we have here is the invention of a modern model of a family unit in which the man supports the family and the woman is solely a housewife, but in traditional Japanese life it wasn't that way. Everyone worked, wives and mothers as well, and the birthrate was high. I recommend a trip to southern Japan and you'll see that in farming it is still that way."

In today's Japan, as opposed to Scandinavian countries, according to Ueno, there is no option to educate children in public frameworks that relieve women of their roles at home. Indeed, Japan's expenditure on education as percentage of its GDP is amongst the lowest in the OECD countries. Conservative politicians don't want to raise taxes to create a full welfare economy. On the other hand, they abhor immigrants who would change the country's character. There is thus no inexpensive childcare provided by foreign or immigrant nannies, and for the same reason there is no housework help provided by foreign workers, as there is in many parts of the US.

What she describes is a conservative-capitalist trap—women suffer discrimination in the labor market, men must work unreasonable hours that keep them from being active parents, and women have no job security. There is no willingness to allow mass immigration, which would make raising children and keeping house easier. Most single mothers live in poverty. Taxes cannot be raised to pay for a welfare state that could support families. Women are caught in the middle—they are expected to be mothers without support and to accept a role in a labor market in which they are not treated as equals. How many women readng this are thinking to themselves that that situation is not limited to the Land of the Rising Sun? Many, I'll bet. Ueno's account is just a case study of what women and families in typical capitalist societies experience. In other words, it's not a story about Japan. It's about all of us.

CHAPTER 12

"Humankind Is the *Titanic*"

. . . A time when the advantages of remaining childless make most people feel [that even] a single child is a burden.

—PLINY THE YOUNGER, FIRST CENTURY CE

The world's fertility rate has halved since 1950, and continues to decline.[1] Many of the world's countries are now experiencing a "baby bust": not enough children are being born to maintain current population levels.[2] Economic constraints and lifestyle changes are not the exclusive reasons why fewer children are being born—there are also physiological causes. World population growth is projected to end by 2100, for the first time in modern history. The number of people in Europe, Asia, and South America will start shrinking slowly, and then more quickly before the century ends, while Africa will remain the only continent with high birthrates. Together with rising life expectancy, the consequence is that global median age will reach forty-two by the end of this century. In 1950 it was twenty-four.

The declining birthrate is not often considered in the context of globalization and the market economy. Yet it has sweeping implications. It is occurring against the background of a fractured world order. Failing to replace ourselves will affect the next generation far more than the ascent to power of radical or populist forces. Regardless of politics, these demographic trends herald a shift in the composition of societies that will affect all aspects of life. The decline in the number of working-age people and young voters will have a major impact on industrialized economies, their politics, and their culture, which once sanctified childrearing. Since the Industrial Revolution, societies and market economies have been structured on the premise that populations will continue to increase, and that this is closely connected to economic growth. Pension schemes, tax policy, job markets, and core political values are largely based on that projection.

If the dip below the replacement level of 2.1 continues, and even more so if it drops even further, there will be a need for a major recalibration of economic policies and institutions. It will be a crisis for capitalist-consumer society. But to brush off the gravity of the problem by observing that a declining population will be good for the environment, or to just reconsider economic policy so as to take the new demographic situation into account, is to miss much of its significance. Such attitudes rush to address the problem's repercussions while ignoring its more profound significance. What does it mean when societies begin to shrink? What does it say about the individuals living in them, their families, about happiness and health? The fertility crisis indicates how unsustainable the current global order has become; it is having an impact not only on the economy and politics but also on people's *capacity* for forming families.

SINJUKU PARK IS A FINE PLACE TO VIEW THE CHERRY BLOSSOMS in Tokyo. When the trees bloom, Japanese families gather around them for the traditional flower-viewing and to photograph the stunning pink blooms. The grass remains yellowish after the cold days of early spring, but the park bustles with children. I'm there with Mei (née Kyotani) and Asi Rinestine, a Japanese woman and Israeli man, both in their early thirties, who met and married in Japan and were expecting their first child. We spread a blanket on the lawn and they explain what makes being a new parent so stressful in Japan.

Asi has a local business and Mei works for a large corporation. Beyond her management job, she also models in the advertisements for some of the products that her company manufactures. She has an excellent regular job that provides full benefits and employment security. Asi is animated and anxious about the approaching birth, while the self-assured Mei glows.

We speak about her work. "Before I got pregnant, I was working from eight a.m. to like twelve midnight," she tells me. She'd take the last train home. "I would sleep like four to five hours," she recalls. She would then get up at 6:00 a.m. and go to work. Asi says she worked between seventy and a hundred hours a week.

I am taken aback.

"It's a samurai thing," Mei says. "Bushido. You don't bitch about it." Bushido is the Japanese warrior code of honor, and it remains the foundation of the ethos of loyalty and sacrifice that are expected at workplaces as well.

Mei says that when she told her colleagues she was pregnant, "people were positive, but some people do think I kind of let my career go. Because when you have a baby you work less hours, so of course I cannot make those seventy hours a week, right?" And there are other problems. Mei explains that "the ideal mother in Japan is a housewife. So if you use services like nannies they will think you are a lazy mother. People want you to be perfect, a supermom. You want to work? You have to do all the housework and also your job, as well as taking care of the kids."

With such demands at work, Asi says, it's hard to find time to spend together, or for intimacy. "Our parents, and your parents, too, they would come home from work at five or six, watch some TV, and then at nine o'clock what are you going to do?" He and Mei laugh. "Sex and children were big parts of life. For us now, 24/7 there is something on, 24/7. Good stuff on. Like internet, mobile phones, mobile devices, we are connected twenty-four hours and live in a city that doesn't sleep. And you can live a full, entertaining life with no children, with no sex."

Apparently a lot of people feel that way. While since 1914 the human population has grown by about 6 billion, the trend will break by the end of the current century.[3] In South Korea, Poland, Spain, Romania, Slovakia, Russia, Italy, Germany, and Greece, for example, the birthrates are similar to or lower than Japan's. In 2018 the general fertility rate in the United States reached a historic low of 59 births per 1,000 women, a drop of 15 percent since 2007.[4] Nor is this the case only in developed countries. In Iran the fertility rate in 1985 was an average of 6 children per woman. Today the number stands at 2.1. Globally, in 1960 each woman gave birth, on average, to 5 children. Today the figure is 2.45.[5] The only continent on which the birthrate remains high is Africa.

Europe is already experiencing a very real crisis. Spain's birthrate, which is even lower than Japan's, combined with the economic crisis and emigration, caused that country's population to shrink between 2012 and 2018. Portugal's population has been declining since 2010 and is projected to be 10 percent smaller by 2050.[6] At the same time, urbanization is accelerating globally, so societies are changing in other ways even as they produce fewer children. In Spain's autonomous region of Galicia, for example, some 1,800 villages lie abandoned because of the steadily contracting population and the move into the big cities. In Spain as a whole there are almost 3,500 such villages and towns.[7] It is striking that, in 2017, the leaders of Italy, Germany, France, and Britain were all childless.

We are witnessing "the globalization of fertility behavior," according to the late John Caldwell, one of the world's leading demographers. The fertility decline that began at the end of the 1950s took place in countries in which 80 percent of the world's population lived; "the range of populations involved in the decline was unpredicted and unprecedented," Caldwell wrote. He stressed one point—it is the first time in history that a change in the birthrate has occurred on a global rather than a regional scale. "It seems likely," he maintained, "that such near-simultaneity was the product of the same forces everywhere."[8]

The forces involved are the Industrial Revolution, the information revolution, globalization, the ascent of liberal values, rising educational levels, urbanization, and the changing position of women.

Declining birthrates around the world are not weakly linked to globalization—they are one of its salient manifestations. More than ever before, the most significant decision that humans everywhere make, to bring children into the world, is affected by global factors. With interconnections growing more potent, that should hardly be a surprise. It is possible that the drop in the birthrate will prove to be temporary and will end when our species reverts to more sustainable numbers. But even if that happens, the population will be older than it has been in the recent past, older on average than it ever has been before.

Between 2000 and 2016, life expectancy rose by five and a half years on average.[9] European sociologists talk about a continental gerontocracy, rule by the elderly. Already, the median age for Europeans was forty-two in 2018. A report issued in 2019 by the Bank for International Settlements argued that "the political divide of the future will be over the elderly protecting their social safety net and the working age population their real post-tax incomes."[10] Three factors—declining fertility, rising longevity and health expenses, and a decline in workforce participation—threaten the dream of retirement that was one of the fundamental promises of the postwar market economy. The values and policies of the age of responsibility resulted in large part from the postwar surge in the birthrate. The baby boom increased the pool of workers, and the children of that generation produced and consumed at levels that brought about half a century of prosperity. The boomers who created it expect to live in comfortable retirement, but that age is over.

The revolt against globalization and its values is in part a reflection of generational tension. Parents feel that they have worked all their lives and have earned the right to political and economic stability and, especially, their pensions. Young people see their taxes paying for the older generation's needs, and are finding jobs scarce because older people enjoy tenure and are in no hurry to retire. Furthermore, the jobs they do find do not offer the security that their parents enjoyed. In 2013, Japan's finance minister, reacting to the longevity of his country's senior citizens, said that they should "hurry up and die."

A Mammal Society in Crisis

The substantial drop in the world's birthrate is a fateful phenomenon of our time.[11] For many readers, it might sound like good news. After all, the world is overcrowded, and even if fertility declines, it will continue to become even more packed with human beings in the short term—the population is expected to balloon to 10 billion before it stabilizes, according to current trends. The rising human population exacerbates catastrophic climate change, the destruction of natural habitats, a collapse of biological diversity, and the pollution of the atmosphere, hydrosphere, and lithosphere. Braking the multiplication of the human species would seem to be vital for the planet's future.

But a shrinking population is also a daunting challenge for *Homo sapiens*. We have no proven model for how to maintain a modern society with a declining number of members. Germany, for example, is considered a stable industrial nation, but its birthrate is almost the same as Japan's. The Federal Statistical Office of Germany projects that, by 2030, the German workforce will shrink by 5 million, and that by 2060 it will be only three-quarters of what it is now.[12] A study by Bertelsmann, a German foundation, shows that, to compensate for the aging of the population and declining birthrate, that country will need to welcome in half a million immigrants a year.[13] Given the current unrest over the refugees who have arrived from the Middle East, the chances that Germany will agree to do so are nil. If birthrates continue to fall, in 2060 there will be only 1.8 German workers for each pensioner. The government will need to raise taxes or reduce social benefits, purchasing power will decline, and jobs will dwindle. The capitalist consumer culture will not be sustainable, and the economy—in Germany and elsewhere—will head into a stagnation similar to or worse than that of the Japanese economy during the last twenty years.

The prospect of decline is fodder for the far right, which fervidly seeks to terrify the public with prophecies of impending catastrophe. In 2017, Germany's extremist AfD party put up posters depicting a pregnant white woman. The caption was: "New Germans? We'll make them ourselves." In 2018, Matteo Salvini, leader of the far-right Italian Northern League, asked, "Are we as a country facing extinction? Unfortunately, yes."[14] Centrist parties in democracies are so wary of touching reproduction issues that they have left a vacuum for nationalists, who have a lot to gain from the fertility crisis. As rich countries face declining births, they need immigrants to sustain growth; growing numbers of immigrants feed the fears and perceived threats that voters feel, prompting them to support nationalist-xenophobic parties.

CONSERVATIVES, RELIGIOUS AND OTHERWISE, TEND TO SEE the fertility crisis as a portent of doom. According to George Alter and Gregory Clark, "New products and new lifestyles in the growing metropolitan societies created by the Industrial Revolution expanded choices. Wealthy families responded by consuming more of these new products and services instead of producing children."[15] While Clark and Alter present this as an observation, for conservatives it is an indictment.

Lord Jonathan Sacks, Britain's former chief rabbi and a prominent Jewish religious leader in Europe, warned in 2016, "The contemporary historian[s] of ancient Greece and ancient Rome saw their civilisations begin their decline and fall," he told *The Telegraph*. "Both the Greeks and the Romans attributed it to falling birthrates because nobody wanted the responsibilities of bringing up children. They were too focused on enjoying the present to make the sacrifices necessary to build the future."[16] He cautioned that "Europe is going to die because of this . . . Europe can only maintain its population by unprecedented levels of immigration."

Tradition-touting leaders, from the Roman Empire to religious leaders of the twenty-first century, are apt to blame the declining birthrates on hedonism. But there is no empirical data to back up this claim. Research demonstrates that people are having fewer children because they are making rational decisions about education, jobs, and family planning. For instance, more education for women correlates with their having fewer children. A study that examined fluctuations in 70 countries over 130 years, beginning in 1870, found that education "has been the main socioeconomic determinant of the demographic transition."[17] When girls begin to receive elementary-school education—that is, six years of schooling—it results in a decline of from 40 to 80 percent in the birthrate. A study conducted in Nigeria found that each additional year of education for girls leads to a decline of 0.26 in the number of children they eventually have, on average.[18] Conventions about family size change more rapidly if women learn to read and write. In those states of India where the illiteracy rate is declining, more and more people believe that a smaller family is happier.[19] Women have been empowered by the weakening of repressive patriarchal norms, followed by increasing use of contraception and legal and practical access to abortion.

The economist Gary Becker has suggested that the desire to have a child should be treated as a demand for a good, which might bear a price.[20] When the economic value of an education increases, there is an incentive to invest in giving a good education to fewer children rather than investing a small amount in the education of many children. When a society shifts from an agricultural to an industrial economy, it becomes more worthwhile to invest in quality. As women join the workforce, having children bears a double cost—not just the cost of raising them but also an opportunity cost, in the form of the money that the woman could have earned during the time she devotes to her children.

Another driver is falling child mortality. Some researchers claim that there is a causal connection between declining child mortality and parents' decision to have fewer children.[21]

To put it another way, people have many children because they think that some of them will die. When they become convinced that the health system and their lives are stable and safe, their inclination is to reduce the birthrate. Before the modern age, more than a third of children died before the age of five, and families, especially in the developing world, sought to "insure" themselves against the prospect of being left with no children at all by producing more of them, a phenomenon that the research refers to as "child hoarding." Poor countries experience a temporary jump in the birthrate when diet and sanitation improve, because it takes time for people to realize that, since more of their children are surviving, they need not have so many. When that sinks in, the country's birthrate declines.

Educated women earn more, and their families enjoy a higher standard of living. Contraceptives are available, and more women can make their own decisions about their bodies. Child mortality is decreasing. These are all positive explanations for why, in societies in which the average family once had five to six children, families now average two to three children. But what happens when a society begins to shrink slowly as it suffers from a sputtering economy and faces a grim future because its workforce is diminishing?

And even if the economy nevertheless maintains growth and good living standards—is a slowly vanishing society healthy, or successful?

First, a distinction needs to be made between a decline in the birthrate and a long period in which the birthrate remains below the replacement rate. The difference between that rate, 2.1 children per woman, and 1.7 children per woman is of vast significance for the survival of a species.

I use zoological terms deliberately. According to the projections of demographers and sociologists, the decline will almost certainly accelerate. It seems reasonable to ask the same questions about human growth that population biologists do when studying a mammalian society that does not breed effectively in nature or in captivity. Why does a population stop reproducing? Is it because it is not successful in obtaining the resources it needs? Is it because of physiological problems caused by, say, pollution or climate change? Is it the result of excess stress? Does the hierarchical structure of this mammalian society make reproduction more difficult? Is there some outside factor that puts pressure on these social structures, causing population decline?

The reason most often evoked is that more men and women are postponing having children to their thirties and forties, when conception and pregnancy may be more difficult. But the data indicates that both sexes are also facing fertility problems. A meta-analysis conducted by the Hebrew University of Jerusalem together with Mt. Sinai Hospital in New York, published in 2017, found that between the years 1973 and 2011, the sperm count of men in Western countries declined by about half. Based on data from 185 peer-reviewed studies, published over four decades, it also demonstrated that the degeneration of human semen is accelerating.[22] There was not enough data from Africa or Asia, but a series of studies published in recent years indicate similar numbers there as well.[23] According to the meta-analysis's lead author, Dr. Hagai Levine, a professor of epidemiology at Hebrew University, the findings are an "urgent wake-up call."[24]

The study did not examine the causes of this phenomenon, yet previous ones have connected decreased sperm counts to exposure to chemicals, pesticides, smoking, mental stress, and obesity. There are many other theories as well—perhaps there is a connection to the presence of electronic devices around men, or to the slight increase in global temperatures brought on by climate change, or perhaps even to some as yet unidentified substance in the environment that has an extreme impact on fertility. "Right now the human species is like the *Titanic* a moment before the collision, or maybe already after," Levine says.[25] He adds that human fertility is sensitive and easily disrupted, in both men and women. A small aberration, such as stress, can cause changes to the endocrine system, which in turn can lead to defective development expressed in adult life. A reduced sperm count correlates more closely with having had a mother who smoked while pregnant than with whether the man himself smoked at the age of eighteen.

Levine suggests that social changes may also affect the human endocrine system, not just pollutants, although he stresses that this is speculative. "That is true with regard to animals, so perhaps it applies to humans as well," he says. "It could be that the very fact of living in a city, the social interactions typical of a city, where you see an enormous number of people but your connections with them are different than the interactions that characterize tribes or extended families or small groups, affect us hormonally, and therefore have an impact on our sperm count. That is another thing we need to look at."

He says that such problems become critical for public health only after they reach a tipping point. That would be the case if parents manage to get pregnant with difficulty but their children are unable to do so at all. A recent study compared the sperm count of young males and their fathers; it found the fathers, who were in their fifties, had better sperm quality. "Imagine that height or IQ were to plummet by 50 percent in just a few decades," says Levine. "We have a biological marker for survival that has seen a similar dramatic shift." Recent studies have shown that poor semen quality is a marker of increased risk of morbidity and mortality. In any case, he has no doubt that the change is long-lasting. "Even if we turn the ship around now, there is no assurance that we will see good results. The decline may continue for a long time forward."

The Handmaid's Tale

It sounds something like the opening of a season of *The Handmaid's Tale*, the television series based on Margaret Atwood's novel of the same name. The novel is usually read as an allegory of the status of women in contemporary society, but as the birthrate falls it no longer seems merely allegorical. Meimanat Hosseini Chavoshi, an Iranian-Australian demographer, was arrested in Iran in 2018 for publishing work about the steep decline in the number of births in that country. Iran's spiritual leader, Ali Khamenei, is concerned about the decline in fertility. The government thus not only arrests demographers (Chavoshi was released in 2019) but also encourages polygamy, makes divorce more difficult, artificially inflates the price of birth control pills, and permits child marriage. None of these measures has boosted the birthrate significantly in the past decade, but they befit the Iranian patriarchal theocracy, and that of Atwood's Republic of Gilead. In Hungary, Viktor Orbán's nationalist government trumpets "Hungarian babies" and "Christian values" while trying to increase birthrates (the current fertility rate is 1.5 children per woman) by means of housing subsidies and three years of parental leave.

Democratic societies are reluctant to interfere with or to attempt to influence the reproductive choices made by individuals. Linking such choices to national interests has major moral repercussions and evokes the specters of totalitarian and authoritarian regimes. Indeed, childlessness is not ethically wrong, and the replacement level cannot become a legal or moral demand to couples in a liberal society. Nevertheless, governments are so worried they are now going into the matchmaking business. The birthrate in South Korea is the lowest of all the developing countries, less than one birth per woman (0.98),[26] and since 2006, the government has invested more than $130 billion in a desperate attempt to raise it.[27] It pays women for each birth and compensates them for some of the costs. The country's universities offer elective courses in sex, love, and healthy relationships that require students to date classmates. The state subsidizes local matchmakers and social and cultural activities at which singles meet each other, offers bonuses to couples with babies, and conducts an educational campaign to promote births.

None of these policies has been particularly effective. When asked why they are not having children at all, or more children, South Koreans usually cite the financial burden. Here's one example: families in Seoul spend 16 percent of their income on private tutors for their school-age children. Young Koreans are renouncing courtship, marriage, and childbirth. They refer to themselves as the *sampo* ("giving-up") generation.

Rulers throughout history have tried their hands at birthrate policies, aimed at enlarging or reducing their populations or specific parts of them. Policies other than mass murder have generally failed. The Roman Empire faced declining birthrates among its elites over the course of centuries, which by some accounts was a factor in its fall. In 18 BCE, the Emperor Augustus promulgated a law that provided incentives to Roman citizens who married and had children, and imposed sanctions on those who did not. The law, *lex*

Iulia de Maritandis Ordinibus, sought to promote the purity of the Romans by forbidding members of the nobility to enter into marriage with harlots or former slaves. It failed. "Childlessness remained the vogue," the Roman historian Tacitus wrote a century later. It's not known why the birthrate in the Roman heartland was so low, given that the people who lived there enjoyed relative security and stability. Some historians of ancient Rome speak of the collapse of the family as an institution, and of a decadent culture. Others suggest that lead leaching out of cookware poisoned the Romans and caused sterility. Some even blame the aristocratic practice of taking long baths in very hot water, which these scholars allege was detrimental to the quality of the semen the men produced.[28]

There is no lack of current examples. When Japan was under military rule during the 1930s and during World War II, the government encouraged citizens to have children in order to supply soldiers for the empire. The effort failed. After the war, when Japan was devastated and poor and the government did not pursue pro-natal policies, the country had a huge baby boom. When the Nazis came to power in Germany, they were concerned about the low birthrate, but despite a range of government plans, some of them brutal, the birthrate in 1938 was low, about the same as the average world birthrate today.[29]

The consolidation of the centralized nation-state, along with modern technology, might seem to be an opportunity to shape reproductive decisions. The one-child policy of China would seem to be a successful example of such social engineering. The Chinese government believed that the main way to put the country on the path to rapid industrialization was to reduce the size of its huge and poor population. It forbade couples in large cities to have more than one child; parents elsewhere, about half of all those in the country, were allowed to have two children. The punishments for violating this edict were severe, including imprisonment, sterilization, and forced abortions. Modern history has seen no other such carefully planned and aggressive attempt to shape human demography.

It was a draconian policy, but the party cited strong reasons for it. Four-fifths of the country's inhabitants lived in severe poverty. It was common wisdom that the root source of the country's problems was poverty, and poverty was caused by limited resources that were being spread thin over a growing population. During the four decades it was in force, the birthrate did decline steeply; the one-child policy was repealed in 2016 because the birthrate had gone down so much. In 1970 the average Chinese woman had 5.7 children. Today the number is 1.6–1.7, and in 2020 the country recorded the lowest number of births since 1961, although the population has grown by about 680 million since that year. China now suffers from a "demographic hole" because of the decline in the birthrate, and the implications for the workforce and economic growth are daunting.

Actually, it's not clear that the number of births contracted because of the government's policies and the sanctions it imposed. A comparison of Chinese society and other societies with similar birthrates shows that the downward trend during this period was similar. While China used aggressive means to reduce the number of births, Taiwan, where the government actually wanted its population to grow, also experienced a sharp decline. Other models of fertility and the birthrate show that "in other countries without a one-child policy the birthrate also declined, and it declined below the level predicted for China."[30] If that is indeed true, then the Communist Chinese government set up a huge, brutal, but ultimately ineffective system to oversee and limit births. The same thing happened, sometimes to an even greater extent, in countries whose governments refrained from intervening in what happened in their citizens' bedrooms.

An unintended and dreadful consequence of the Communist Chinese policy was that it spurred the revival of the custom of murdering baby girls. That did not happen only in China—female infanticide was a familiar practice throughout the ancient world. "I am still in Alexandria," wrote a Roman stationed in Egypt to his wife in the imperial capital. "I beg and plead with you to take care of our little child, and as soon as we receive wages, I will send them to you. In the meantime if (good fortune to you!) you give birth, if it is a boy, let it live; if it is a girl, expose it."[31] That meant abandoning her and leaving her to die.

The estimate is that some 200,000 baby girls were murdered in China each year during the 1980s, and hundreds of thousands of others were abandoned. As medical interventions became more accessible, parents sought to prevent the birth of girls, for example by ultrasound examinations to determine the sex of their fetus, followed by abortion if it was a girl. Much the same thing happens in the Indian subcontinent. In China the result was a major imbalance in sex ratios. Among inhabitants of the People's Republic under the age of twenty-five, there are 115 males for every 100 females; in the country as a whole there are 34 million more men than women.[32] This may have a destabilizing effect on all of Asia. Societies with large male majorities are apt to have violent internal crises that can deteriorate into war.[33]

Political leaders around the world are alarmed by the low birthrate not simply because of underpopulation and the fiscal disaster it might generate. Children enable a society to grow and prosper, but more significantly, they are foundational to the human experience. The disappearance or diminished presence of children in a society is an omen of a troubled future. A perceived threat to family as an institution can engender political tensions, leading to violence and fundamentalism.

The Village of the Dolls, 2015

Nagoro is a village "on the edge." That is a Japanese expression that means "on the verge of extinction." To reach it, you have to fly from Tokyo to the smallest and least populated of Japan's four major islands, Shikoku. From the airport we drive far into the mountains. It's a beautiful but difficult route. Sparse snowflakes melt over the windshield, which passes traditional quiet villages. The higher the car climbs, the fewer people we see.

Eventually we reach a small village on a large river. It is very cold, but the sky is clear. In recent years, Nagoro has gained fame in Japan as the "Village of the Dolls," and the very first view of the homes and fields there shows why. The dolls, which look something like scarecrows, can be seen everywhere. They wait in groups at bus stops. They rest under trees. Wearing hats against the sun, they grasp hoes in fields. One doll sits on a large boulder on the riverbank, holding a rod and fully equipped with fishing gear. Their faces are handcrafted, and they wear real human clothing. It's a remarkable, almost intimidating, thing to see.

Ayano Tsukimi, sixty-six years old, is the dolls' creator. She returned to her parents' village many years ago and observed its decline. The elderly were dying off. Most children had left. The last birth occurred in 2001. Tsukimi remains there with her father; there are twenty other inhabitants of the village. Her father is eighty-six years old; during our visit to Nagoro he tends their small potato patch.

Tsukimi says she decided to make dolls in the image of the real people who were now gone. Some 350 of them are scattered through the small valley where the village sits. Many of them can be seen in the local school. We cross a bridge over the river to reach the building that is the center of the community in any Japanese village. Today it stands empty of actual human beings.

Tsukimi assiduously cleans the school and makes sure everything is in good working order. She brings a key to unlock the chain on the door. The school is of generous size, well equipped, with a large playground that any school would be envious of. Dolls of teachers and pupils populate the classrooms. Some of the staff sit in the teachers' room, enjoying imaginary tea. The swings in the playground are empty, though they sometimes sway slightly in the breeze. Tsukimi says that she misses the sound of children, but that, on the whole, she does not think that her installation of dolls grows out of loneliness. It's "just nice" to make them, she says.

After visiting the school, we wander along the village's main street, parallel to the river, to see more of the all-so-human dolls that Tsukimi has carefully fashioned. She does not speak much. She advised us in advance to bring food with us because there are no restaurants or gas stations anywhere within a radius of several hours of driving. When she sees that we have not brought anything, she invites us to the modest home that she and her father share. It's a low structure, full of pictures. It is also warm, built around a cast-iron charcoal stove topped by silvery kettles. Taking refuge from the cold outside, we sit on the floor around the stove, and Tsukimi serves us simple and tasty local food—boiled potatoes spread with miso mixed with mountain flowers, and *onigiri* (rice balls) sprinkled with a yellow spice, and of course a *tamago* (omelet).

A few weeks after my visit to this empty Japanese village, there was a disaster in the Mediterranean Sea. Up to 700 people drowned in a matter of a few days as they tried to make their way by boat and raft from Libya to Italy. The Italian navy rescued some, taking them, exhausted and traumatized, to Lampedusa, a speck of an island in the central Mediterranean that is part of Italy.

One element of the global refugee crisis is that the birthrate remains high in Africa, meaning that the continent has been grappling with a baby boom over the last decade. For every 100 adult workers in Africa, there are 73 children aged fifteen and under who need to be sheltered and fed.[34] Many Africans are thus impelled to head to the West, where they hope to find a better life and earn a decent income. It's not only a matter of improving their standard of living—it's the best way to survive. In 2050, four out of every ten of the world's children under the age of five will be Africans. They will live in places with insufficient infrastructure and faulty roads, in patchwork cities built by colonialists.

There is a bitter incongruence between a virtually empty Japanese village with a wonderful, well-appointed but vacant school that is filling itself up with dolls and the Africans who, seeking a new future, take their lives into their hands in rubber rafts because they are so desperate to find a safe place to live. Globalization drives and accelerates both—the economic pressures felt by Japanese and other modern societies that have contributed to the birthrate crisis, and in Africa the rise in the standard of living and the population explosion.

In a world of unrestricted globalization, Nagoro would not remain empty. True, migrants seldom head for rural areas, preferring, at least at first, city centers. But quite often their arrival causes the established inhabitants of the countries they settle in to move further out to suburbs and peripheral areas. If population movements functioned efficiently, tens of millions of Africans would resettle in Europe, and people from Bangladesh, China, and Vietnam would flood the Japanese islands. If societies cannot grow by means of local families, globalization commands, the growth will come thanks to immigrant families.

But the Japanese do not want to live in a global state, with a world government telling them that the voice of children will again ring through their empty villages—except that the children will be black, or Chinese. On the other hand, they want their pensions to support their long old age—the longest in the world—while only having 1.36 children per woman.

Since I visited Nagoro in 2015, the Village of the Dolls has become a popular tourist site. It's not hard to understand why—it is a great backdrop against which to take pictures of a decaying civilization. A light snow fell as we sat in Tsukimi's small house that afternoon. We drank gentle tea around the stove; the dolls stood silent in the fields.

A few months later, hundreds of thousands of refugees from the Middle East began to cross the Mediterranean, fleeing war and failing economies. Europe's refugee crisis began.

CHAPTER 13

Faces of Exodus

Why are you here?
"You know why."
Where do you want to go?
"Anywhere."

—A CONVERSATION WITH LILAN, SEVENTEEN, A SYRIAN REFUGEE

SUMMER 2015

It's late afternoon. I'm standing on railroad tracks in Hungary, where the line crosses the border with Serbia. Hundreds of refugees stride along the ties—women with headscarves, in heavy, dark dresses, bags in hand; young men in t-shirts, smartphones bulging from the back pocket of their jeans; girls with their hair tied back and hoop earrings, toting backpacks as if heading for a trek in the Himalayas; middle-aged men with dark sunglasses, well-trimmed mustaches, and small potbellies; and children, countless children. Some of them clutch the hand of parents, brothers, or sisters. The parents or family members lift the exhausted ones into their arms or hoist them onto their shoulders. All of the walkers who have the strength plod on, their eyes glassy and vacant. Most are not hungry or ill, and few have obvious signs of injury. But, when asked, many will reveal scars on their bellies, backs, or feet, marking where a bullet entered and exited, where shrapnel shattered a finger, where a burn left a scar that they hope will fade, but which probably never will.

It's easy to spot where they come from. The Iraqis for the most part look well-off. They have the latest cell phones; some sport designer clothes. They are the best packed and best equipped of the migrants, with all the gear they need for the trek along the tracks. The Afghans are the most ragged and clearly the poorest. But the majority of the refugees are Syrians. Unlike most of the others, they look anxious and alarmed.

These are people who left their homes in Baghdad, Raqqa, Aleppo, or Damascus a couple of weeks before. In most cases, after crossing the sea between Turkey and Greece, they rode into the heart of Europe, by bus or train, and hitchhiked rides. Now they've been barred from the trains. For the first time on their journey since sneaking across the Syrian border into Turkey, they must continue by foot. The heat is excruciating, the humidity oppressive. Artifacts of people's lives are scattered along and beside the rails: an English translation of a diploma from Damascus University, a bundle of diapers, candy wrappers, empty water bottles, and redundant winter clothing.

Just an hour before, the refugees had crossed the border into Hungary on foot. After trudging a few miles more, they were apprehended by the police and put into a makeshift detention center on the spot. Exhausted, they sprawled in a green Hungarian field. The police gave them bottles of water, but there was no shade. For the refugees, shade is a treasured commodity—almost as much as water or food.

A cheerfully red wheelchair stands nearby. Who put it here, alongside the train tracks? How could anyone have pushed it through the field? And what happened to the Syrian, Afghan, or Iraqi it belonged to?

The refugees ask no such questions. They sit and wait. Perhaps they are fantasizing that a big bus will materialize and take them onward—after all, Hungary is merely a waystation on their trek northward to Germany or Sweden. A woman and her teenage daughter walk over to me. "We will die here," begins the girl in English, "we are so tired." Her mother adds: "We want to go to Germany." The daughter explains that "Merkel is a good woman, she loves the entire world."

The people on the grass know their location only in the vaguest way. What is important to them is that they are at a point along their route. For them, Hungary is just another country on the journey. But the Hungarian police have their orders: they are to stop everyone, and then take them to a detention camp, where they'll be fingerprinted. So Hungary's ultranationalist government has dictated in its resolve to deter refugees from crossing through the country. Its leaders hope to score political points at the expense of these foreigners.

The refugees fear fingerprinting. It can be used to prove the obvious—that they did not arrive in Germany or Sweden directly, but instead via another European country. The migrants believe that, by law and international treaty, they can claim refugee status only where they first arrive from Asia—which in most cases means Greece. Any subsequent country they enter, such as Germany or Sweden, has no obligation to give them that status. While this is technically true, in fact Germany and other countries were at the time granting asylum to refugees from Syria even when they knew that they had entered Europe elsewhere.

But the Hungarian policemen know nothing of such fine points. Two of the younger officers invite the refugees to play soccer with an improvised plastic ball wrapped in white tape. "Like soccer?" they ask the refugees. One of the Syrian boys is wearing a Ronaldo jersey. The policemen start to mark off a playing field and to gather the children for a game. It's a naive gesture; it would be perfect for a saccharine film with a corny-sweet scene about how sports can break down barriers. But in the context of the refugees' dusty journey, the game has a sinister subtext. It means: you are not going anywhere soon. You'll be here for a while, so let's play and act like friends.

Almost as one, the refugees get up and start walking again along the railroad tracks. "*Yalla ya shabab*," they mumble to one another. "Let's go, guys." They simply disregard the young Hungarian policemen in the field. They march quickly, urgently, fearing that the Hungarians will soon follow and detain them.

A bit farther on, a small group of refugees stops and looks back anxiously at the policemen. As I approach them, a dog appears out of nowhere at my side. Most members of the group take a seat on a knoll beside the railroad tracks. Two teenagers, a boy and a girl, stand and gaze at me in silence. Dressed in jeans and a striped shirt, the girl makes furrows in the soil with her sneakers. She has a treble clef tattoo on her neck, and her left eyebrow is pierced. The young man at her side is wearing a Nike warmup jacket. I asked the girl her name. She's Lilan, she tells me, and she's seventeen, from Aleppo. The fellow with the Nike jacket remains nameless. Lilan says he is her cousin.

The conversation is as sharp as a knife being drawn across someone's neck.

"Can you tell me where you came from?"

"From Syria. We passed through Turkey, Greece, Macedonia, Serbia, and now we're here. The journey is very hard." She left the rest of her family, including her siblings, aged sixteen and twenty, behind in Aleppo.

"Why are you here?"

"You know why." There's a sardonic smile. The war somehow lives in that smile.

"Where do you want to go?"

"Anywhere."

She asks me what awaits them now that they have walked dozens of kilometers. I tell her that the Hungarians will soon try to stop them. The young man at her side curses, walks away from us, and sits down next to the train track. "May God help the Syrian people," he mutters.

"What do you want to do in Europe?"

"I just want to study. And victory," she says.

"What do you mean by victory?"

"Syria, freedom, victory, stay, food, water, shower, everything. It's really simple. But we need it . . . I want everything. Everything."

The conversation peters out under the heavy burden of hope. Here is liberal Syria, with a treble clef inked into her skin, gone astray along the Hungarian-Serbian border.

"You want a life."

"Yes, an absolute life."

Lilan and her small group resume their trek along the tracks. My cameraman walks alongside, filming. We lose contact with each other, and there's no cell phone reception. So suddenly I'm unable to direct the video shoot, and the people I'm watching are no longer the material for a television report. I can just be with them.

A gap opens in the procession of people on the tracks. Those who have passed by keep walking; those still to come are still far away. I stay put, searching in vain for my cameraman. I see a boy, perhaps five years old, on the tracks. He's sitting alone. His clothes are neat and clean. I look at him and his eyes are blankly focused on the trees nearby. All parents know this expression. It means utter fatigue. For a moment I panic; is he completely alone?

Then his parents appear. Both are staggering, about 800 meters behind. His father approaches, and now, as he draws near, I can see he's carrying an infant in a sort of baby sling tied around his neck. "*Yalla, ya ibni*," the father says to the exhausted boy, "Let's go, son." As the father walks past me, I realize why he insists that the boy get up and walk on his own. A second baby hangs in a sling on the father's back.

The boy refuses to get up and the father stands over him, his strength spent. Suddenly, an older boy, about ten, appears from behind. The father, with a sigh, removes the baby from his back and places it in the arms of the brother. Then the father takes his weary son into his arms.

My immediate thought is, where's our car? Maybe we can change our route and take the family to the Keleti train station in Budapest, which is where the whole convoy is headed. But the refugees continue to walk on and then dwindle into black dots bobbing along the railroad tracks through the green expanse, and I remain alone, feeling absolutely powerless.

New Wars

I have already referred to the global challenge posed by the birthrate crisis. The solution would seem obvious—to welcome and assimilate the roughly 80 million refugees, asylum seekers, and displaced persons.[1] According to the UN, we are living through the worst refugee crisis since World War II.[2] During the age of responsibility, political leaders and their citizenry were prudent compared to today. In that period of relative stability, there were far fewer displaced people. During the 1990s, with the collapse of the Eastern bloc and the outbreak of war in the Balkans, the number of refugees began to rise significantly.

Within eight years, the number of displaced persons doubled, from 20 million to over 40 million people; by 2019 it had reached 79 million.[3] Millions more have since joined this fellowship, at a staggering pace. Most recently, much of this surge has been due to the conflict in Syria. Approximately two-thirds of the Syrian population, more than 10 million people, have fled their homes since civil war broke out in 2011. Some migrated outside the state's boundaries—3.5 million live in Turkey, for example—but others became refugees in their own country.[4] Even so, the Syrian civil war, with its enormous political repercussions, is only part of the international humanitarian crisis.

The current refugee crisis stands out because it is not the consequence of a global conflict. In the 1940s, tens of millions of people lost their homes and wandered in search of safe havens. But they were displaced in a brutal world war. Despite the economic crises of the 2000s, the world has arguably become more secure and prosperous during the past two decades. In fact, the data shows that people are much safer from conflict today than they ever have been before.[5] We have not experienced another world war, a great depression, and COVID-19 has yet to cause mass displacement. The number of interstate wars is at a historic low. There is almost no country in the world that is engaged in war with another country. That is nearly unprecedented in the modern era. So why have so many people been displaced?

Notably, less than a third of the displaced people are refugees who left their countries. Such refugees numbered about 26 million people in 2019. Over 45 million are internally displaced persons—that is, they remain within the borders of their country but have been forced from their homes on account of war, hunger, or expulsion. They find themselves in gigantic camps or in impoverished neighborhoods on the outskirts of big cities and cannot return to their homes.[6]

The huge increase in the number of internally displaced persons is the result of internecine strife and the loss of freedom of movement across borders. The latter almost goes without saying: today, states have more ways to keep people from crossing their borders than they have ever had in the past. There are so-called smart fences equipped with movement sensors, closed-circuit television, satellite technology, mobile phone networks that can monitor people's locations, biometric passports, and retinal scans. It has become much more difficult to flee a country as an unregistered asylum seeker.

Increasingly, it is not interstate wars that are driving people from their homes. More and more, people instead leave because the societies they live in are collapsing. "New wars" is a term coined by Professor Mary Kaldor to describe the nature of conflicts in the world after the fall of the Soviet bloc.[7] The contending forces include both state and non-state actors; crime and human rights violations are common and political control more crucial than actual geographic possession of territory. In the post-bipolar world, Kaldor notes, conflicts are less often over ideology (Communism versus liberal capitalism, for example) and more because of ethnic or religious identity.

Simply put, wars these days are less about what you think, or what your country demands of another country, than about who you are. A person's opinions can be changed through persuasion or coercion, but religious and ethnic identities obviously are less fluid. One is born Hungarian or Romanian, Kurd, or a Muslim or Christian Arab. An identity-based dispute is very hard to resolve. The examples that prompted Kaldor to write her book were the wars in the former Yugoslavia, where the adversaries aspired to establish ethnically and religiously homogeneous states; some were willing to engage in genocide or ethnic cleansing to achieve that goal. The state that pursues such a policy is not concerned about the political views of the people it is cleansing from its territory. It seeks to obliterate the adversaries' legitimacy and identity altogether. Sitting in a coffee shop in Dubrovnik, many years after the end of the Serbian-Croatian war, I spoke with a former Croatian army officer. "These people were not here when the Renaissance or humanism happened," he said of the Serbs. "Having no history of their own, they resorted to mythology! Their entire history is a myth!"

The contemporary implosions of societies have been attributed to new wars, failed states, and hybrid conflict. Whatever the term used, the context is always globalization. When crude oil sold for $100 a barrel, Venezuela could afford the populism and corruption of Hugo Chávez and his government. When global prices dropped in 2014, the state began collapsing, forcing more than 3.5 million Venezuelans to flee poverty and seek refuge in other countries in Central and South America.[8]

Global economic factors also accelerate the lethal tango of state and non-state actors. The Mexican drug war is ostensibly a conflict between the state and drug cartels. Actually, it is more a war among the cartels over money, power, and market share, as the government cracks down on them by arresting their foot soldiers and confiscating their narcotic merchandise. Most of the war's victims are cartel soldiers or civilians. The government effort in Mexico is also the result of political pressure from the country's northern neighbor, which seeks to contain its long-running and worsening addiction crisis. The US is also a source of weaponry used by both the drug merchants and the Mexican army.

In this conflict there are no set battles and no clear and enduring winners or losers, at least not so far. Economic interest is the predominant motive. A World Bank study found a correlation in Mexico between income inequality and an increase in murder rates during the drug war.[9] Then-president Felipe Calderón launched the drug war in 2006. Since then, at least 120,000 people have been killed, and tens of thousands more have disappeared and are presumed dead.[10] That makes it one of the deadliest conflicts in the world, and it has virtually nothing to do with nationalism, tribalism, or religion. The most potent force in a world of globalization is supply and demand. The drug war is a byproduct of a desperate attempt to eliminate the supply, while demand only grows.

Globalized interconnectedness not only sets the material conditions of conflicts and their motivations but can also transform their nature. The Syrian civil war did not break out because of the divide in the Muslim world between Shiites and Sunnis but as a people's uprising against an oppressive regime. As it escalated, state actors joined forces with non-state groups and turned it into a proxy war that conformed, to some extent, to religious divides: Sunni forces fought against the Assad regime, which has its power base in the Alawite community, and which has the support of Iran, the regional Shiite power. Money was raised and fighters recruited using religious identity and a call for a jihad against fellow Muslims. As the Syrian conflict became more global, the values fought for became more local, historic, sectarian, and tribal.

Since contemporary conflicts include non-state actors, such as crime syndicates, financial institutions, corporate lobbyists, the media, and of course official international institutions, they no longer accord with the state system built after World War II. Globalization has created a world in which conflicts cannot be confined within sovereign or geographic borders. As a rule, there are no "sterile" wars, devoid of external involvement. The dynamics of global interconnectedness turn regional confrontations into black holes, powerfully sucking more elements into them. What started as an internal Syrian struggle morphed into a regional conflict, and then one involving multiple global actors. The world in which countries could have internal, distinct, and relatively confined conflicts has died, if it ever existed. Indeed, attempts to confine conflicts to local theaters have arguably failed. The Obama administration decided to keep its feet out of the toxic swamp of the Syrian war. They had plenty of good reasons for doing so. They saw little economic or strategic value in Syria, and believed that America's ability to bring about liberal democracy in the country was extremely limited given the evolving sectarian nature of the conflict. The administration was probably correct in thinking that the United States could gain little from intervention. In different, less interconnected times, the policy might have succeeded. In the current era, you cannot bypass the swamps of regional upheavals. It sucks you in with a vengeance.

Where do the refugees come from? Syria is in first place by a lot, followed by Venezuela, Afghanistan, the nascent South Sudan (which has collapsed into tribal warfare), and Myanmar (which has conducted ethnic cleansing against its Muslim minority).[11] All of these countries, besides Venezuela, which is collapsing as a result of its repressive regime, sanctions, and failing economy, are in the midst of long and brutal civil wars, and the number of internally displaced persons is very high. The breakdown of communities and states is a fundamental phenomenon in international life in the twenty-first century.

The UN Charter guaranteed the sovereignty of the international organization's member states. At the time, states most feared hostile occupation and imperialism and sought to preserve national self-determination. Today, their main challenge is to endure as cohesive political communities—that is, to exist.

There are no effective international mechanisms for intervening in states in the throes of intense internal conflict, nor are there rules or procedures for halting the slide of countries into collapse. Indeed, the international system is built on precisely the opposite paradigm—states are sovereign and thus generally shielded from outside interference. While the World Bank and the IMF might be able to assist countries in economic strife, they do not pretend to supply an encompassing approach to a state experiencing a multifaceted crisis. The doctrine of humanitarian intervention is not generally accepted, and even if it were, it is invoked only in the most severe cases, such as genocide—and even then, only rarely. Countries are not supposed to meddle in the "internal affairs" of other countries.

As states fail, they set off a social and political chain reaction that defies attempts to control it. When the Syrian civil war broke out, millions of people were expelled or fled from their homes. They sought refuge in the coastal cities or deep in the Syrian desert. Another three and a half million people made their way to Turkey. A million and a half arrived in affluent Europe, with its population of about 740 million. The era of prosperity has camouflaged the weakening of states. Weak states, mostly in the global South, fragment under internal and external pressure, sometimes forcing people from their homes in the process. These displaced people seek a haven elsewhere in the collapsing state, or in neighboring countries, which are often in dire straits themselves.

The chain reaction can be triggered deliberately and artificially, and that has happened in Syria. First, President Bashar al-Assad ordered his forces to forcibly conscript young Syrian men. This included abductions carried out by special gangs directed by the government.[12] The regime then extended the term of mandatory military service, practically indefinitely. Many Syrian refugees cited forced conscription as their reason for fleeing the country. Another new edict was that draftees could defer their service by paying a fine of several hundred dollars. Men who paid this fee were issued passports, not an easy document to obtain in a dictatorship like Syria. But in 2015 the regime suddenly made the process much briefer and easier. Passports were even issued to Syrians who had left the country illegally. Up to this time, refugees from the country generally made a dangerous and lengthy journey to Libya. There they tried to cross the Mediterranean in rubber rafts or decrepit boats, together with African migrants, to Italy. The presence of millions of Syrians in Turkey encouraged local smugglers to upgrade and lower the price on a shorter and a much safer route—the crossing from the Turkish coast to the Greek islands in the Aegean Sea. The cost of getting to Europe plunged. It cost a Syrian about $6,000 to get to Europe via Libya, but only about $3,000 to make the trip via the fifteen-mile link between the Turkish port of Bodrum and the island of Kos.

The news that the gates to Europe had opened quickly spread throughout the Middle East. Thanks to social and other media, vital information about how to make the journey to a new future could spread even in the midst of the hellish Syrian civil war. Young Syrian men faced a stark choice—they could stay in Syria and risk death or forced conscription into the army of a reviled dictator, or they could pay to defer their army service, obtain a Syrian passport, and leave the country in relative security.

Many of the refugees claimed that the mass emigration was a case of ethnic cleansing, intended to empty the Syrian heartland of its Sunni population. The goal was profound demographic change in favor of Assad's loyalists. In 2016, General Philip Breedlove, NATO's supreme allied commander in Europe and commander of the US European Command, gave testimony before a congressional committee that did not receive the attention it deserved. He maintained that the Assad regime was driving out its population not only as part of its war effort but also to wreak political havoc in Europe. "Together, Russia and the Assad regime are deliberately weaponizing migration," he declared, "in an attempt to overwhelm European structures and break European resolve."[13] Breedlove cited the regime's bombing of Syrian civilians, with Moscow's backing. He said that the bombing had "almost zero military utility. [It is] designed to get people on the road and make them someone else's problem. Get them on the road, make them a problem for Europe, to bend Europe."

Demographically, the refugees are negligible as a proportion of Europe's population, but they have triggered a huge political reaction. An increasing number of voters and politicians in Western countries are painting immigration as an imminent threat. It's an easy argument to make, as people in industrialized countries have been encouraged to believe that migration can be controlled forever. In fact, attempts to limit human movement are experimental, new, and probably not sustainable. The Syrians I met on the Hungarian border faced the prospect of death at home, which they could not and did not accept. Their journey was a revolt against the current world order. They sought, as seventeen-year-old Lilan asserted, "victory."

Less developed countries bear most of the brunt of the refugee crisis. Turkey has taken in more than twice as many refugees as the entire EU. One day, while my crew and I continued to document the arrival of boatload after boatload of refugees to Kos, we stopped for lunch at a restaurant on the beach, five miles from the closest point on Turkey's coast. At the next table next sat a middle-age Turkish woman sporting a designer handbag. We asked her about the impact of the refugee crisis on her side. "I'm from Bodrum," she said, "and about two months ago, an entire Syrian family moved into my backyard and set up a tent."

"What did you do?"

"I called the police to get rid of them," she replied. After hearing the details of her complaint, a police officer replied: "Ah, they are personal guests of our esteemed president, Recep Tayyip Erdoğan." She hung up and dialed again, hoping to speak with a police officer who wouldn't play games with her. Another police officer answered her call. When she again reported the squatters in her yard, he had an answer ready. "Those people? They are guests of President Erdoğan. No one told you?"

I asked her what she did after that.

"I didn't do anything, I understood that it's the policy," the Turkish woman shrugged.

CHAPTER 14

An Experiment and Its Costs

Everyone has the right to leave any country, including his own, and to return to his country.

—ARTICLE 13(2) OF THE UNIVERSAL DECLARATION OF HUMAN RIGHTS[1]

My grandfather, Joel Shastel, was a short, silent man. On Sabbaths he dressed in heavy suits totally unsuited to the Mediterranean climate, baffling me and my brother as we watched him return slowly from synagogue. We never had a real conversation with him. When we grew up, we realized that he seldom had real conversations with anyone. Sometimes he would give us candy, but mostly we were scared that he would shout at us to stop playing, not to run through the fields around our house or to climb the ruddy loam hills that lay past them. He demanded that we be careful all the time. We shouldn't fall, we shouldn't be wild, nothing should happen to us. Caution was paramount. When my father was a teenager, he rebelliously put a shelf up on the bare wall of the tiny house he and his two brothers grew up in. What a ruckus my grandfather caused over putting a hole in the wall, the trouble it involved, the danger of putting up a shelf.

He lived his life in dread. The most tragic part he expunged from his mind, leaving only three cryptic words, "*Ha-Nazim, yimah shemam.*" "The Nazis, may their name be obliterated." Calling down oblivion on the name of a persecutor is an ancient Jewish curse, and for Grandpa Joel it contained everything—his beloved wife and son, who were murdered by the Nazis; the disappearance of other members of his family transported to death camps; and the annihilation of the culture and environment in which he grew up. It was also his curse at having had to see all that happen from afar, without any power to intervene. "*Ha-Nazim, yimah shemam*," he would mutter, and then fall silent, and the conversation would expire even before it began.

He was born in 1905 in Bialystok, Poland, to a poor family of that city's large Jewish community. What we know of his life we have put together from letters and official documents; he told us almost nothing. He apprenticed as a tailor, and at the age of twenty-five married Rejzle Winokur, five years his senior, in a ceremony conducted by the city's chief rabbi. He served in the Polish army before his marriage; a photograph shows him to have been an uncommonly handsome young man, with large eyes and lips. In the photo, he looks determined and full of hope, but we never saw him look that way.

Some members of his family emigrated to America, as he also wanted to do. Palestine was another option. Unfortunately, Britain had just instituted major restrictions on Jewish immigration to the country, which it ruled under a League of Nations mandate. The world was sealing its borders against immigration. Three years after his marriage, the couple made a bold decision. Joel would travel to Palestine on a tourist visa good for three months and violate its terms by staying on illegally. From there he would make arrangements for the rest of the family to come. He left Poland in 1933; his travel documents show that his journey was a long one. He went through Czechoslovakia, Greece, and an undesignated Arab country before arriving in Palestine. The Syrian refugees I spoke to followed a similar route, but in reverse, from the Middle East to Europe. The paths of flight did not change, only the circumstances and direction.

The year he arrived in Palestine, the Nazis came to power in Germany.

My grandfather settled in a farming village north of Tel Aviv, staying in touch with his wife and son by letter. He kept trying to obtain a "certificate" for them—an immigration permit—but without success. Certificates were in short supply because of the British restrictions, and were allocated by functionaries of the Jewish Agency. He was a laborer and illegal infiltrator into Palestine. The security situation was dire in any case. Three years after his arrival the Arab Revolt broke out; the Arabs were incensed at the British policy of permitting Jewish immigration and settlement in the country. The troubles continued until 1939. That same year Bialystok was conquered by the Germans, who then handed it over to the USSR as part of the Molotov-Ribbentrop pact. Two years later the Germans launched Operation Barbarossa against the Soviet Union. On the first day after the Germans retook Bialystok, they set fire to the city's central synagogue. Eight hundred of the city's Jews burned to death inside. And that was only the beginning.

One can only imagine what my grandfather went through, mad with worry and fearing for the lives of his wife and son, his parents and family, as he remained trapped in Palestine. The documents tell the story. He knocked on the doors of consulates. He used what little money he had to pay professional letter writers to compose petitions and pleas. He paid a notary to translate and certify his birth and marriage certificates. He joined political parties, hoping that their functionaries could help. In November 1942 the world learned that the Nazis were carrying out systematic mass murder of the Jews. It was not persecution, or pogroms, but rather a deliberate campaign to annihilate the entire Jewish people. My grandfather's despair is encapsulated in two brief documents we found after his death. The first was a letter he sent to the Jewish Agency; written at its center in large letters was the word "*hatzilu*," "save them!" The second was the reply, stating that the representatives of the Jewish Agency "understand his difficult situation" but that there was nothing they could do.

Other documents indicate that he may have tried, unsuccessfully, to make his way back into war-torn Europe, where the gas chambers were already under construction, to be with his family. If that is true—we have no way of knowing for certain—he failed even at this attempt at suicide. The vast majority of Bialystok's Jews were murdered in Treblinka, Majdanek, or Auschwitz.

Following the war and frantic searches that turned up nothing, he married Fortuna Tabach, from the ancient Jewish community of Beirut. She was a strong-minded and exuberant woman, very unlike him. They had three sons, one of whom became my father. My grandmother fell ill and died young, another tragedy for a man who had already lost everyone he had.

My grandfather was a man whose family was ensnared between borders and malign evil. Then he, too, was trapped, bereft of citizenship, without a country, knocking on foreign doors, begging for mercy, powerless. Not only did he fail to rescue his wife and son, he could not even die with them.

The Birth of an Experiment

Homo sapiens' distinctive capacity for migrating and adapting to new habitats has allowed the species to spread and flourish globally. History is to a great extent the story of mass human movements, for example the biblical story of the Exodus from Egypt, the invasion and reshaping of the Roman empire by tribes from northern Europe, William the Conqueror's conquest of England in 1066, the spread of Islam and the early Arab conquests, and the European invasion of the Americas.

Our age—paradoxically—is turning into an anomaly. Today's globalization amplifies political, economic, and cultural exchange among nations and individuals. But there is a kind of black hole at the center, exerting a force that impedes human movement. While information, capital, and goods flow much more easily and with much greater security than before, the movement of people is more severely constrained.

Starting in medieval times, kings and lords realized that controlling entry and exit was fundamental to power, and to the ability to manage their lands. Even within their realms, rulers sometimes sought to prevent population movement into large cities, out of fear that it would destabilize their regimes. Beyond formal slavery, which entirely deprived people of freedom of movement, the upper classes in Europe, Asia, and Central America maintained systems of serfdom and peasantry that tied farmers to the land as parts of a single property. It was not until 1861 that Czar Alexander II freed Russia's serfs—some 23 million human beings.

With the inception of the colonial era, similar attempts were made to control the populations of imperial lands in the New World, Africa, and Asia. Spaniards were permitted to enter their country's colonies only with a license that proved that they were "Neither Jews nor Moors, nor children of such, nor persons newly reconciled, nor sons or grandsons of any that have been punished, condemned, or burned as heretics or for heretical crimes. . . ."[2] An opposing model took form in

Britain, which sometimes preferred to use its new possessions as penal colonies or to permit religious minorities to settle there, as in the case of North America and Australia. In the ancient and modern worlds, expulsion of people from their land, in essence forced migration, has always been an established practice. The Jews are among the few that survived as a community despite suffering this fate again and again. First after the fall of the ancient Hebrew kingdoms, and then again after the loss of Jewish sovereignty in the first century AD, the Jews have wandered, seeking new homes from which they were then again uprooted.

The Jew was the ultimate refugee and the perpetual stranger. "From place to place the homeless Jew wanders in ever-shifting exile," writes Prudentius, a Christian born in the fourth century CE in a Roman province that is now part of Spain, "since the time when he was torn from the abode of his fathers and has been suffering the penalty for murder, and having stained his hands with the blood of Christ."[3]

After being expelled from Judea, Jews were expelled from France, England, Spain, the Rhineland, Austria, Lithuania, and the list goes on. In some of these cases, they were allowed to stay on condition that they convert to Christianity, but for the most part they refused, preferring to leave. It was a world in which monarchs could employ violence against Jews and other minorities, but the minorities could still flee their persecutors to other realms.

When the Jews were expelled from Spain at the end of the fifteenth century, many took refuge in neighboring Portugal. King John II accepted them on the usual terms that rulers who gave refuge to Jews demanded—payment in gold, and in advance. The six hundred wealthiest families paid a large sum and received permanent residence permits; the rest paid smaller sums and were allowed to stay for eight months. When the deadline passed, they were given the option of paying special taxes or becoming slaves. According to some testimonies, those who sought to leave also had to pay at the borders.[4] Refugee camps sprang up not far from the Spanish border, and epidemics broke out. As historian Francois Soyer documents, many Portuguese officials did all they could to keep foreigners out of their cities and towns. The king tried to protect the newcomers; his officials told local authorities that they lacked the authority to prevent the refugees from settling in their jurisdictions. In short, it all resembled today's refugee crises.[5]

When it turned out that some of the Jews were unable to pay the taxes imposed on them, the king instituted a child separation policy cruel even by medieval standards—he ordered the abduction of Jewish children, making them slaves by royal decree. Estimates of how many children were kidnapped vary from the hundreds to the thousands. It is well documented in Portuguese archives as well as in Jewish sources, which include accounts of mothers casting themselves under the hooves of the king's horse to plead for mercy. The children were handed over, as slaves, to a Portuguese aristocrat and sent to São Tomé, an island off the West African coast. Nothing is known for certain about their fate, but the assumption is that most died within a short time.

The Jewish people were subject to many cruel persecutions of this sort. But it was in Portugal that the story took a dramatic turn that demonstrates what happens when the possibility of escape is cut off. John II's policy was to grant residence permits; his successor, Manuel I, sought matrimonial union with the Spanish royal house. The pious Spanish sovereigns had already expelled the Jews; King Manuel realized that they would not agree to unite the two dynasties unless he did the same. According to some accounts, the Spanish made this a precondition for the marriage.[6] Whatever the case, five years after the Spanish expulsion, in 1497, Portugal decreed that all Jews must leave its territory. But, unlike Spain, and unlike with almost all other such expulsions in the medieval era, Portugal neither wanted the Jews nor wanted to allow them to leave, because of the large role they played in the country's economy and its merchant class.

The king thus forced them to stay and compelled them to convert to Christianity. Judaism was banned, and Jews were prevented from fleeing communally. Synagogues were shuttered, sacred books burned, and the forced converts were forbidden to sell their land and homes. Thousands of children were forcibly baptized and systematically taken from their families. According to a few sources, some Jewish families elected to commit suicide with their children.[7] Some of these "New Christians" continued to practice Judaism in secret, and in the end managed to escape. The rest were brutally incorporated into Portuguese society. In 1506, during a mass pogrom in Lisbon, thousands of suspected Jews were slaughtered, many of them burned at the stake. A year later, bowing to pressure that followed the massacre, the king allowed the New Christians to leave, ten years after ensnaring them. It was a temporary measure; Manuel then asked the pope to establish the Holy Inquisition in Portugal. Decades later, it would still be rooting out and murdering secret Jews. Within a single generation, one of the most important Jewish communities in Europe had been obliterated.

The Jewish experience in Portugal at the cusp of the fifteenth and sixteenth centuries teaches a universal lesson. History is rife with terrible deeds, but they are made worse when people and communities are not allowed to leave. In the long run, personal and communal survival require the ability to migrate. The very lives of Jews and many other tribes and peoples depended on the ability to move.

IN ANCIENT GREECE, BEING ABLE TO GO WHERE YOU WISHED was one of the four components of freedom, and marked the difference between slave and freedman.[8] In much later times, when the world still lacked rapid communication, it was difficult to create and enforce standards of permitted and forbidden movement. Passports were royal documents that, beginning about the fifteenth century, kings granted to their emissaries. But ordinary people did not always need such documents to move from one country to another. It was not that kingdoms and empires advocated free international movement—that was hardly the case. In an attempt to stop Protestant emigration, King Louis XIV of France forbade his subjects to leave his kingdom without a permit, an early form of the passport; he also required permits for travel from one part of France to another. Medieval Britain required that travelers embarking by sea from the island present a special document. Formally, "papers" or passports were often required by royal orders. Yet, in his illuminating book *The Invention of the Passport*, John Torpey explains that this kind of documentation functioned in a much murkier world. "Passports had a notorious propensity to go 'lost,'" he notes, "in which case replacements were to be secured in the area in which the traveler then found him or herself . . . Passport restrictions were a nuisance for many, to be sure, but administrative laxity and the well-meaning assistance of a variety of benefactors frequently made a mockery of the state's use of documentary controls as a means of regulating movement."[9]

Kingdoms and principalities had only a limited ability to prevent the entry of people into their domains, except when the numbers were large and involved entire communities.

The newcomers had no guaranteed legal rights and were not asked about their citizenship, a concept that did not emerge until the eighteenth century. Neither were they received with open arms. Sometimes they were killed when they crossed a river, or starved when they were refused entry into a walled city. But when a community faced a calamity and fled, it did not encounter along the way the effective technological barriers that, for example, today's asylum seekers from Central America confront when they seek to cross into the United States over its southern border.[10]

It's an important point. Just like today, rulers wanted absolute control over who entered and left their dominions; but they lacked the technological ability to do so. An exception appeared in the nineteenth century, and it produced one of the most important migration movements of the contemporary age. Before World War I, the United States did not have any substantial barriers to immigration, "reflecting a tradition of laissez-faire labor mobility that dated to the colonial period," writes Mae Ngai of Columbia University.[11] The millions of Europeans who immigrated to the United States in the nineteenth and early twentieth centuries sometimes came with passports or sometimes did not; they arrived without visas and without any guarantee that they would be allowed in, though relatively few were rejected. As time passed, Congress imposed more and more restrictions, barring anarchists, prostitutes, and Chinese laborers, among others. But generally, people who arrived, were not obviously sick, and could prove that they had some money, or relatives who could support them, who could walk without a pronounced limp and who were not feeble-minded, entered unimpeded.

The Great War was the turning point that prompted countries to institute closed and policed borders, as a security measure.[12] States grew stronger, and passport control at border crossings became a fixture of international travel.[13] The Paris International Conference on Passports, Customs Formalities, and Through Tickets of 1920 set uniform standards for passports,[14] thus laying foundations for the greatest experiment in human history—a global regime of effective borders that would isolate people and prevent them from entering countries without permission. The new passports included photographs, for the first time providing a relatively effective means of identifying people at border crossings. The United States Immigration Act of 1924 set, for the first time, quotas for immigrants of different races and from different countries, ending, for all intents and purposes, free entry into the United States. The institution of international passport standards made this easier.[15]

This system seems obvious to us today, as it did for our parents, but in fact it was revolutionary. It has changed the way people lived, survived, and flourished over previous millennia. Between 1820 and 1930, more than 30 million people immigrated to the United States, remaking the republic.[16] They were German, Polish, Irish, English, Dutch, Slav, and many others. This huge wave of migration transformed the US from an agriculture-based country of 10 million inhabitants into a superpower in waiting. It was a unique moment in which free movement was sanctified. Now it represents an age long dead, when starving Irishmen and persecuted Jews could board a ship and sail into a new future. It was a nation of immigrants formed not at passport control but rather from the tired and poor and huddled masses that Emma Lazarus, in her poem "The New Colossus"—inscribed on the Statue of Liberty's pedestal—welcomed into the New World.

Lazarus was a Jew of Spanish-Portuguese descent. Her family is thought to have fled Portugal for Brazil. Only in hindsight do we comprehend that the age of great migration was the last twinkling of a dying sun. By the twentieth century, nation-states were already modern and powerful enough to stop migration to their territories. Limitations on human movement became a global phenomenon not on account of a systematic ideology but rather because of advances in technology. Governments and sovereigns wanted that power, but had lacked the means. Then, suddenly, they could, so they did.

The Jews were the first to feel the brunt of the new restrictions, and in the most severe way. When the Nazis came to power in Germany, they initially pressured Jews to leave. Many did. But the immigration quotas instituted by the United States and Britain, and by the British administration in Palestine, caused a humanitarian crisis in the 1930s. Jews tried to do what they had always done—flee and find refuge elsewhere.

But the countries of the world fell in love with the godlike power they had gained and which gave them control over the physical location of all people. It well served the Nazis' ultimate aspiration of giving birth to a new world ruled solely by the Aryan race.

The Third Reich's murder machine began to devour Jews, as well as homosexuals, Roma, the "mentally unfit," political opponents, and any other groups the Nazis reviled. The story of my grandfather's family was that of all of those minorities: human movement in reaction to crisis, once natural, was now blocked, the result of the new experiment.

The vast majority of Europe's Jews were exterminated by the Nazis and their allies. After the war and the Holocaust, the international community agreed on conventions and norms aimed at preventing a repeat of the genocide. Yet, while nothing as monstrous as what happened in World War II has occurred since, people continue to be caught between the evil that is forcing them from their homes and the walls keeping them out of any place they might hope to find new ones. Sometimes they die on the wall itself.

Global Citizenship

What's new isn't the mass movement of people across borders but rather the successful attempt to limit that movement during the last hundred years. The general expectation in the West is that the great experiment in border controls will continue forever, even though it is unsustainable for many people and ethnic groups. The world is in convulsion, as people who must move face off against barriers to movement never known before in human history. British comedian Russell Brand has a routine that puts it in a nutshell:

> "Immigrants! Immigrants! Immigrants! . . ."
>
> You know that an immigrant is just someone who used to be somewhere else.
>
> "Ahh! Have you always been there?"—
>
> "No, no, no,—I used to be over there."
>
> "Ahh! Keep still! I can't relax with people moving around. Keep still on this spherical rock in infinite space. Keep still on the spherical rock with imaginary geopolitical borders that have been drawn in according to the economic reality of the time. Do not pause to reflect that free movement of global capital will necessitate free movement of a global labor force to meet the demands created by the free movement of that capital. That is a complex economic idea and you won't understand it. Just keep still, on the rock."[17]

The definition of an immigrant as "just someone who used to be somewhere else" is far from the way large portions of the public understand the word. It disregards cultural differences, and the disparity between people from the Middle East and those who have lived their lives in the suburbs of London. Those dissimilarities can be attributed to religion, social position, economic state, colonialism, and common historical memories, which are the fabric of any community. Such differences have a very significant effect on the way that the public thinks about and responds to immigration. A community's character is determined, first and foremost, by the individuals who compose it, and when the composition of the community changes, so do many other aspects of the public space.

Brand's agenda is universalist. He assigns little importance to the sentiment of national self-determination, or to traditional or religious values. This type of universalism jargon provides useful fodder for politicians. In 2016, Theresa May, then prime minister of the United Kingdom, declared that "today, too many people in positions of power behave as though they have more in common with international elites than with the people down the road, the people they employ, the people they pass in the street. But if you believe you're a citizen of the world, you're a citizen of nowhere. You don't understand what the very word 'citizenship' means."[18] This may sound like an incisive statement of how human communities work, but it's not. It is another attempt, much like those we saw in the chapters on nationalism and fundamentalism, to create a diametric opposition between a one-world concept of global community and being connected to a local identity. Cosmopolitanism is a flexible and open concept. Cosmopolitans can be traditional and proletarian; they are not only the stereotypical alienated people who fly business class.[19]

There's another problem with the distinction May makes. Half the globe's people—indeed, half her own country's citizens—feel that they are citizens of the world. At the same time, they are certainly citizens of a particular country. The World Economic Forum's Global Shapers Annual Survey, which polled tens of thousands of young people from all over the world in 2017, asked them to name the primary component of their identity. A plurality said "human," and the second most popular reply was "citizen of the world." Together, these accounted for 60 percent of the responses. Identification with a particular nation came in third, with a religious identity lagging far behind.

When the respondents were classified by income level, it turned out that young people of low or high middle income were most likely to identify as humans or citizens of the world. Among the poor, only 40 percent identified this way; the very wealthy also scored lower than the middle class.[20] In chapter 2, I cited an international survey conducted for the BBC over the course of many years, as part of which inhabitants of eighteen countries were asked to say what they thought of the statement "I see myself more as a global citizen than as a citizen of my country." In 2016, 47 percent of those asked in Britain agreed or agreed strongly with this revolutionary idea. The same was true of 54 percent of the Canadians surveyed, 59 percent of the Spaniards, and 43 percent of the Americans.

What was perhaps surprising was that respondents from developing countries, where there are high rates of extreme poverty, and which are often in conflict over their borders, were the ones most likely to agree with the statement. Seventy-three percent of the Nigerians surveyed saw themselves as citizens of the world, along with 67 percent of the Indians and 70 percent of the Peruvians. In the big picture, there was a virtual tie between those who identified themselves primarily as citizens of the world and those who identified primarily as citizens of their countries.[21]

That's astonishing. After all, people who identify as citizens of the world are actually signing on to something that doesn't exist. For them, citizenship is concrete and has clear legal manifestations and ramifications—an official language (or more than one), a flag, an army, and, mainly, a political system using its power to indoctrinate others into national citizenship. Yet with what ease they seem to be willing to adopt an ambiguous concept—global identity—with no legal foundation. Isn't it clear why all local powers are in panic? Hell hath no fury like a nation-state scorned.

GLOBESCAN, THE PUBLIC OPINION RESEARCH FIRM THAT performed the survey for the BBC, has been conducting this same poll since 2001, but it is only in the most recent one that the self-styled citizens of the world achieved parity with those who identified with their own countries. The increase is due largely to the developing world, which is enthusiastic about globalization—it is Nigerians, the Chinese, and the Brazilians, for example, who are pushing forward the idea of global citizenship. In contrast, in the seven European countries surveyed there has been a decline over the years in the percentage of people who believe that a person can view himself as a citizen of the world. In 2017, only 30 percent of Germans chose the universalist option, a decline of 13 percent since 2009. In a more and more oppressive, nationalistic Russia, only 24 percent see themselves as citizens of the world.

Surveys of this sort should be taken with a grain of salt, but in this case the data correlates with major developments in the West—Britain's departure from the European Union, the reluctant response to the Syrian refugee crisis in most of Europe, and the new political map of the West, suffused with xenophobia.

What we are witnessing is desertion. It is Europeans and Americans who introduced the modern ideas of supranationalism and universalism. As the nonwhite world, China, and the global South board the boat of globalization, the people of the developed countries are jumping ship. It raises the question of whether their erstwhile support for globalization was not actually a way of maintaining the global North's dominance. When they saw that globalization was becoming truly global, they began to sign out. When it turned out that globalization was emancipating the world's Others, who were by the sweat of their brows climbing out of fourth class and demanding to steer the ship, entire groups of people preferred to get in the lifeboats and leave.

In their haste, many Europeans and Americans are abandoning the lessons of World War II and their own history, not to mention simple common sense. Europe, after all, has been one of the great beneficiaries of international trade and immigration—think of Germany rebuilding its economy after World War II with the labor of Turkish guest workers, or the free movement of Europeans within the EU, which helped Britain enjoy increased efficiency in its service industries.

The United States, the most successful nation of immigrants in history, has been taking in ever-fewer refugees since the 1980s, and many fewer under Trump.[22] In October 2019, for the first time since records began to be kept, the US took in zero refugees. A celebratory moment for people such as Republican congressman Steve King, who declared in 2017 that "we can't restore our civilization with someone else's babies."

In his final speech as president, Ronald Reagan spoke of the magic of immigration and its importance to the American experience. "Thanks to each wave of new arrivals to this land of opportunity," he declared, "we're a nation forever young, forever bursting with energy and new ideas, and always on the cutting edge; always leading the world to the next frontier. This quality is vital to our future as a nation. If we ever closed the door to new Americans, our leadership in the world would soon be lost."[23]

Is it really true that the United States owes its power to immigration? Some say that is more of a romantic conception than a political fact. Either way, the concept has been cast into the trash heap by Trumpism. In the summer of 2019 Trump went so far as to adopt the rhetoric of the Ku Klux Klan and other racist groups, not just opposing immigration but calling on four congresswomen from families that had immigrated a generation or two ago to go back where they came from. "These places need your help badly," he said. "You can't leave fast enough." In saying this, Trump rejected the fundamental principle of citizenship, a liberal construct born out of the notions of progress. Citizenship unites people of disparate backgrounds in a single national community, as equals. Trump is instead presuming that the primal identities of one's parents and family taint you genetically and forever—even if you were elected to represent the people of an American congressional district. It is at this point that opposition to immigration morphs into nativism, and nativism into racism.

The desire to tighten borders to the point of sealing them entirely is an attempt to regain a capacity that has been largely lost, or at least weakened, under globalization. The attempt is spreading, and as it does it sometimes loses its anchor in reality. When I walked with the Syrian refugees,

I saw watchtowers and high fences rising up over the border between Hungary and Serbia. An energetic Hungarian police officer gave me a tour along the barrier, proudly displaying its advanced technology. “Soon,” he told me confidently, “we will be able to block them all.” None of the refugees I spoke to had the least interest in settling in his country.

CHAPTER 15

Rivers of Blood

We will go to Germany, study and work there. Happy life, Inshalla.

—SHAWQI ABOUDAN, SIXTEEN-YEAR-OLD SYRIAN

It was the time of the Syrians. The shores of the Greek island were covered with thousands of life jackets, the last relics of their flight. In the early afternoon, on a beach packed with bikini-clad women and men streaked white with sunscreen, my cameraman and I watched a rubber raft make landfall. About twenty stunned Syrians disembarked. "Where are we?" they asked in Arabic. When we told them that they were on Kos, one of them, holding a young girl to his chest, began to leap into the air and shout, *"Al-Yunan, shukran, al-Yunan!"*—"Greece, thank you, Greece!"

The bathers, most of them Europeans who had bought package tours, found themselves in the midst of a humanitarian drama—and they lived up to the moment. Many approached and offered the exhausted refugees water, fruit, and food. Some of the vacationers picked up the young children and bore them along the shore to the main road.

These newcomers were exhausted but exultant, full of gratitude to the beautiful and calm sea that had not sunk their raft; to the Greek navy that had not intercepted them; to the beach that had provided them with a harbor; to God, who brought them there; and even to the Israeli journalist who stood on the sand and greeted them.

On the promenade along the sea I met Riyadh Biyram, a tall man with a big smile and square glasses. He and his family hail from Kobanî, a Kurdish city that had been besieged and then largely leveled by ISIS during the war. Biyram promptly handed me a business card that identified him as a computer programmer, and offered his contact information; he specified that he uses both WhatsApp and Viber. Despite being without money, a passport, and a permanent home—like most of the refugees, he lived in a tent on the street—Biyram was already prepared for his first job interview. Probably no refugee prior to the current wave has made a point of being accessible on all possible networks, posting pictures of his journey on social media. "I am a software engineer, and I left because Syria is completely destroyed," he told me in English.[1] "We decided to leave because there is no possible future there. We did not want to delay, because maybe in another month or two Europe will close its borders, and then we would have been unable to leave."

Biyram's younger brother, Osman, who had been studying construction engineering at the University of Aleppo, stood next to him. "My brother is very smart," Biyram said. "If he had stayed in Syria he would have had no chance of completing his studies. Why stay in Syria? If we come here, maybe he can become an engineer, a doctor, a teacher, or some other important profession. In Syria he would have been at most a day laborer. That is the thinking that led us to make this journey." His optimism was infectious.

On the promenade, a few hundred meters to the north, about 200 people were demonstrating in front of a police station. The protesters spoke Persian rather than Arabic, and were rhythmically, desperately shouting "Iran! Iran!" at a few weary Greek policemen. "They aren't giving us papers because we are from Iran," an unshaven man who looked about thirty years old explained to me. "We have nowhere to sleep, there are no bathrooms, and they aren't letting us continue on to the continent." The Greeks had not yet decided what to do about the Iranians because, unlike the Syrians, they were clearly not war refugees. "They executed my brother by hanging," the man told me, clenching a hand around his neck. "If we go back to Iran they will hang me."

Suddenly a man with short blond hair approached, wearing a white shirt and striped vacation trousers. He was clearly not Greek; his accent was north European. He began to shout at the crowd: "Don't you have any respect for us? Don't you have any respect for this country? Don't you have any respect for the laws and police office? And you want to come to Europe? What are you doing, unbelievable, no respect for nothing!" Then he turned to me. "You see the garbage, they were welcomed to Europe, we give them everything, food, money, housing, and no respect for nothing!" he griped. When one of the Iranians tried to pantomime that they had no water or food, the blond man imitated his gestures and hooted like a monkey.

"What would you do if you didn't have any water or food?" I asked him.

"If it is your country, you would kill them, expel them!" he shouted at me. "And here we can do nothing. They destroy the economy, destroy Kos, destroy the citizens." He turned back to the demonstrators and shouted at double speed: "Thank you! Welcome to Europe!"

Such hostility was relatively uncommon in mid-2015, but it was a prelude to what has since become much more widespread. The rash of refugees from the Middle East came on top of a long flow of refugees from Africa. There is hardly a country, especially in Europe, in which anti-immigrant feelings have not grown stronger, sometimes in the form of political parties dedicated to preventing refugees from coming in, sometimes in the form of existing parties or movements seeking to tap such sentiments. An Ipsos survey released at the end of 2017, conducted in twenty-five countries, found that while one out of every five people surveyed asserted that immigration had a positive impact on his or her country, twice as many claimed that the opposite was true.[2] About half of those surveyed said that there were too many immigrants in their country.

The Poor and Immigrants

Immigrants are taking our jobs, nationalists almost always charge. And there are indeed studies that show that an expansion of the labor force pushes down wages among workers who lack anything other than basic skills. That makes sense in the short run—when more workers compete for jobs, employers can pay less. But most studies show that it doesn't take long for the positive effects to appear. The consensus among economists about immigration is, overall, positive. It creates more demand, drives innovation, enhances labor efficiency, and boosts productivity, all of which compensate for the short-term problems.

The country taking in the immigrants will incur high costs at the start—for example, the taxes immigrants pay do not cover the costs of the education and health services that they receive (in part because immigrant families are often relatively large). An International Monetary Fund paper shows a direct correlation between the number of immigrants and a rising standard of living in society as a whole. The correlation was not dependent on the skill level of the immigrants—both those with high and low skills lead to increased labor productivity.[3] According to a calculation from 2017, if the United States were to take in 8 million immigrants a year, it would enjoy annual growth at a stunning 4 percent rate.[4]

One of the fundamental problems in industrialized countries is aging populations, which require higher expenditures on health and on care of the elderly, and consequently a severe actuarial crisis. Immigrants themselves tend to be young, and they generally have more children than the established population, making the population as a whole younger. The second generation, those born to immigrants in the new country, often makes a major contribution to society. In the United States, the average income of this first American-born generation is the same as that of the population at large, with a lower poverty rate and a higher percentage of academic degrees.[5]

Immigrants have enormous motivation to get ahead, and they change in dramatic ways as a result of settling in a new country. A 2009 study showed that a Mexican worker in the United States earns two and a half times what he would have earned in Mexico, in terms of purchasing power. A worker from Haiti makes ten times as much, and one from Nigeria fifteen times as much.[6] With the help of the institutions, infrastructure, higher education, and personal security that an industrial country provides, these people have a high chance of succeeding and thriving. Even unauthorized immigrants in the US, those who have difficulty gaining access to all the rights and services granted to the local population, make a decisive contribution to the economy. A recent study pegs it at $6 trillion a decade, and shows that legalizing these immigrants would increase their contribution to American private-sector product by more than half.[7]

■

THE QUESTION IS NOT WHETHER ADVANCED ECONOMIES BENefit from immigrants, but *who* in the country's population is benefiting. In the American case, where economic inequality has been on the rise in recent decades, it is a critical question. An analysis of the most comprehensive American report of recent years on the economic consequences of immigration reveals who is benefiting the most:[8] immigrants themselves, and the rich. Immigrants, of course, enjoy a significant rise in their quality of life and earning power. The entire immigration surplus, meaning the increase in wealth to the local population, is quite modest—only $54 billion, just 0.31 percent of the overall increase in income.[9]

But what is the effect within the local population? George Borjas, a Harvard economist, is perhaps the skeptic cited most often. His take is that local poor and immigrants compete with each other for jobs, driving down wages and thus costs for firms, transferring massive amounts of wealth from lower-income earners to the top percentiles of the American economy. The arrival of immigrants hurts the poor, composed of prior immigrants and the American-born poor who lack high school educations. According to one of Borjas's studies—one often debated by his peers—a growth of 10 percent in the labor supply as a result of immigration leads to an average decline in wages of 4 percent among competing workers of similar educational levels.[10] Borjas maintains that immigration is actually an asset for corporations. So while immigration caused wealth overall to grow by $54 billion, he contends that it also led to a transfer of wealth from workers to corporations totaling half a trillion dollars.[11]

Other studies, for example of the effect of the sudden entry of labor migrants from the Czech Republic into a region of Germany, have shown that it led to a modest decline in wages for local workers and a significant drop in employment levels for older local workers.[12] Yet the IMF paper cited at the beginning of the chapter maintains that an increase in the percentage of migrants in the job market leads to per capita income growth among the lower nine deciles of earners and in the top decile.

"Rivers of Blood"

This debate among economists gives a respectable veneer to the real clash over immigration, which is not about immigration's effect on wages and growth. It is actually about identity and the character of society—and demographics is crucial. In 1970, the percentage of American inhabitants who had been born overseas was 4.7 percent. By 2017 it had reached 13.6 percent, the highest since 1920.[13] In Britain, the percentage of the foreign-born has more than doubled since the 1980s.[14] The picture is the same in Germany—within three decades, the number of noncitizen foreign-born residents has doubled. The sense that immigration has risen and that the countries of the West have changed is not the delusion of the disgruntled people who the then-candidate Obama famously said "cling to guns and religion." The proportion of foreign-born people has risen in all these places for two principal reasons: first, tacit consent among the political classes that legal immigration is vital for the economy and should be expanded, and, second, a significant rise in the number of unauthorized immigrants. In the US, the number of the latter has tripled since the 1990s.[15]

These are comprehensive changes in the makeup of society and therefore its identity. They have profound implications for culture and people's sense that they are in control of the societies they live in. Too often, liberals trumpet data showing that the major resistance to immigration is to be found in places largely devoid of any immigrants. It must, then, be irrational. It is a patronizing argument. People don't need to live in New York or London to sense that their country has changed. Its public culture, which streams into living rooms on television and social networks, reflects the new demographics. Physical distance from these new members of society can make the threat they seem to pose to identity seem more dismaying, not less.

The political discourse about these issues was silenced by the mainstream political parties in almost every industrialized country. Economic rightists wanted massive immigration for the sake of economic growth; the left believed that immigrants would become its voters. For the well-off sectors of these societies, immigration had only advantages; after all, the indigent immigrants, with their foreign culture, did not live next door to the rich, nor did they compete with them for jobs that could be filled by people without a high school education. And the rich did not run into the newcomers in their schools or in the line at the public health clinic. What the upper classes saw was that immigrants were boosting the economy and bringing down the prices of services, while exerting virtually no political influence. For both the mainstream right and the mainstream left, support for immigration was morally attractive. It also looked good.

Yet, as the immigrants came, social time bombs were ticking. In European countries, which never fostered an ethos of immigration, there were severe problems integrating the newcomers and their children. In 2012, a third of French youth whose parents were born in Africa were unemployed, and almost a third dropped out of school without earning any sort of diploma. That was twice as much as French youth with no immigrant background.[16] In 2015, the median income for immigrants in France was 14 percent lower than that of the population at large.[17] Even immigrants with the same qualifications as the established population had trouble finding work. They were socially marginalized, the victims of racism and discrimination in the allocation of resources.[18]

That's not surprising. Western Europe is the homeland of white Christian homogeneity. Canada, the United States, and Russia all have higher minority populations than Western European countries do. For centuries, Europe specialized, more than any other place in the world, in the purging of ethnic and religious minorities—by segregation, expulsion, and sometimes annihilation. Perhaps it is possible to transform the Europe of the European Union into a continent of immigrants, but it can't be done while ignoring history, and without an earnest and courageous public conversation. When mainstream politicians like Britain's Enoch Powell used incendiary rhetoric to criticize immigration, they were denounced as xenophobes or racists.

Powell was probably dog-whistling when he offered, in 1968, a grim prophecy: "I am filled with foreboding; like the Roman, I seem to see the River Tiber foaming with much blood." Yet a dialogue about immigration was very much needed. With extremist Islamic terror cells operating in Europe for two decades now, the European right today maintains that the predictions Powell made in his "Rivers of Blood" philippic have come to pass. Some of the terror attacks Europe has suffered in recent years were perpetrated by immigrants or the descendants of immigrants. Such were the attacks in London in 2005; the massacre at a Jewish school in Toulouse in 2012; the murder of Lee Rigby, a British soldier, in 2013, whose assailants tried to behead him on a street in the heart of London; the attack on the Paris offices of the French satirical weekly *Charlie Hebdo* and a kosher supermarket in 2015; the series of coordinated terror attacks at the Bataclan theater and other sites in suburban Paris that same year; the truck attack on the Nice promenade in 2016, in which eighty-six people were killed; a similar attack at the Christmas market in Berlin later that same year; and the car attack in Westminster, London, just a few months later, in 2017.[19]

The attacks perpetrated by immigrants and their descendants have served as a powerful weapon for those who have long argued that the loss of European homogeneity would undermine the continent's security. Loss of a sense of personal security is what drives the core opposition to immigration, not loss of jobs or a decline in wages. The far right tried for decades, without success, to gain public support with slogans like "immigrants are stealing our jobs." Xenophobia spiked only when terror attacks made people feel unsafe. Elites have constantly underestimated the political impact of people's fear that foreigners want to kill them, and the fact that a sense of vulnerability can persist long after the actual threat has passed.

Many people in the West maintain that there are far more Muslims around and within their communities than there really are. When the Ipsos survey asked a representative sample of the French population what percentage of their country's population was Muslim, the average guess was 28 percent.[20] The true figure is about 9 percent.[21] Belgians, Canadians, Australians, Italians, Americans, and many others have the same mistaken perception.[22] People see the Other everywhere, even where he is not.

This distortion is a great opportunity for the violent extreme right, and for populist nationalism of the type promoted by the Alternative for Germany (AfD) and Matteo Salvini in Italy. The political discourse has degenerated and is delimited by widely held falsehoods, well-heeled elites, and economic interests—so much so that extremists are able to kick down the rotten barriers that once confined them and their ideas to the fringe.

The Aboudan Family

Summer 2015. I find Shawqi and Shahed Aboudan sitting on a wharf, their legs dangling over the water, gazing at the blue sea. They are a brother and a sister from Aleppo, the city that the Syrians themselves call Haleb. She is fourteen and he two years older. They stole over the border into Turkey by night and then crossed the sea to Greece in a rubber raft. Now, this morning, they are waiting for more dilapidated refugee boats to arrive, hoping that their parents will be on one of them. Shahed flashes me a warm smile under her long, dark hair, and Shawqi keeps swinging his legs enthusiastically.

Both are beautiful and so optimistic, naive, and confident about their future that it is hard to believe that they have just fled a war-torn country. Shawqi recounts their journey: from Syria to Gaziantep in Turkey, then to Istanbul, "where nothing went right," and from there to Izmir, "where everyone slept on the street," and finally to Bodrum, where they boarded their boat. But they were intercepted by the Turkish police, who took them back to Izmir, where they again took to the sea under cover of night. He tells the whole tale with a smile, as if he is regaling me with the story of a school trip and not a perilous escape.

They speak of their everyday lives in Aleppo as war raged—when it was okay to go out into the street, when it was too risky. Shawqi says he was once detained at an ISIS roadblock because his haircut was, he says, "too fashionable"; the Islamists suspected he was a Christian. Shahed tells me that her school closed after it was bombed by Assad's forces, and that she wants to be a pediatrician. "I want to be an engineer," Shawqi says. He assures me that they will reach Germany. "We will study and work there," he predicts. "Happy life, *Inshalla*."

Winter 2015, near Frankfurt, Germany. This time I meet Shawqi and Shahed with their parents and siblings, whom they found not long after we met on the wharf in that refugee-packed town in Kos. They are at an old German military barracks that has been refitted to house refugees. The Aboudans are delighted with the warm welcome they have received from the Germans. According to one poll, more than 30 million Germans donated food, clothing, and money to the refugees who arrived from the Middle East. The family has begun German lessons sponsored by the government, a response to the earlier failures to integrate immigrants. They tell me that their home in Aleppo was bombed by the Syrian air force before they fled.

"And then you decided to leave?"

"No," answers Shawqi and Shahed's father, Abdallah. "Then we decided to rebuild it."

"It was destroyed a second time," Shawqi interjects, his eyes sparking. "My brother hid under a table, that's how he survived."

"So then you decided to leave."

No, they say. They moved to another place in Syria, along the coast. But it wasn't safe there, either. Shawqi added that his mother still had shrapnel in her chest from an explosion. We are all silent for a short moment. "It's a new beginning for us," Shawqi interjects. "We will forget the past and begin a new life."

The conversation illuminates something that has almost been forgotten in the fervor of the West's acerbic polemics about immigration. People usually leave their homes and homelands only when they have no other choice, when all other options have been exhausted, whether the home is in Syria or El Salvador. Immigrants are generally painfully aware of the heavy price they'll have to pay. They will be foreigners for the rest of their lives, and will have to struggle with a new language that they will likely never learn perfectly or be able to speak without an accent. They will descend many notches on the social ladder, be at least somewhat alienated from their new culture, and will at times long for their homeland and distant past. Most will never be able to realize their full potential. Emigration is not a fashionable choice in Syria, or in Venezuela or South Sudan. Poor but stable countries, where people feel things are going in the right direction, do not produce many refugees and asylum seekers.

Summer 2019. This time Shawqi is on Skype.[23] He and his entire family now speak German with one degree of proficiency or another, an indication that the government's policy of teaching refugees the language and culture has been at least partly successful. "Nothing is impossible," Shawqi says. "It was difficult, but we succeeded. We learned." Some of the teachers were excellent, others less so, he remarks, but, on the whole, they were nice, "not like in Syria, where teachers sometimes beat you." One of his teachers in Syria once struck him, he recalls, for wearing jeans to school. "There's nothing you can do about it there."

His father is working as a bus driver and his mother in the kitchen of a local school. Shawqi himself has completed high school and is working at a catering company. The family lives in a rental apartment and is saving up money to purchase it. Shahed, who had wanted to be a doctor, has decided on pharmacy school instead. Her little sister aims to study law. Shawqi, who had wanted to be an engineer, has decided to begin learning photography next year. He says that the journey to Germany changed his thinking and made him realize what he really wants.

"I want to show people how I see the world, to work in television or social media," he said. His parents are fine with it: "They say I should work in whatever I want, the most important thing is to be happy, just do something." The Germans, he reports, "are very nice, but not all of them. A few of them, they hate us, I don't know why, they don't want any refugees in Germany. But these are only a few . . . in the east side of Germany there are [a] few Nazis, not a lot. I didn't meet them in my city, but my friends had encounters with them. The police are very good and nice, and German people here are sweet. They say, the Germans, we are all humans . . . and we are all the same." He feels very lucky. "It's like a dream to be here. This was our dream to be here, to have peace and a good life."

When he says "a good life," I think of Aristotle. With its extremes of great progress and misery, it is migrants who understand best what the good life is. That makes sense. In our circumscribed world, very few persecuted people manage to cross borders and find new homes. When they do, the chains of poverty or conflict that bound them break. The escapees create a new world for themselves, and might experience what Aristotle spoke of as the mark of living a good life, eudaemonia, human flourishing.

THE MEDIA, TRADITIONAL AND SOCIAL, OFFERS FEW STORIES of immigrant success like that of Shawqi's family, instead focusing more on the difficulties of absorption into a new society, the accelerating hostile responses of communities, the rise of the extreme right, and the threat of terror. More than

1.4 million asylum seekers arrived in Germany from 2015 to 2018. They came because of intrastate wars in the Middle East, climate change that caused political volatility, and as the result of tribal, ethnic, and religious animosity. Among them were a very small number of militant extremists who entered Europe in the guise of war refugees.[24] Attacks and other violations of the law, the best-known of the latter being the mass sexual predation on New Year's Eve in 2016 in Cologne and other German cities, provided an opening for the extreme right—whose presence has increased. With it has come severe violence against Muslim immigrants and minorities in general.

In the sixty or more years between World War II and the 2008 recession, and in response to the Holocaust, German society has fought racism with determination. The effects of the recession were limited; the German middle class was not economically weakened and threatened as countries elsewhere in the West were. But the sudden momentum of mass migration and the 2016 incidents were enough to reverse previous gains and fan the embers of German racism.

Dueling Revolts

On June 23, 2016, less than a year after refugees began crossing into Europe, the EU suffered one of the heaviest blows in its history: the people of the United Kingdom voted in a referendum to leave the European Union. The vote, probably the most concrete assault on globalization since the 9/11 attacks, stunned the media and political elite, both in Britain and the rest of the West. The referendum campaign had been tempestuous and acrimonious, and devoid of classic British moderation. A member of Parliament, Jo Cox, who advocated remaining in the EU, was shot and stabbed to death by an extreme rightist, who shouted as he killed her: "This is for Britain. Britain will always come first."

There were complex historical and political reasons for the results of the referendum. They are connected to deep currents in British conservatism dating back to the government of Margaret Thatcher, to the love-hate relationship between British voters and the EU, the leadership vacuum in both the Conservative and Labour Parties, and the economic crisis of 2008, which intensified the sense that the country's power structures were vacuous and incapable of solving the real problems faced by average people. The EU was vulnerable in Britain both because Britons have traditionally viewed themselves as being apart from and superior to the European continent and because the EU itself operated as a top-down bureaucracy with very limited popular support.

None of this guaranteed that the Brexiters would win the referendum, but in political contexts there are sometimes factors that tip the balance in a particular direction. Brexit's supporters had one decisive advantage—they opposed immigration, a position with more widespread support than ever. Free movement within the EU had been a constant source of political tension in its wealthiest states, even though those countries benefited from the influx of educated workers from Central and Eastern Europe who were willing to accept low wages. The combination of legal immigration from other EU countries and the entry of Middle Eastern refugees effectively armed a political bomb that detonated with Brexit. The Leave campaign made opposition to immigration a central plank in its anti-European platform. A month before the vote, a report showed that Britain held a record for immigration, having taken in, in a single year, 330,000 immigrants from the EU and elsewhere.[25] From the moment this number was made public, the debate was no longer about immigration itself, for or against, but rather who could stop the alleged "flood." The Remainers suddenly argued that it would be easier to control immigration from within the EU, but that looked like a patchwork of hypocrisy.

The very fact that such extensive attention was given to immigration was a victory for the populist right that sought a British exit from the EU. Nigel Farage, the leader of the anti-European Independence party (UKIP), displayed a provocative poster depicting a long line of hundreds of refugees from the Middle East who had been photographed during their journey through the Balkans. Emblazoned on the poster in red letters was the legend BREAKING POINT and, under it, the tag line "The EU has failed us all."[26] The poster was widely condemned; some compared it to Nazi propaganda. Yet in the age of social media, condemnations have one substantial meaning: more engagement and exposure.

The focus on immigration seems to have worked. One study found that 73 percent of those who said they were "worried" about immigration voted for Brexit.[27] Researchers who accompanied voters to the polls reported that the indicator that best predicted the way people voted was their attitude toward foreigners—much more so than holding right-wing opinions on economic, security, and other issues, or their age.[28] Some, like Boris Johnson, who led the Brexit campaign and who became prime minister in December 2019, argued that the best way to fight xenophobia was to leave the EU because "if you take back control, you do a great deal to neutralise anti-immigrant feeling generally."

Like many things Johnson said in the campaign, it sounded like nonsense in real time, but after the referendum it turned out to be dangerous as well. The British police reported a significant jump in racist attacks and hate crimes in the lead-up to the referendum, and even more so afterward.[29] In the month following the referendum, hate crimes increased by 41 percent,[30] and the rate has continued to rise. According to figures provided by the London police force, hate crimes have gone up by 15 percent each year since the vote, in comparison with the period prior to the referendum.[31] A study conducted before and after the referendum claims that minorities complained much more about hostile comments masquerading as humor, anti-immigrant rants, and online abuse after the vote than before.[32] Nationalism is an autoimmune disease. When it breaks out, it is very difficult to restore the body politic to its previous balance.

THE REVOLT IS MULTIFACETED. WHILE SOME IN THE MIDDLE classes of the West are revolting against the liberal inclination toward migration, the migrants are revolting against the global order of tightly closed borders. Confrontation is inevitable.

At any given moment during the last decade, millions of migrants and refugees were waiting for a chance to cross borders. A German government report that leaked to *Bild* in 2017 estimated that more than 6 million people were waiting to cross the Mediterranean Sea into the EU.[33] Since 2018, more African migrants are making their way to Central and South America, hoping to continue on to the United States. If something doesn't change quickly, the climate and biodiversity crisis will only increase this massive movement of humanity from the most vulnerable parts of the world to calmer, more stable, and more temperate places. For the hundreds of millions of people who live in poverty-stricken countries, the choice to emigrate is the most rational and effective way to save themselves. If they are parents, it is also a basic human urge—to protect their children.

These migrants will eventually land in other countries, countries that are other people's homes. I don't necessarily mean that as a metaphor. Recall the woman from Bodrum, in Turkey, who had refugees camping out in her backyard. Determining who will be let in or rejected at the border is not just an internationally recognized right of sovereign states; deciding who is an insider and who is an outsider is fundamental to the nature of a national polity.

The old international order grants countries the right to prevent changes in the composition of their populations through sovereignty. Can this formal right shield them from the vicissitudes of history? From the moral challenge of millions of human beings fleeing death, oppression, and starvation at home?

Nationalists invoke the fall of Rome, which they attribute to the influx of foreigners that caused its decay and fall. Yet these incursions led to the birth of modern Europe. It's a concrete example of how migration both alters and advances countries and civilizations.[34] History's answer to the question "can communities keep change out?" is a resounding no. But the judgment of history does not take into account border controls and barbed-wire fences. This experiment has the potential to stymie or delay history.

Even when migrants successfully cross borders, the new countries they come to may themselves be in crisis. They may be contending with fundamentalism, or grappling with economic inequality, or suffering from a birthrate crisis. The entry of refugees is a powerful disruptive force and may signal a state's weakening relevance for the communal fabric of its people. If a politician cannot prevent the Others appearing uninvited, speaking a foreign language, what is he worth anyway. Yet for some politicians, migration may present an opportunity. The first nationally televised ad released by the

Trump campaign after he won the Republican Party's nomination in 2016 included footage of Syrian refugees and an ominous warning that Hillary Clinton would allow outsiders to "flood" America.[35] In his second televised debate with Clinton, Trump said: "People are coming into our country, like, we have no idea who they are, where they are from, what their feelings about our country is [*sic*] . . . this is going to be the great Trojan horse of all time."[36]

Exit polls in the three states that decided the presidential race that year—Michigan, Wisconsin, and Pennsylvania—asked voters what they saw as the most important issue facing the country: foreign relations, the economy, terror, or immigration. Clinton had a small advantage over Trump among those who thought the economy was paramount. In all three states, among those who maintained that immigration and terror were the gravest issues, Trump beat Clinton by a large margin.[37]

Elites and power centers in the global North have sought to make immigration a national consensus, or to enable it to happen under the radar without public discussion. But when the middle class began to feel that its identity was under threat and that its personal security was menaced, the attempt exploded spectacularly. With immigrants holed up in—to use Trump's image—a Trojan horse at the gates, many citizens wanted to bolt the gates shut. And then to build a wall.

CHAPTER 16

A Subject of the Empire Speaks

MARIANNA, PENNSYLVANIA, JUNE 2016

We've been driving back and forth over the town's narrow, puddled streets for an hour, searching for the Quigley family's home. Torrents of rain plunge from the gray skies of coal country. Night has already fallen, the visibility is horrible, and our GPS has stopped working.

Just as we're about to give up, we find it.

Inside the wooden house, Jessica and Joel's two children frolic in the warm living room, full of family photos. They play with each other and the smartphones in their hands at the same time. On the couch, alongside Joel and Jessica, sit Joel's parents, Joel Sr. and Carolyn. I came to hear who they planned to vote for in the coming presidential election.

On the face of it, the question was pointless. All the polls showed Hillary Clinton holding on to a consistent lead in Pennsylvania, a state that no Republican presidential candidate had won since 1988. The Republican and Democratic functionaries I had spoken to in Washington had all stated unambiguously that, given the polling data, Trump could not win. That was the buzz everywhere. When he declared his candidacy, the *Huffington Post* announced that it would cover him in the entertainment rather than the political section. I read articles with titles like "Relax, Donald Trump Can't Win,"[1] and "The argument for Clinton in 2016 is that she is the candidate of the only major American political party not run by lunatics."[2] An influential voice in the American media told me, off the record, "Don't believe it. Things are still going to happen. The Republicans will liquidate him from inside. This man will not be their presidential candidate. Believe me, you have no idea who I was just on the telephone with." To me, an outsider in America, the Trump campaign looked like an eccentric billionaire's gimmick.

My mind was changed by listening closely to what he was saying rather than fixating on the outrages he was committing. Trump offered a brutal attack on the world order. Often, he backed up his attacks with outright lies and incitement. But a repressed and powerful truth about the dark side of globalization showed through the cracks in his rhetoric. When he blustered about the "rigged system" and free trade, or when he raged against political correctness, it sounded like an exercise from a course in international relations and political economy: "Write five pages about the ways in which a nationalist-populist candidate could gain support in America because of the defects of globalization and its effect on the labor market and local culture."

Television journalists research a story before filming it.[3] By the time I head out with my camera crew to conduct an interview, I already have a good idea of what the story's protagonists will say. Marianna took us all by surprise. I expected Jessica and Joel to tell me that they were uncertain about whether to vote for the Democrats, the party they had always voted for. But I was totally unprepared to hear the magnitude of the disillusionment that these hardworking people voiced about the American dream. It went far beyond the question of whether to vote for Trump or Clinton. In the past, such cynicism was the preserve of Communists and the extreme right, but you would never have heard it in the living room of a typical middle-class family in Pennsylvania.

Sodden American flags on the house across the street flapped in the wind and rain. The Quigleys, an all-American family, told me at length why they were sick of America. It was the first glimmer of what was about to happen. The presidential election was turning into a referendum on the crises I have chronicled in my previous chapters—defective globalization, immigration, international trade, employment, and security. Hillary Clinton wanted the election to be an affirmation of a universalist America's brilliant future, but instead it centered on the sins of the past and the betrayal of large swaths of the middle class. In Marianna, I realized that America had passed a tipping point. It was no longer what we, those living in the distant outposts of its empire, thought it was.

IN A WORLD DOMINATED FOR DECADES BY THE UNITED States, everyone has his own image, dream, idea, or fear of America. My image of America took form fourteen years before I went to Marianna, a few months after the 9/11 attacks. I was in the press corps accompanying Israel's prime minister, Ariel Sharon, on a visit to President George W. Bush in Washington. The second intifada was raging in Israel and the Palestinian territories; the US had already invaded Afghanistan and was preparing to invade Iraq. Sharon was resisting American pressure to negotiate with Yasser Arafat, the unchallenged leader of the Palestinian people. Warm relations with Bush were vital for Sharon's grand plan, ultimately successful, to prevent the establishment of a Palestinian state even as he swore that he actually supported it. His strategy was to always agree to negotiate but never to speak to Arafat, whom he profoundly loathed. To survive in the Middle East, Sharon once told me, you need to say one thing, think another, and do something else entirely.

In 2002, Washington was emerging from the shock of the al-Qaeda attacks. It was shrouded in the deceptive fog of the war on terror. The neoconservative faction among Bush's advisers was taking the lead; their long advocacy of an assertive use of military might to project US leadership had become official policy. It was a portentous moment to be in the American capital, and it was my first experience there as an adult. When I stepped out of my hotel onto Connecticut Avenue and wandered the streets of Washington for hours on end, I felt something that it is hard to explain to Americans.

It was a sensation totally unrelated to current events or politics. A general feeling that there was order in the universe, that suddenly everything made sense, and that I had arrived at the power center from which that order and logic emanated. I was in the capital of the empire I lived in; I felt like a young man who had grown up in Iberia in the second century must have felt when he arrived in Rome and saw for himself its grandeur, and its decadence. Throughout history, people have been drawn toward the center of gravity from which their world was ruled; after all, in the chronicles of human civilization through the first half of the twentieth century, the essential political unit was the empire.

We who live in the West's far-flung provinces see the markers of empire everywhere, even though the United States is not an empire in the full historical sense. We know, rationally, that the dollar is the world's fundamental currency; that the US wields power, hard and soft, in all sorts of ways, all over the world; and that American culture, in particular US film and television, are all-pervasive. But it's only when you physically arrive in Washington or New York that you suddenly feel, rather than just know, that a most salient part of our lives comes from America. You understand that our political, economic, and even aesthetic discourse at home is really this distant culture dubbed in any one of a number of languages.

It comes with the force of an epiphany, and with no sense of inferiority, that the United States broadcasts continually into your present life and your history. Pay attention, the message says, make no mistake about it, you are the subjects of an empire and Americans are its rulers. For me, it all comes together in Washington. It happens as I take in the sights—as I ascend the steps of the Lincoln Memorial and turn to its south wall to read the passage from the Gettysburg Address inscribed there: "government of the people, by the people, for the people." Many rulers and regimes promise that, and liberal democracy is no longer confined to the United States, which may not be its best incarnation today. Yet, this is the wellspring of that fairness. And it is incised in stone. No one else will ever sound like Abraham Lincoln; his assertion of "government of the people" will always be primal. A subject of the empire encounters this reference point and feels just like someone who wakes up past midnight with the solution to a long-forgotten riddle.

A seated Lincoln gazes out from his memorial. It is a rare pose for statues of historic national leaders and heroes. He emanates a serene might. This serenity in the use of power, its naturalness, is a modern and exceptional innovation of the American empire. It is not a mercantile empire that avariciously exploits plantation colonies, nor a land-based power that seeks to expand into new territories to despoil them and bestow their riches on its upper class. Rome turned from republic to empire. Britain set out to acquire an empire with eyes wide open. America was born to be an alternative, "an empire for liberty as she has never surveyed since the creation,"[4] in Thomas Jefferson's words.

The United States expanded its territory steadily, by a number of means. It oppressed, displaced, murdered, and brought about the annihilation of native populations with characteristic European brutality; it based its economy on a widespread system of slavery that continued well into the nineteenth century. Everywhere it fought, it stayed. While it often handed power over to local inhabitants, American military forces and economic interests always remained. It has an armed presence in more than seventy countries and territories and is the most powerful military power in history. As it faced off against the Soviet Union during the Cold War, America supported dictators, brutal kings, and drug dealers, fomenting coups and murdering and torturing dissenters who opposed the regimes it put in place, everywhere from Iran to Honduras. All means were legitimate in the campaign against the rival empire—and both of them were imperialist powers, even though both Washington and Moscow used the term "imperialist" as a pejorative.

But, alongside wielding power, and unlike every other empire in history, it seeded the places it went with liberal values. American military occupations of Germany and Japan were temporary, and during them the US promoted human and civil rights, and of course free market economics, in a way never done before in history. In doing so, it became an empire for the democratic age. Fittingly, Lincoln in his memorial sits surrounded by his words about civic freedom. His hands lie on armrests on which are engraved fasces, the bundle of wooden rods borne by Roman magistrates and symbolizing that they held the authority (*imperium*) to compel others to obey. At the Lincoln Memorial, they represent the power of the Union and the need to preserve it after the Civil War. That is the message of this American landmark. But what does Lincoln's *imperium* say to people like me, who live outside the United States? One possible answer: here there is liberty, but also the power to cause others to obey, at one and the same time.

The hierarchy is clear. Let's say you're waiting outside the Oval Office to be allowed to enter for a brief photo-op between your prime minister and the president he has made a pilgrimage to. The White House staff explains the rules precisely, and removes from the close-packed line anyone who comes close to violating them. It is clear to everyone that this pilgrimage is a routine event, like that to pay obeisance to the emperor of China in, say, the fourteenth century. When you enter the room, you see President George W. Bush sitting, cool and composed, beside your elderly prime minister, a man who commanded the armored battles that saved your country during the Yom Kippur War. Your leader is perspiring a bit, and anxious, and it's very clear who holds the power and can wield it with serenity.

The point of view is a provincial one, of course, but since the 1950s half the world has been in one sense or another a province of the American empire; since the 1990s, far more than half. Globalization expands and deepens these relations, but there is no equality. The United States created the sense that it is the source of abundance for a divided, frightened, and poor world. For example, have Americans grasped how the free refill is the essence of the sweetness of the American experience for foreigners?

It makes no difference whether the foreigners come from Africa, struggling to improve a standard of living; or Eastern

Europe, which still remembers the stinginess of Communism; or efficient Western Europe, which at times enjoys a higher standard of living than Americans do, or from the Middle East. The first time they get a free refill of their coffee or soft drink is a revelation. As a child, I remember the joy and wonder of the concept. Why, I marveled, doesn't everyone just stand there and continually refill their Coke cups over and over again, without end? The simple answer is—because they've had enough. And if they want it, they can take it. You don't need to be poor to be surprised by America's plenty and sense of security; you just need to be a person from outside the United States. The allure of the American dream is manifested in the free refill and summed up in the trust the cafeteria or gas station has in its customers; and the customer's fundamental decency means that he will not abuse that trust beyond reason. It is also found in the capitalist insight that encouraging a consumer to consume one product, even at a loss, will often lead him to buy others. It signals that consumption is fun. Of course, it comes along with deadly excess calories, environmental damage, and overindulgence as a lifestyle.

In my native language, the word "America" has long been an adjective. "How was your vacation?" you ask someone. "America" is the reply. "How's the new car?" you ask a friend. "America" is the answer. A similar expression is *lehiyot large*, the second word taken from the English, meaning to be generous like in America. Generous in the form of large reclining armchairs (for watching television) sent from America on large ships as if they were idols transported from the capital of Assyria to temples in the far-flung corners of the empire. Large is also those huge buckets of popcorn that you can buy at the movies, or a steak topped with two jumbo shrimp and butter.

The Victory of Moderation

Although the United States involves itself in many conflicts, and its model for doing so has disturbing, even horrifying, cracks in it, the US consistently comes in at or near the top in every measure of "soft power."[5] The United States is not France, which offers the world gourmet food; nor is it Britain with its royal family, Oxbridge, and the Beatles. The US is an assertive superpower but nevertheless remains popular.[6] Its aura of power derives not only from a yearning, on the part of those outside the US, for the American dream and American products. There is also the promise of protection to distant provinces. In the Middle East, for example, the knowledge that war is always on the horizon and can be brutal when it breaks out comes along with the tacit presumption that any war will not drag on for long, because it will be brought to an end swiftly by American (or, in the past, also Soviet) intervention.

Israel's security policy is founded on the assumption that at one point or another in any armed conflict, an American decision will be made and an order will be given that will bring an end to the fighting. As such, the Israeli army's combat doctrine has always been based on the premise that it had limited time to fight and to conquer territory, and that it would have to stop when the superpowers (both at first, now the only remaining one) told it to hold its fire and cease advancing. Indeed, the Soviet Union and the United States used their might to halt hostilities, and occasionally to draw cease-fire lines down to the inch.

Washington's position at the helm has not always been welcomed in parts of its empire and within the societies of the countries under its influence, but it has had a mitigating effect. Decision makers in these countries could resist domestic warmongering by claiming that they were under US pressure, using Washington as the ultimate excuse.

A given policy might incur Washington's disfavor; another option might gain us the backing of the world's strongest power. Egypt pursued the latter strategy when, at the end of the 1970s, it signed a peace treaty with Israel, trading its alliance with a declining USSR for one with the USA. What was true for the Middle East was even more relevant for the nations of Western Europe and, after the collapse of the Soviet Union, for the entire continent.

In 1956, the United States made it clear to France, Britain, and Israel who called the shots in the world it had begun to construct. It (along with the Soviet Union) forced the three countries to withdraw completely from the Sinai Peninsula and the Suez Canal, which they had seized from Egypt in a joint operation. From this point out, and despite the protests (from France in particular), the countries of the West realized that major strategic decisions required American consent. When, in 1963, President John F. Kennedy went to West Berlin after it had been blockaded by Communist East Germany and proclaimed "*Ich bin ein Berliner*," the message to America's allies was "you are not alone."

That doesn't mean that Americans were immune from error and destructive folly, or that they did not participate in wars that involved mass killing, as they did in Vietnam. They were fully aware of their power and believed that it was their unique historical and even God-given role to use it to fight Communism and spread freedom. As they saw it, this made the United States of America different from any other country of its time and of all past times. It was a doctrine—adopted as early as the 1840s—that came to be called American exceptionalism.

When American presidential candidates swear fealty to American exceptionalism, it sounds to foreign ears like a mystic initiation ceremony. When President Obama made a remark that was taken to be a disavowal of America's imperial role, he faced strident resentment for it throughout the rest of his term.[7]

America's view of itself as a city on a hill, with a moral duty to lead and to take action, is nothing new. The exceptional thing about American exceptionalism is how unexceptional it is, historically speaking. In the first century BC, not long after Rome intervened in a civil war in my country and installed a vassal king named Herod, Virgil preached Roman exceptionalism to the youth of his country: "Never forget that this will be / Your appointed task: to use your arts to be / The governor of the world, to bring to it peace, / Serenely maintained with order and with justice, / To spare the defeated and to bring an end / To war by vanquishing the proud."[8]

YET HERE'S THE THING. THIS SENSE OF SECURITY, THE UNDERSTANDING that the public is not at the mercy of local politicians because they must report to a higher authority, America, was something that only some people *outside* the United States benefited from. There was no external force to guarantee America's own security or to scold its leaders if they acted irresponsibly. Americans never received friendly advice from a stronger ally not to go to war because it would end in disaster. An American president never had to take a phone call from Washington, an event that every other leader in the world dreads. No American leader ever had what we here in the provinces have had—someone who keeps us from acting on our worst instincts. At the same time, the fate of America, like that of all empires, became inextricably tied up with the dominions it rules from afar, and with the responsibility, or the imperial prerogative, that it took upon itself. As Reinhold Niebuhr wrote in his classic *The Irony of American History*, "A strong America is less completely master of its own destiny than was a comparatively weak America, rocking in the cradle of its continental security and serene in its infant innocence . . . We cannot simply have our way, not even when we believe our way to have the 'happiness of mankind' as its promise."[9]

But the US rejected Niebuhr's approach. The pivotal role it played in the twentieth century's two world wars, followed by its success in overwhelming its rival superpower, made its leaders feel invincible and indispensable, and often naively ambitious. Thomas Friedman, the noted author and *New York Times* columnist, once told me in an interview: "People love to make fun of America, love to make fun of our naivete . . . But American naivete is extremely important to the world. If we stop standing for rights, for privacy, for gender equality, for basic decency, you change the whole world. And if we go dark, the world goes dark."

We spoke in the spring of 2016 in his room at the paper's Washington bureau. A few months later, Donald Trump was elected president.

The Imperial Project

Globalization as we know it today spread through the Western world as Christianity spread through the Roman Empire. It changed the way people communicate and trade with one another much as the great naval empires of the fifteenth and sixteenth centuries did. Globalization encompassed and brought under its dominion more peoples and territories than any great power ever did, becoming the most successful imperial project since the dawn of the modern age. And it was made in America.

The Bretton Woods agreement and the institutions it established, along with the Marshall Plan, were enunciated at the beginning of the Cold War with the USSR. The desire for a world open to international trade and connected by strong reciprocal ties grew out of the state of the American economy at the end of World War II. In the 1950s, the United States produced about half the world's goods,[10] but it was home to only 6 percent of the world's population. Someone had to buy all the radios, automobiles, soft drinks, and all the other products produced by America's efficient factories, assembly lines, and bottling plants.

The United States invested more than some $12 billion (equivalent to $170 billion in 2020 dollars) rebuilding Europe after the war. That was the equivalent of 4 percent of its annual GDP at the time.

In his book on the Marshall Plan, Charles L. Mee recounts how newspaper offices throughout Europe, and in Britain in particular, were briefed in advance on the substance and significance of the speech in which the secretary of state laid out the plan.[11] In contrast, the administration did its best to ensure that the speech would be covered as little as possible within the US. American reporters were not invited to what was dismissed as a "routine commencement speech." Reporters who nevertheless sought to explain just how dramatic the event was—so the *Washington Post* later recounted—were rebuffed by their editors. Afterward, the State Department informed them that nothing new had been said.[12]

It's a wonderful example of how the subjects of an empire received promises of security and economic backing that the empire's own citizens heard about only later and in a vague way. That's typical—the empire was not built on honesty with the public. The truth about America's policies was located at the empire's frontiers, not in the information it supplied to its citizens. President James Polk lied to Congress in 1846 because he wanted to expand America westward by means of war with Mexico. In 1940, President Franklin Roosevelt, while preparing the country for war, told the American public that "your boys are not going to be sent into any foreign wars." In 1961, President John Kennedy told the American public, just a short time before the Bay of Pigs operation, that the US had no plan to intervene militarily in Cuba.

Europe's rehabilitation was leveraged for the creation of strategic alliances and provided the basis for NATO.[13] When the Berlin Wall fell, the United States promoted the Washington Consensus, which prescribed privatization, open capital markets, and the elimination of trade barriers.

"We are the indispensable nation," declared Secretary of State Madeleine Albright in 1998. She meant that the United States alone had the power to guarantee the world's security and prosperity. In the decade that followed, America tried to achieve long-term settlements in sometimes centuries-old epic conflicts, such as in the Balkans, Northern Ireland, and in the Middle East. Sometimes it succeeded.

Three years after Albright's pronouncement, al-Qaeda attacked, and it turned out that the world was not so secure and that fundamentalism was posing a profound challenge to the liberal order. Americans' inner sense of confidence turned out to be an illusion. Faced with this strategic surprise, the United States launched two wars—in Afghanistan, against the Taliban regime that had supported and protected al-Qaeda, and then, in a massive act of deception and self-deception, in Iraq, against the regime of Saddam Hussein. The US, which had always avoided establishing a classic empire that rules other nations, began to get accustomed to the idea and to see it as a military necessity and perhaps even as manifest destiny. No longer was it an empire that held the fasces and proclaimed "government of the people, by the people." It just held the fasces.

It was a dramatic departure for globalization, the handiwork of the United States after the war. Empires that rule by force over other peoples, wars between "civilizations," closing borders—all these were profoundly out of line with the open world and economy that America fashioned after 1945. Desperately seeking personal security, the fundamental requirement of every polity, the Americans fell in love with an anachronistic version of power. The historian Niall Ferguson proclaimed that the US "is an empire . . . that dare not speak its name," and charged that, because it was "an empire in denial," it was endangering the world's security.[14] "American Empire (Get Used to It)," the *New York Times Magazine* headlined a piece, and it was followed by more of the same.[15] A senior adviser to President George W. Bush told the magazine *Foreign Policy* off the record, "We're an empire now."[16] In December 2003, Vice President Dick Cheney sent out a Christmas card bearing a quote from Benjamin Franklin: "And if a sparrow cannot fall to the ground without His notice, is it probable that an empire can rise without His aid?"[17] When a US vice president sends out a Christmas card with an epigraph of this kind, it is not simply a literary act. At the time, American forces were spending their first Christmas in Baghdad. President Bush thought it necessary to publicly deny his country's imperial ambitions, declaring that "America has not empire to extend or utopia to establish." There is much debate over the costs of these two wars, but the lowest sum cited is $1.6 trillion,[18] and most estimates put it at close to $3 trillion.[19] A study in 2018 estimated that the total cost was $6 trillion, if the pensions of the war wounded over their lifetimes were factored in.[20]

The US was warned by some of its allies to stay away from these wars. But no American president who spoke softly and carried a big stick warned it that "if you, a power that was attacked, respond with an expensive war, you will find yourself deeply in debt, investing less in infrastructure and other drivers of growth, with spiraling defense expenditures and internal economic crises. That is a classic recipe for decline." No greater power was there to block Americans and save them from themselves.

Never Here

The empire replaced effective soft power with ineffective hard power. The wars took place far from American soil. In the meantime, the United States itself plunged into its deepest recession since the Great Depression of the 1930s. As usual, the price was paid by the shaky middle-class people like those I met in Marianna, Pennsylvania, in the heart of what was once the Steel Belt and is now the Rust Belt, a few months before Trump won the biggest political upset in American history.

At the beginning of the twentieth century, Marianna was a model of American industrial achievement. Its inhabitants lived in modern stone houses with indoor plumbing. Immigrants from Central Europe, Russia, and Italy flocked there, thanks to that era's free immigration. The town was built close to three coal mines run by the Pittsburg-Buffalo Company, which had a reputation for operating the most advanced and safest mines of the time. Its position at the technological forefront brought President Theodore Roosevelt there for a tour in October 1908, accompanied by European coal experts. Forty-four days later a national disaster struck—an explosion killed 154 miners.

In the town's small but well-kept library I met Joe Glad. He was eighty-nine years old, with an erect posture and probing eyes; he wore a baseball cap. Stroking a large chunk of soft coal, the type that was used in steel manufacturing, he said it was "good clean coal." Joe began working in the mines two years after the end of World War II. His father brought him into the job and taught him the lore that a miner needs to survive. For example, if you see a bit of dust, hear a creaking noise, and see a mouse running past, "you better start moving with the mice."

When he began to work, he and his father loaded coal. Each of them loaded about three tons of coal in a day's work. For each ton they received 93 cents.[21] He shows me his large hands. On some mornings, he recalls, he couldn't open and close them "because of the calluses. So Dad used to take me to the spigot, turn the hot water on, you get them loose, and let's go."[22]

Jeremy Berardinelli, a borough councilman, opens the doors to the old buildings at the mine, which closed in 1988. It's a majestic ruin, impressive in its size, with a scarred beauty. He shows me the last heap of coal that came out of the ground. "They left it here in case we ever put up a museum," he said. No museum was ever built. Glad takes me to where he was hired, the superintendent's office. He is moved to tears. "It hurts, it hurts. Not only me. Good people who can use jobs," he says. "I'm old, I'm gonna die, I got a year or two. What's the young people gonna do?"

Joel Quigley is one of those young people. He used to work in a coal mine near Waynesburg, not far from Marianna, and then it closed, like the rest. In the Quigley living room, the conversation begins with coal, continues with coal, and ends with coal. The question of whether it will bounce back, why it is needed, who its enemies are. Jessica, Joel's wife, has stayed home with their children for seven years, but now she needs to find a job because "he's not making enough money." Joel says that he has to work "twice as hard and longer days just to make half" of what he brought in as a miner.

Referring to companies that ran mines in the region, he says that "they're blaming coal not booming for their bankruptcy. But really they put themselves in bankruptcy." His father, Joel Sr., who worked as a miner for thirty-two years before retiring on a pension, attributes it to "corporate greed." His son agrees: "Your CEO's still got all their big money." They sound apologetic when talking about Donald Trump. "Trump's going to try to put the working man back to work," says Joel Sr. "So what are you supposed to do?" Joel's mother, Carolyn, makes clear she's not fond of him: "I do not like him, but if he's going to get us back to work—I'm voting for him."

"You know," I say, "that the leaders of this country say it's impossible to save the coal industry. If the mine has closed, there are other jobs. We won't subsidize coal. It's a free market. Maybe that's America, I don't know, I'm not American." They laugh for a moment. Joel says that, if that's the case, he doesn't understand the logic of sending American coal to China "for their power plants to run and shut ours all down." Indeed, the US exports coal to China.

They say they have never before voted Republican.

"And you'll vote for Trump?"

"We might have to." We all laugh.

"Do you think this is the greatest country in the world?" The living room shakes with their protests: "No! No!" Joel turns to his father. "Maybe America was when your Dad was growing up, but now?" Jessica adds that America is "embarrassing." Out of everything they've said, that's what grabs me. Their country embarrasses them. They say that children are starving while other people "live off the system." It was only after Trump's victory that I understood the extent to which such perceptions had permeated America.

According to Jessica, "Nobody helps their own, it's never about here. It's always what we can buy cheaper overseas, what we can do in other countries, *never here*." She stresses each syllable of the last two words.

There is something obvious in her claim that "it's never about here." It points to the neglect of rural America, "flyover country," and the gap between America's flourishing urban centers and the rest of the country. What is less obvious is that there is no "here" here. The world has become so global, so connected, that nothing is really here.

I'm from one of those other countries the Quigleys talk about. We benefited from the safeguards provided by the age of responsibility. No matter who was president—Clinton, Bush, Obama—or whether he pursued wise policies or foolish ones, America was always around.

Yet while the power of American empire survived in the far-off outposts, its engine, the American dream, lost its spark. Like an old Japanese soldier walking out of the jungle decades after his country's defeat, we were possessed by something long gone. We were talking, hoping, dreading America, but the promise in its heart was already lost. Trump said as much, with apparent masochistic glee, at the very start of his presidential campaign, when he declared: "The American dream is dead."

The Quigleys voted for him in 2016. They had to.

CHAPTER 17

"My Mother Was Murdered Here"

The American dream was nowhere more alive than in Detroit, where it died. I am on an urban expedition on a small bus, driving among abandoned buildings. My companions are an eclectic bunch of travelers fascinated by ruins—a Japanese man, a young German hipster, an American couple from Florida. The tour began with the traditional American ceremony—the signing of a form releasing the tour operator from all liability in case of a crime or an accident in the derelict buildings we will enter, the abandoned schools that have turned into drug dens, or the ruined choirs of dusty churches.

The bus stops at the Harry B. Hutchins Intermediate School, a huge and impressive structure located on a sad street. When it first opened, a local newspaper proclaimed that it "makes full provision not only for the academic, but also the physical and vocational education of children . . . as well as forming a community center."[1] That was in 1922, when officials came from all over the United States to learn about Detroit's superior school system and mimic its success back home.

When we go inside, our guide warns us to use flashlights and not to pick up anything from the floor, especially needles. Water drips from walls riddled with holes and furrows—metal scavengers have dug out whatever they could. The guide shows us the locker rooms—the showers were built with blocks of imported marble. Detroit was the Silicon Valley of the early twentieth century, and no luxury was spared for the city's children.

Beyond the locker rooms are two large, deep swimming pools, now empty; one was for boys and the other for girls. There are also workshops where students learned electric work, printing, carpentry, and auto mechanics. Natural wood closets, custom-made, are still in place; sunlight from an inner court floods the unused classrooms. "It's good neutral light for photographs because of the clouds," the guide says. One of the tourists picks up an old shoe and places it on a rotting desk. The walls are covered with faded graffiti. Cameras flash.

Detroit's population in 2018 was about 670,000.[2] It's a huge city of nearly 140 square miles, as befits a city built for the automobile industry. In 1950, Motor City, as it was called, had a population of 1.8 million; it has since dropped by 65 percent.[3] In 2010 there were 53,000 abandoned homes and at least 90,000 vacant lots, most of them the sites of houses that had been burned or demolished. People are scarce in some of the neighborhoods. Many streets are occupied by a few lonely houses with empty lots between them, where homes once stood.

Detroit was not like a gold rush town of the nineteenth century, with an economy based only on mines that would eventually be depleted, killing the community. Neither was it depopulated by an earthquake, famine, or war. It dwindled in size because of the reversal of the river of globalization—instead of flowing from east to west, it flowed west to east.

The city's bad fortunes are emblematic of the brutality with which the American dream was shattered. Detroit suffered from flawed urban planning, racial inequality, dependence on corporations, a vicious circle of poverty and declining tax revenues, crime, a drug epidemic, and botched renewal initiatives. In recent years, after it officially declared bankruptcy and began a rehabilitation process, its downtown has begun to revive and is enjoying an impressive surge of construction. But as of 2018, Detroit still ranked third among US cities in murders.

St. Margaret Mary Catholic Church stands locked and barred on what seems to be a forsaken street. A vehicle stops next to us. Two white women are inside, Sharon Probst and her daughter. They have come on a family roots tour to this old neighborhood, and saw us in front of the church. Sharon, in her seventies, was married here in 1963. "The church looks so different now," she says, in flat midwestern tones. She leans on her cane and points at her daughter. "She was baptized here."[4] She asks me to help her make her way past the overgrowth into the building.

The interior is covered in graffiti, and everything metal has been stripped from its walls. A piano, rotting and swollen with moisture, stands on its side. In the sacristy hangs a white gown, gray with dust, like a cheap prop in a horror film. Sharon stands on the chancel, not far from the altar, under two golden wings. She taps her cane hard on the floor and says that the church was once full of people.

"Sooner or later you had to leave," she says. "You couldn't trust anybody anymore. Times got hard, really hard." There is a moment of silence, and then she adds: "My mother and stepfather were murdered in this neighborhood, in their house . . . December 6, 1974."

I asked her if it happened during a burglary.

"Yes, people next door," she replied. "They shot them twenty-two times apiece. They stabbed and killed the dog first, because Lady was protecting the house. And when they came home, they killed them. My mother was talking to the police when they shot her. You hear the phone drop and her fall on the floor. And then my stepfather was down in the basement and when he came up they were right there, and they shot him."

She tells me about the atrocity without drama. An old newspaper clipping I found afterward relates that Lady, the dog, was given a hero's funeral by the children of Clifton and Lee Ledbetter, the victims.

IMPRESSIONS FROM AMERICA:

2008, Grant Park, Chicago. I am walking with the masses who are leaving Barack Obama's victory rally. They are so elated that they seem to be walking on air. I look at my BlackBerry; just an hour earlier, I had managed to get through the security perimeter around Obama's podium and take footage of him coming onstage. A black woman, about seventy years old, also leaving the rally, asks me, "What have you got there, son?" I show her, and we both laugh as if we share a secret. When I reach my hotel, suffused with the evening's euphoria, two white men in jeans and sweatshirts get into the elevator with me and my companion. My friend makes a joshing comment about the election; they seem indifferent, but within a tenth of a second something happens, like an animal going on the attack. One of them comes up very close to us in the elevator and shouts: "Fucking foreigners, don't you be messing with our electoral system!" His friend joins in. They reek of alcohol and I realize that they think we voted for Obama. We flee the elevator.

An abortion clinic in Ohio, a few weeks before the 2016 elections. Dr. David Burkons, the doctor who runs it, feels like he's being hunted down. He is almost seventy years old and says that he is the youngest abortion doctor in the state. No one else wants to do such a reviled and dangerous job. To the two men who are permanently protesting outside the clinic he is a baby murderer; he receives death threats. A young student has come for a medication-induced abortion. "I'm making the right decision," she tells me. "It's clear to me." The men outside, wearing baseball caps and holding lurid signs, wait for her to emerge. "It's murder, it's Auschwitz," they harangue me. "It's a holocaust." A young black woman is in the waiting room. She desperately needs an abortion but does not have enough money. The price is $425. She asks if there is government funding. The startled doctor chokes back a bitter smile. There definitely is not government funding.

North Carolina, in Ron Baity's church in Winston-Salem. The members of his devoted congregation are ready for the Sunday service. Baity likened gays to maggots, and warned against same-sex marriages in the language the Bible uses to describe Sodom and Gomorrah. "You think Ebola is bad now, just wait," he has said.[5] It's Father's Day; screwdrivers are being handed out as gifts to all males. I gaze at the pile. They look so forlorn, those screwdrivers at the preacher's church door, that I can't help but break out laughing. The presidential election of 2016 is less than two months away. Outside, Rev. Baity tells me that transgenderism is a mental disorder and that his America is still alive and kicking. "Don't bury us yet," he admonishes me.

Earlier that week, just sixty miles away in Charlotte, I watched Erica Lachowitz, a transgender woman, violate

North Carolina's so-called bathroom bill,* which Rev. Baity had fervently promoted. She entered the women's bathroom in a trendy local produce market along with her eight-year-old charming daughter, Alice, her head deep under her mother's shirt. I invited them to an ice cream parlor. At a table, Alice told me about one of her friends. "I've known her since kindergarten and I always went to her birthdays and she went to mine. But when we were in second grade she didn't invite me." She turned to her mother. "When she didn't know you are a transgender she always invited me, but since I told her, she hasn't really been speaking to me."

A Radical Moment

A society's radical moment—when the previously unthinkable breaks out of the cage of convention and becomes an option—lies hidden in these snapshots. The election of Barack Obama was such a moment, touching on Americans' deepest-held feelings about citizenship, social mobility, and race. But some felt that their country had been stolen from them. Many, like Donald Trump, rushed in with conspiracy theories to fill the gap between what they thought should have happened and what had in fact happened. The birthers, who insisted that Obama had not been born in the US and was thus ineligible to be president, were a manifestation of the fact that many Americans could reconcile themselves neither to his candidacy nor his victory. They needed to claim that Obama was

* The Public Facilities Privacy & Security Act, commonly known as House Bill 2 (HB2) or the "bathroom bill," was enacted in 2016. It compelled schools and public facilities to bar people from using bathrooms that do not match their birth gender. It also prohibited municipalities from instituting their own antidiscrimination policies.

a counterfeit president, an imposter, elected on the basis of a sinister lie.

Radical moments entail the loss of capacity for making reasonable predictions of how social life will look in the near future. Political questions are answered in absolutes, with gray areas ignored. In today's America, it's not clear if people of color are striding toward more equal opportunity or are instead caught in a discriminatory maze built by the establishment and maintained by a toxic status quo. Just as their votes are rendered meaningless by gerrymandering, so the powers finagle to limit their right to vote and to receive an equal share of public resources.

The transgender mother I met in Charlotte does not know if she lives in a state that is advancing toward LGBTQ equality or the opposite. Rev. Baity cannot predict whether he will soon again be able to tell people whom they may have sex with or which bathrooms they can use, or whether his time is past. Millions of Americans do not know whether Obamacare will expire in a few years or exactly the opposite: Will it be replaced with a comprehensive public health system? Inner-city residents don't know if they will experience the renaissance people have been promising them for two decades, or whether their cities are doomed to crumble like the ruined houses of Detroit's east side. No one knows if the United States will continue to protect, to some extent, women's freedom to control their reproductive rights and their bodies, or whether the Supreme Court will reverse *Roe v. Wade* and allow states and localities to put women or medical practitioners on trial for imperiling the lives of their fetuses, as has already begun to happen.[6]

These uncertainties are not so much a matter of ordinary political divisions; they profoundly touch on how Americans view their country, whether they think it should remain a global power with global responsibilities or whether it should retreat and avoid entanglements and conflicts overseas that cost blood and treasure. Another division is about the difference between an America based on an empowering liberal vision and one in which white supremacy leaches into mainstream politics. And there's the debate whether America should turn toward the socialism that has captivated the hearts and minds of many young people or remain capitalist.

Take the Money and Run

America's dynamo has long been its reliable promise of prosperity. But it is running out of steam. After World War II, the free market system and the global trend toward lowering trade barriers unleashed a wave of globalization that brought prosperity to the US.

When the war ended, almost 40 percent of America's nonagricultural labor force was employed in manufacturing. Globalization was essential for a country with superior industry and access to markets. But when the world began to recover from the war and the economies of countries like France, Britain, Germany, and Japan grew stronger, America's relative advantage in manufacturing began to erode. In 2015, only 9 percent of American workers were employed in that sector.[7]

This is not only a result of automation and rising productivity. In a globalized world, competition grew fiercer all the time, but the US acted as if Detroit's Big Three automobile manufacturers, General Motors, Ford, and Chrysler, were a force of nature, just like the wind and the sun. Productivity in the rest of the US increased at a much faster pace in other areas than in the automotive industry, but its workers enjoyed higher salaries.[8] The subject of the Steel Belt decay is replete with misconceptions that blame trade and free markets. Many of Pennsylvania's shuttered steel plants do not, in fact, stand empty because of imported steel, but because jobs and manufacturing moved south within the United States, where salaries were lower and labor protections less developed.

Imports nevertheless did play a critical role in the decline of American industry. Manufacturing in countries with lower labor costs saved corporations money, gave consumers more disposable income, and raised the country's standard of living overall. Few Americans were willing to give up those benefits. When I asked the Quigleys of Marianna if they would be willing to pay thousands of dollars more for their TVs and smartphones so that manufacturing could remain in America, they laughed.

David Autor of MIT studied the ways in which international trade decimated US communities. He found that at least a million people lost their manufacturing jobs within the space of a few years in the 2000s; the jobs they subsequently found were lower-paid and offered much less security. He showed that small communities throughout the US never recovered from the shock of the massive rise in Chinese manufacturing that began in 2001, the year China joined the World Trade Organization.[9] In an interview in 2017, Autor explained that the conventional wisdom that workers could move from sector to sector, at the very most by accepting a cut in pay, was false. "People don't move really readily," he said. "They have skills that are specific to their industry, they have attachments to their jobs, it's wrapped up in their identity. And then the shocks, because they're so geographically concentrated, they're highly, highly disruptive."[10]

The disruption takes the form of a drop in wages for women and men who were already working in low-paying manufacturing jobs; a lower birthrate; a lower marriage rate;

a rise in the number of children born out of wedlock; and a rise in the number of children living in extreme poverty. The economists Angus Deaton and Anne Case of Princeton describe "deaths of despair"—that is, the higher death rate for white men of middle age, generally between twenty-five and sixty-four, from suicide, drugs, alcoholic liver disease, and cirrhosis.[11] The most dramatic manifestation of this in the US is the opioid epidemic and its tens of thousands of victims. The broader trend is of the rising death rate among white American men of working age.

So international trade and imports have taken away the jobs: that makes it out to be a zero-sum game. In the words of Donald Trump, in a speech he gave in Detroit prior to the 2016 election, "The skyscrapers went up in Beijing, and in many other cities around the world, while the factories and neighborhoods crumbled in Detroit."[12] But that's only a partial explanation. For example, in the 1990s American automobile companies improved their efficiency in the face of challenges from Japanese and European competitors. American GDP and productivity continued to rise, yet the loss of manufacturing jobs accelerated. The loss of jobs in the 1980s was related more to automation of production lines. Considerable data shows that technology is responsible for most of the evaporation of American manufacturing jobs between 2000 and 2010—up to 87 percent of them.[13]

In other words, international trade is not the main reason for the loss of American jobs in recent decades. Instead, it was, at first, robotic arms and enhanced automation. It's easy to blame Mexico or China; for now, at least, it's hard to demonize robots.

Globalization means that no society can be an island. No society can really control the forces of supply and demand, unless it is willing to become North Korea. Even the wealthiest of nations, the one that gave birth to globalization in its current form, cannot escape this truth. The same processes that enrich a society as a whole can impoverish broad populations within it—and, from a purely global economic perspective, that is of no consequence. The fact is that imports from China raised the standard of living of middle-class Americans; that benefit outstrips the value of the jobs lost since 2000. The modernization of production lines is brutal but natural.

The issue is not that this happened but *how American elites responded to it.* They could have upped investment in infrastructure and education, and accepted that higher taxation of their wealth would be necessary to pay for a technological leap forward. Another response could have been protections for small and new businesses, the classic way in which Americans have always set themselves up and moved up the economic ladder. But none of that was done. In fact, since the 1970s the number of American small businesses has halved.[14] One of the reasons is that large companies and chain stores have exploited economies of scale to fill the niches that small businesses used to grow in.[15]

But it gets even worse. Corporations and those who own and run them have promoted policies and goods that knowingly cause injury to individuals and society as a whole, not just in the US but throughout the world. Oil companies were cognizant as early as 1977 of the scientific work showing that fossil fuels were causing climate change. That's eleven years before it became a public issue. The oil companies even performed complex experiments to test climate models predicting the consequences of greenhouse gas emissions.[16] They used this knowledge to disseminate disinformation and spent tens of millions of dollars to counter scientific findings on the climate.[17] The oil companies acted in much the same way that American tobacco companies did in an attempt to suppress the enormous health risks posed by cigarettes.

In America, the most egregious example of an established pattern of lucre and lies was the aggressive marketing, using misleading data, of OxyContin, an opioid medication manufactured by Purdue Pharma.[18] OxyContin is one of the main catalysts of the opioid crisis that has taken the lives of tens of thousands of people.

These are exploitation schemes, and even if the ones I have cited are exceptional in their callousness, that does not let the rest of the rich and powerful in America off the hook. At the very least, they have focused on reducing their tax burden and raking in capital gains. They have taken the money and run (to tax havens, for instance). A study from 2017 by Thomas Piketty and two colleagues showed that the bottom 50 percent of the population continued to earn in 2014 the same amount, in real terms per capita, that it had earned in 1980—only $16,000 (before taxes, in 2014 dollars). And add to it that the upper 1 percent's income per capita more than tripled during this same thirty-four years, reaching $1.34 million.[19] As a corollary of this, the share of the US national income earned by the bottom 50 percent has fallen by half.[20]

The wage and income level of American workers is a painful and controversial subject. Economists have been grappling for years with an ostensibly simple question—have these wages risen, and if so, by how much? By one measure, from the mid-1970s through the mid-1990s, the average hourly wage paid to American workers stayed exactly the same. But according to other accepted measures, it plummeted.[21] By one estimate, the average weekly earnings of production and nonsupervisory employees peaked in 1978 and has never returned to that level.[22] Even when the average wage began rising in the 1990s, it was at a low rate compared to what it was in the past. According to a Pew Research Center paper, today's average wage has the same purchasing power that it did forty years ago.[23] The problem is that, using a different computation of inflation, wages, including those of the middle class, actually rose significantly. Given that the standard of living has improved, some economists have contended that wage stagnation is a myth.

One way to resolve the apparent paradox is to look at household income. After taxes and after factoring in the value of entitlements like food stamps and Medicaid/Medicare, the income of the American middle class (broadly defined as the second through eighth income deciles) has risen by 47 percent over thirty-seven years, equivalent to a bit over 1 percent a year.[24] The previous generation of middle-class Americans did much better. It enjoyed annual income growth twice that figure or more;[25] nine out of every ten of those born in 1940 enjoyed an income significantly higher than that of their parents.[26] Income inequality has grown significantly worldwide since 1980, but much less so in Europe than in the US.[27]

The trampling of the American middle class would not have happened had decision makers and congressmen not turned a blind eye to what was going on. They permitted this party because they were its guests of honor. From the mid-1980s, outlays on political campaigns and lobbying increased many times over in the US. A legislative case study showed that companies can get as much as a 22,000 percent return on their lobbying dollars.[28]

American leaders saw that GDP and productivity were rising, on the whole, while at the same time the stock market indexes swelled. But such aggregate figures hide local political abscesses, ones with dire effects on the future health of society. A study by Autor shows how areas with high trade exposure are displaying more political extremism, both of the right and the left, and replacing mainstream elected officials with radicals.[29] Beginning in the 1980s, and all the more so after the fall of the Berlin Wall, Americans were promised, in the words of Ronald Reagan's 1984 campaign commercial, that "It's morning again in America." In fact, for many, night was falling.

Becoming Self-Aware

I sit down, in Waynesburg, Pennsylvania, with some of the people for whom night fell. A few hours earlier I had tried to get into Emerald Mine, which had provided the area's livelihood until it closed in 2015. But the guard at the gate told me to go away. Some 4,000 people live here, half of them with an annual family income of up to $44,500, which is $15,000 less than their state's median income.[30] They say in the town that young workers can expect to earn $14,000 a year, less than the average income in Mexico. About half the population lives under the poverty line.

We are in a crowded room with fluorescent lighting and spartan furnishings. Across the table from me sit local government officials and leaders of the United Mine Workers of America. When I said that the closing of a coal mine is a purely economic matter, they were outraged. "The government wants to close them," said county commissioner Blair Zimmerman. "They are destroying communities. They don't care what happens to us. We have problems with alcohol, drugs, family problems. The husband who used to be at home, to help with homework, help his wife, has become a truck driver." Long-distance truck driving is the highest-paying work that laid-off miners can hope to get. It pays about half of what they earned at the mine, and they hardly ever see their families.

All the assembled are well versed in efficient and clean Danish and Japanese power plants that run on coal.[31] They claim that the natural gas industry is funding environmental and political campaigns against the coal industry. That is not a conspiracy theory. *Time* reported that, between 2007 and 2010, the Sierra Club, America's oldest environmental organization, received more than $25 million from the American natural gas industry and the problematic and polluting oil shale industry. Among other things, this money funded a "Beyond Coal" campaign that explicitly called for closing coal mines.[32] The gas companies wanted to donate at least another $30 million, the Sierra Club reported, but at that point it decided to cut ties with them.

Ed Yankovich, one of the union's leaders, speaks with lots of exclamation points. "We have to go to other occupations, correct?" he says. "That isn't there, it isn't anywhere! It's nowhere here in Appalachia! You can start in the top of the Appalachian mountain chain, which is probably Maine, and go all the way to Alabama, and you tell me where that is! It doesn't exist. Kids in here, in Greene County, in Appalachia, they're not stupid, they're smart. We can train people into the new high-tech industry, but there is no door to knock on tomorrow morning and say, 'Here I am, ready to move to this high-tech industry job'—because there are no jobs here! And no one is bringing one in the foreseeable future that I know of."

He leans back in his chair. What is mostly evident is his sense of hurt pride when he asserts that the children of the people in the room are not stupid, and are simply not being given a fair shake.

■

Hillary Clinton made an appearance in Columbus, Ohio, in March 2016 and presented a plan to replace coal- and oil-fired power plants with renewable and clean energy. "I'm the only candidate who has a policy about how to bring economic opportunity using clean renewable energy as the key into coal country," she declared. "Because we're going to put a lot of coal miners and coal companies out of business, right?"

Her plan involved the investment of $30 billion in coal mining communities. But no one paid any attention to the plan. The sound bite was that she wanted to put "a lot of coal miners and coal companies out of business." She would later describe this as the worst mistake she made in her campaign.

The Trump campaign jumped on it. In May, Trump held a rally in West Virginia, a coal mining state. He put on a hard hat. He squinted and pursed his mouth and pantomimed wielding a coal miner's shovel. The crowd went wild. "You watch what happens if I win," he shouted. "We are going to bring those miners back; you're going to be so proud of your president here. You are going to be so proud of your country."

The coal miners I spoke to, across from the shut-down coal mine in Waynesburg, heard two lies. One was Trump's lie about reviving the expiring coal industry. Yet they also heard a broader, darker lie from the proponents of globalization. The coal miners knew that no government program could save them, and that no high-tech plant would fall from the sky. They knew that Washington had long ago forgotten them. Clinton needed the coal miners' votes, but she was also convinced by the arguments of environmentalists, who wanted coal to be superseded by clean energy. For these miners, Clinton's promises were background noise, like rain pounding on a rusting coal car in an abandoned town. And it's not that they believed Trump. Nearly everyone I spoke to stressed that they did not like Trump; some loathed him. "We had to decide between liars here," said one of two laid-off miners I found in Waynesburg, at the gate to the mine that once employed them. "And we knew that Trump would at least try." The social media echo chamber often characterized Trump and Clinton as being equally dishonest, but fact-checkers consistently showed that Clinton was more truthful. Trump's promise to the miners to reverse the mighty currents of the market economy was empty verbiage. Between 2017 and late 2020, the country's coal-fuel capacity underwent the worst decline of any single presidential term.

These people lost their good jobs. Their savings levels went down by double digits, their share in the national income plummeted, their unions faded into irrelevance, and their children on no account enjoyed equal opportunity. The fear of descent into poverty is a powerful political force that fuels nationalism. The members of the middle class, who had seen their net worth implode in the crisis of 2008, looked down in horror at the bottom half of American society.[33] At the same time, the "other countries," to echo the Quigley family, came to be perceived as threats to their identity and their lives. And that was far more important than their stagnant wages.

And so a new consciousness was born. "*Skynet begins to learn at a geometric rate. It becomes self-aware at 2:14 a.m., Eastern time*," reports the Terminator in the science fiction movie of that name.[34] The people I met had long been mute tools used by a thriving America. Yet through access to information, the rise of social networks, and growth of a globalized awareness, blue-collar workers have become aware of their true situation. The coal miners in Marianna and Waynesburg engaged me on shareholder dividends, corporate taxes, education policy, global economy, and energy markets, displaying expertise and knowledge that was once not widespread in small-town Pennsylvania. As they learned at an exponential rate, they had no intention of missing the opportunity offered by the radical moment ignited after the 2008 crisis. From their perspective, this was not, and is not, a "backlash" or a "populist wave." Rather, it is an attempt to fundamentally alter the vectors of American power and priorities. Having no faith in society's institutions, from Congress to the media, they opted for revolt.

CHAPTER 18

The Anti-Globalizer

After the great victory I had a vision in the middle of the night. In my vision peace and calm prevail. Clear skies stretch to the horizon above fields in bloom. There are no borders, no barbed-wire fences. In my vision people are working in the field and in the factory without hatred, without fear, they're working together, regardless of nation, religion, race or gender, because everyone is working towards a single goal, they're all working for me.

—HANOCH LEVIN, *SCHITZ*, 1974[1]

At the end of 2018 I participated in a gathering of scholars and pundits at a secluded estate in Britain. The weather was chilly, but there was port to sip and warm up with. The attendees offered their thoughts on the international situation two years after Donald Trump's election to the US presidency. The Europeans spoke guardedly, with a degree of imperturbability, although this had its limits, with Brexit impending. In contrast, the Americans, none of whom had voted for Trump, voiced such a profound and existential despair that I could hardly help feeling compassion for them. They questioned whether America would ever be able to reinstate the political norms ravaged by the current administration. The damage to America's standing might never be undone; the sense of security that the US had given its allies might never be restored. But I also heard something unexpected—that Trump had been necessary. "He forces us," one of them said, "to rethink our fundamental assumptions. But, even more than that, he has reignited liberal sentiment." As much as they hated him, they gave him credit for addressing issues that everyone else had avoided. "Trump spoke of the wounds that we all knew were there, but which none of us wanted to touch. From immigration to America's place in the world. Now he's touching them—and how," another participant said. The taboos of American political culture had conjured up someone who would violate them all.

Such assessments require going back to the most important players—Trump's voters.

In January 2020 I returned to Marianna, Pennsylvania, to see the Quigleys again. Jessica and Joel had moved into a larger house his parents had given them to live in while they took a cross-country retirement trip. On a hill overlooking the town, it boasted several acres. Joel was working again, as a coal miner, but in a different mine. Jessica had been promoted at the hotel where she worked. Its clientele came mostly from the energy sector. Marianna looked better. The Quigleys mentioned that they no longer had to drive elsewhere to buy their meat—a butcher shop had opened in town. When it got dark, we drove with their kids to get something to eat. Over some really good pizza, they told me about their trip to Disney World a few months before. Did their return to modest prosperity have anything to do with the Trump administration's approach to coal? Not really, they said, but the family credited the president for the general improvement of the economy. "We're starting to head in the right direction," said Joel.

When we spoke by phone before my return visit, they sounded resolute and hopeful.[2] "Trump's done a lot for the country," Joel said. "He's bringing jobs, unemployment's down, but there are some issues with Trump, like the way he talks and stuff. That can be uncalled-for, but businesswise, he's doing the right thing, I think." They both tell me that they're glad they voted for Trump. Jessica has even registered as a Republican. "The economy overall is doing great, people are out making money, spending money, unemployment is down the most it has been since I've been alive," says Jessica. Jessica bemoans the negativism she sees around her and primarily blames the president's opponents. "I think the social media thing is ridiculous, but I don't think it's just Donald Trump," she explains. "I just feel that a lot, maybe, is exaggerated or too much is out there . . . people are in a panic." The Quigleys think the Democrats' criticism of Trump goes too far. According to Jessica, "It's awfully dramatic. They are just trying to pull Trump down." Joel, for his part, is impressed by Trump's desire "to work with everyone." As Joel put it, "He's making an effort."

Joel doesn't like it when Trump "rants and stuff" at people. "I saw, just the other day, he talked about how he's a better man than Jon Stewart," he relates. "Jon Stewart was a TV host. But Jon Stewart fought for the health care for the 9/11 firefighters, the first responders, and he got their health care taken care of. For Trump to come back and say that he thinks he's a better man than Jon Stewart, you don't need to say something like that."

When I thought about what they were saying, and looked again at my notes from my visit to their home in 2016, I realized that something significant had changed. In 2016, the world and "other countries" played a significant role in our conversation, as did America's standing. Three years later, the Quigleys weren't talking about that at all. They were talking only about America. Joel and Jessica said that they will probably vote for Trump again in 2020.

■

THE AGE OF RESPONSIBILITY SET THE STAGE FOR CONTEMPOrary globalization's first act, in which the United States and the West flourished. During Act One, the world was divided in a fairly orderly way. There were superpowers, flourishing urban centers, and labor and environmental exploitation hubs. Inspired by progressive values and with the help of manufacturing and technology, the exploitation hubs began to shake free.

During Act Two, the system grew increasingly unstable in rich countries. Manufacturing and wage growth moved to the global East and South, rescuing millions from abject poverty. Much of both the left and the right welcomed immigration, for economic and social reasons. At the same time, opposition to liberal values increased, and fundamentalism began to push its rock up the hill.

By the end of the act, all the conditions for the revolt were in place. Many sensed that their personal security, communal identity, and jobs were in danger. The middle class, betrayed, drew a pistol and laid it on the mantelpiece. Now the third act begins, and the pace of events turns fast and furious as the revolt breaks out everywhere—fundamentalism, populism, nationalism, the immigration crisis, Brexit, left-wing radicalism, trade wars, and the disruption of the entire global order. Not everyone is in revolt, not even a majority, but many are. Lacking a coherent, systematic ideology, it is an uprising of many incarnations and local contexts.

The current moment is not similar, as some have suggested, to the outbreak of a revolution, for example the Bolshevik revolution of October 1917. Then, a small militant faction was able to seize control of Russia because they were well organized and impelled by a comprehensive and radical ideology. Almost from the first moment, the Russian Communists offered an alternative model of society, based on the theories of Marx and Engels.

We are not at that point. There is momentum for change, but we are just taking the first step into a new age. The answers that the agents of this radical moment offer are about destroying current power structures; they articulate no coherent consensus view of what will replace it.

If any parallel is relevant, what we are seeing is not the October Revolution but rather something more like the revolution that preceded it by a few months, in February. It was then that the czar abdicated, the monarchy was abolished, the soviets (workers' councils) were established, and a provisional government was formed to rule Russia. There was little bloodshed compared to what would come later, in the civil war, but the scent of the impending carnage was in the air. The February revolution so upended Russia's political system that the radicals were able to organize and attack. Lenin was able to board the now legendary "sealed" railroad car in which he crossed Kaiser Wilhelm II's Germany and return to Russia "like a plague bacillus," as Winston Churchill put it, and lead a coup d'état.[3]

Donald Trump is no Lenin. He is merely the man who has fired the pistol that was laid on the mantelpiece in the second act, signaling the dawn of the present age. Trumpism was just the beginning.

The Dark Scion of Globalization

The irony is that, more than any other American president, Trump owes his rise to the very era of globalization that he seeks to destroy. Trump's political star rose after the global financial crisis, and his strongest rhetoric, that which set him apart from the pack of Republican candidates in 2016, was about immigration, jobs, and failed international trade policies.

But it's not just politics, which is a new chapter in his life. Trump is a property developer who built golf resorts around the world, put up high-rises in Manhattan with Chinese steel, sold apartments to Russians, and rebuilt his bankrupt businesses with financing from a German bank. He is addicted to a global social network, Twitter. That is not a meaningless laundry list. The reason a real estate magnate in New York could thrive and promote himself by using Chinese steel, German financing, Russian money, and foreign direct investment in other countries was that barriers to the flow of goods, information, and capital were steadily removed. Trump flourished in a world designed by the architects of the age of responsibility, the people who set up the Bretton Woods system, the World Bank, the United Nations, and the World Trade Organization, all of which he despises. Without the policies on which that world order was based, he almost certainly would not have prospered as he did, or survived his many financial crises. His father, Fred Trump, built a real estate empire that was almost entirely based in the United States. Donald Trump aspired to be global. An enemy of globalization? Trump is its ultimate insider.

Trump has been wildly inconsistent. He was a Democrat and is now a Republican, a supporter of a woman's right to choose who now inveighs against abortion, and an early advocate of the American invasion of Iraq, which he then condemned. But he has been consistent about two things—since the 1980s he has been a vociferous critic of US trade policy and of what he says is America's weakness abroad. For years he fed a hungry media with claims that the rest of the world was taking advantage of America, economically and politically, and that only he, the ostensible author of *The Art of the Deal*, could save it. From the very start, he flourished commercially in the age of globalization, while using his success to attack it.

Trump is a performer for the age of consumerism. You might have seen him on TV tackling the CEO of World Wrestling Entertainment, dragging him into the ring, smearing foam on his scalp and shaving it on live television. He is the man who called a camera crew into his office and joked as they placed a bald eagle on his shoulder and took footage for his new reality show. He appeared in a bright yellow suit in a *Saturday Night Live* skit along with actors dressed as chickens, to promote an imaginary "Donald Trump's House of Wings." The presumption behind such good-natured appearances was always that he was a personal success. His voters never thought of themselves as an audience listening to his speeches but rather as satisfied clients enjoying a great show.

The label "billionaire" is crucial to his image (Trump loves to insert the expression "billions and billions" into just about any context). But there are thousands of billionaires these days. What sets him apart in the American context is that he has fashioned himself as an archetype of the billionaire, devoid of doubt and nuance, just like he is portrayed on *The Simpsons*. He is his own logo, with his iconic hair, orange face, and bombast. The photos of him eating food from Kentucky Fried Chicken or McDonald's on his private jet are designed to show that, in a paradoxical, even weird way, he's a regular guy, a man of the people, an incarnation of middle-class romantic projection. As Fran Lebowitz put it, Trump is "a poor person's idea of a rich person."

Trump, of course, is no regular guy, but he's the available billionaire. Exceptionally for a man of his wealth, he has made himself available to the media with gusto. He's hardly the richest, smartest, or sexiest billionaire around, but he is the only one who became a constant presence in the media and entertainment industry, and as such he has been far more visible than anyone else in his income bracket. Before he was president, he made regular appearances on Howard Stern's radio show, regaling listeners with stories about whichever fashion model he went out with the previous night and his opinion of the various parts of various women's bodies. He was a fixture on late-night television, gave interviews on economic programs, and has played himself in movies and television. Even as president he calls in to radio and television shows unannounced. He is so available, so eager for attention, that the *Washington Post* reported that he would call journalists and identify himself as "John Miller," a publicist and friend of himself.[4] In this role, he would shower himself with praise. "He gets called by everybody in the book, in terms of women," Trump allegedly said of himself in one such recorded conversation. "Actresses, people that you write about just call to see if they can go out with him and things." The avatar noted in particular that "he's got zero interest in Madonna," even though "she called and wanted to go out with him."[5] As Trump won primary after primary, Bill Kristol, a conservative pundit who despises him, fulminated that "the Republican Party is about to nominate as its presidential candidate the Wizard of Oz."

Kristol was right, but his indignation was misplaced. In America, all it takes to be a wizard is to look like one. Trump made it through all the debt restructurings, companies' bankruptcies, and the economic press's scorn for his business acumen and his supposed good name. What saved him in the end, at the beginning of the 2000s, was his reality show, *The Apprentice*, in which he played the ultimate businessman.

Billionaires are at the top of the capitalist food chain, and Trump was the only billionaire who appeared regularly in the American public's living rooms. Once a New York society celebrity, he became a national icon thanks to his top-rated TV show. Of course, it never addressed Trump's business failures, the well-founded claims that he did not pay suppliers and contractors, or his litigiousness.[6] On television, Americans watched a crude and combative robber baron of the people. He ran a capitalist gladiator show in which losers were executed in market economics style, by being told "You're fired!" He fashioned himself into the supreme authority on something that had evaded a large swath of the American public—success. It was the perfect persona for the world of social networks, one that brooked no indifference.

Even now, years after Trump's 2016 victory, his accession to the White House remains, for many, hard to believe. Twenty-two women have publicly accused him of sexual offenses ranging from indecent assault to sexual assault to rape. He was taped saying that "when you're a star, they let you do it. You can do anything. Grab them by the pussy."[7] No other American president has come close to a scandal of this magnitude, but it's just one more precedent he has set. He's the first president who never held public office prior to his election, the first since Richard Nixon not to release his tax returns, the first candidate to have his businesses go bankrupt multiple times. "Donald Trump is a phony, a fraud," Mitt Romney declared in 2016.[8] By any objective measure, Trump is one of the most prolific liars to head a great power in the modern era. But what makes him so important is not only his willingness to lie but his penchant for lying so publicly and baldly that his fabrications are evident to all. None of that kept voters from lining up behind him in 2016 and much of the American public to continue to support him since.

It could not have happened had the American empire not been suffering from severe systemic maladies that were ignored and neglected by its leaders. They denied that the global system was hurting entire communities on the American continent. They denied the rollback of American power around the world, and the growing sense of insecurity that Americans have felt since 9/11. They did not really want to engage with combustible, explosive issues like immigration and identity. They often acknowledged the distress of, say, blue-collar workers, but in response simply recited fortune cookie globalization mantras about positive thinking and offered mystical and tautological prophecies to the effect that America could best any challenge. Such claims were based on a classic inductive fallacy. America would always come out on top because it always had.

Trump's revolutionary message for Americans was exactly the opposite. The feeling of many that the country was on the verge of apocalypse was accurate, he argued. A marketing genius, Trump was touting a novel product—despair. The antagonism to globalization in its broad sense—not just global trade but also involvement in and engagement with the world, and universal values—had become so pervasive that even he, the billionaire from Manhattan, born into a hugely wealthy family, had tapped into that sentiment. His acceptance speech at the Republican National Convention was dark and wrathful. "Our convention occurs at a moment of crisis for our nation," he declared. His speech contained a collection of grim prophecies:

> The attacks . . . the terrorism . . . violence in our streets . . . chaos . . . The crime and violence . . . Homicides . . . victims of shootings . . . illegal immigrants with criminal records . . . roaming free to

> threaten peaceful citizens . . . one international humiliation after another. One after another . . . America is far less safe and the world is far less stable . . . disasters . . . ISIS has spread across the region and the entire world . . . radical Muslim Brotherhood . . . chaos . . . crisis . . . the situation is worse than it has ever been before . . . Death, destruction and terrorism and weakness . . . America—a more dangerous environment than, frankly, I have ever seen and anybody in this room has ever watched or seen . . . [9]

Global Elections

Trump's rhetoric fits in well with the destabilization and insecurity that characterize the age of revolt. The 2016 elections were the most globalized campaign in American history, and they led to his elevation—however tainted—to the leadership of the world's most powerful country.

First and foremost, Donald Trump benefited from an intervention by a foreign power in American political discourse, in the form of the wide-ranging covert Russian operation, aimed, according to the US intelligence community, at sowing discord in the US and boosting his chances of winning. Even if one believes that Russian interference did not tip the election, it can hardly be denied that the Russians carried out an unprecedented espionage operation. They accomplished something that the Soviet empire at the height of its powers never had—they disseminated information to millions of Americans, many of whom became unwitting dupes, effectively promoting an agenda that apparently came from Vladimir Putin's office.[10]

But Putin was not the only player on the field. False information was spread by private operators, many of them acting from abroad, for profit. Cambridge Analytica, a British political consulting firm, harvested personal data and preferences from the Facebook pages of millions of people without their explicit consent and used it to fashion targeted ads for specific kinds of voters. The Obama administration decided not to get involved in the Syrian civil war, but then the US found itself facing off against ISIS. The civil war set in motion an international refugee crisis that had a major impact on Europe. In Britain, it increased support for Brexit; in the US, it boosted Trump's presidential campaign. In a globalized world there are no isolated local conflicts; superpowers that remain passive pay a price.

Globalization has gradually eroded the effective sovereignty of states. That erosion peaked in 2016. Trump may have adopted the slogan "America First!," but no previous presidential election was marked by so much involvement from outside the US, up to and including a foreign power gaining access to American voting machines[11] and the capability to spark demonstrations from afar.

It was an inevitable result of ever-tighter global interrelations. Not only did globalization set the stage. It also fixed the play's rhythm, music, and script. When the stage was set, globalization sent on the players, led by Trump. Trump in turn brought out globalization, which had previously been a back story, and placed it in the spotlight. "We will no longer surrender this country or its people to the false song of globalism," he proclaimed. "The nation-state remains the true foundation for happiness and harmony. I am skeptical of international unions that tie us up and bring America down."[12] Hillary Clinton explicitly defended globalization and warned against isolationism, which she claimed would hurt the United States and its workers. But by this point, the former certainties of the members of America's middle class—job security, community solidarity, and the sense that one's children would lead even better lives—had been replaced by fuzzy holograms.

Trump was conjured up by this simulacrum of reality. A billionaire of the people, he spews hatred for the media but has a symbiotic relationship with it. As his brilliant biographer, Michael D'Antonio, told me, meeting him was like stepping into a scene that had already been scripted by Trump. The texts he recites trumpet American greatness, but the idiom had become detached from its source, from the values and policies that contributed to America's rise. It is appropriate that he broke into America's consciousness via a reality show, a counterfeit that no longer even makes a pretense of representing the real world, much like his faux-Versailles gilded penthouse.

Spectacularly, the fabrication was exactly what many people in places like Greene County, Pennsylvania, craved, places where the American dream had grown hollow. Middle-income American workers now had credit rather than savings, voracious consumerism instead of true financial security, overtime instead of pay raises, junk food instead of nourishment, connection to the world via smartphones, but with a declining chance of enjoying the prosperity their parents had enjoyed. In a world in which real things have been replaced by fake ones, the boldest fake can be king.

The election results looked arbitrary, or close to it. Clinton won the popular vote by a margin of nearly 3 million, but her loss in three states, by a total of less than 80,000 votes, cost her the presidency. Had just a quarter of a percent of Americans been less hostile to electing a woman, had she succeeded in persuading more nonwhites to get out and vote for her, or even if then-FBI director James Comey had held back his now infamous letter about reopening his investigation of her emails, she would have won. But she lost. America had gone so off-kilter that it was prepared to push a man like Donald Trump across the finish line.

Political leaders had curled up and dozed in their down quilts of faith in inexorable progress, free trade, and lax migration laws while the American heartland felt menaced. Then came the red MAGA baseball cap. The revolt against globalization that had begun at the fringes staged a guerrilla strike at the political center of the world's greatest power and captured its top office. Trump is not a coherent response to the grievances against globalization but merely the weakest horseman of a possible apocalypse.

A Nationalist Is Born

A short time before the midterm congressional elections of 2018, Trump flew to Houston to campaign for Senator Ted Cruz. The two men had said horrible things about each other when they competed for the Republican presidential nomination; Trump said far more than Cruz did. He gave Cruz the moniker "Lyin' Ted," since he was, he claimed, "the single biggest liar I have ever dealt with in my life," not to mention "a very unstable person" and "a little bit of a maniac." Cruz said that Trump was unfit to be president, that he was a "narcissist" and "serial philanderer," and that Americans might wake up one morning to discover that he had "nuked Denmark." That's just a small sample of the slurs they traded.

After Trump's election, Cruz, like most other Republicans, fell into line behind the man he had claimed was a danger to the US. And, as president, Trump of course did not want the Republicans to lose Cruz's seat.

So Trump changed his tune about Cruz, whom he now called "Beautiful Ted"; it was as if a medieval pope had granted an indulgence to a wayward nobleman. The evening was festive, with a large and enthusiastic crowd. In his speech, Trump spoke at length about the state of America's economy, and declared that the countries of Europe are "not taking advantage anymore, folks."[13] Then came the part where he explained how he intended to upend the liberal order and globalization in their current form. It was a pretty old, hackneyed trope. "America is winning again. America is respected again because we are putting America first. We're putting America first. It hasn't happened in a lot of decades," he told the crowd. Then he identified the enemy. "Radical Democrats want to turn back the clock for the rule of corrupt power-hungry globalists. You know what a globalist is, right? You know what a globalist is? A globalist is a person that wants the globe to do well, frankly, not caring about our country so much."

Casting people of liberal values as traitors is classic nationalist rhetoric. It sets up a false dichotomy, according to which universal values necessarily contradict local interests and national cultures. The message is stated explicitly: there is an enemy hiding among us. On the American extreme right, the word "globalist" is used in this context with anti-Semitic purpose. Has Trump, whose daughter and grandchildren are Jewish, grasped this? In any event, he has had no compunction about embracing a term that has been toxic in American political discourse throughout most of the last century and into this one.

"You know, they have a word," he said. "It's sort of become old-fashioned. It's called nationalist, and I say really, we're not supposed to use that word. You know what I am? I'm a nationalist, okay? I'm a nationalist. Nationalist . . . Use that word. Use that word."

Trump did not explain that night what he meant by nationalism, and he avoided doing so the next day as well, at a White House press conference, where he reiterated that he is a nationalist, while denying categorically that he knew that the term is linked to racism and white supremacism. As nationalists customarily do, he uses the term to define not what he is but what he is not. He is not one of those treacherous globalists who seek to sabotage America in the name of a secret internationalist agenda.

This prompted, of course, a public debate about nationalism, addressing everything from ethnic or racist nationalism, the classical idea of liberal nationalism as formulated by the Italian revolutionary Giuseppe Mazzini, and the economic nationalism advocated by Steve Bannon. From a historical point of view, it also addressed why the concept had vanished from legitimate political discourse in the US. The debate overshadowed the obvious: Trump had said what he really was. "Nationalism" is an accurate designation for his long, consistent, and deeply troubling record, and it's the only "ism" that he himself has used to describe it.

For years, some observers argued that, in Trump's case, form was content, and style passed for substance. On this account, he spouted empty demagogic rhetoric in an incessant stream of consciousness, composed of narcissistic utterances on Twitter, the only app installed on his phone. In this view, he was at most a dangerous buffoon.

For his supporters, however, he was a populist in the positive sense of the word, pragmatic and resolute, who had no truck with protocol or accepted standards of speech and conduct, a man who advanced in the face of the determined opposition of a moldy political establishment. Even before his victory in 2016, Salena Zito, writing in *The Atlantic*, put it in a nutshell: "The press takes him literally, but not seriously; his supporters take him seriously, but not literally."[14]

In retrospect, he needed to be taken both seriously and literally. Trump has tried to carry out many of his promises, even at heavy cost to American society, to its economy, foreign relations, and to the political culture of the world's most important republic. Whether it is building a wall on the Mexican border, trade wars, or the maltreatment of immigrants and asylum seekers, no one can say that he has gone soft, or become more measured, or adopted established norms. The famous assessment of France's Bourbon monarchs might just as easily be applied to him: as president, he has learned nothing and forgotten nothing. POTUS 45 was the volcano erupting in the midst of his own administration, the source of its frictions and rifts, which he exploits to his own ends.

Trump constantly speaks the language and tropes of nationalism. The idea that he is some sort of classic isolationist, seeking the serenity of an America that dwells alone, has been debunked. He clearly sees half the US population as collaborators in cahoots with a global conspiracy against the country. Trump sees his political opponents as "enemies of the people." These opponents include journalists, immigrants, globalists, Democrats, companies that transfer factories overseas, the Federal Reserve Bank, the late Senator John McCain, elements within the FDA during COVID-19, and many others. His slogan "America First" comes from the movement that tried to keep the United States out of World War II and more than flirted with xenophobia. His economics are classically nationalist, quite distinct from the free market principles of the American conservative tradition. He has sought to block people from Muslim countries from entering the US. He engaged in endless public confrontations with other countries—from Denmark to China—and their leaders. His obsession with keeping out immigrants, sealing America's borders, and building a border wall is symptomatic of an ethnonationalist doctrine of drawing a line between foreigners and "real Americans"—by implication whites who ostensibly share a common history, heritage, cultural values, and identity.

When neo-Nazis and white supremacists marched in Charlottesville, Virginia, and clashed with counter-protesters—ending in the murder of one of them—Trump declared that there were "some very fine people on both sides."[15] He told four congresswomen of color to "go back" to the countries they came from.[16]

This is not a coherent agenda. His administration's trademark is constant improvisation. In any case, his nationalism is flawed. A true nationalist would not be indifferent to, or actually welcoming of, a foreign power's interference in his country's elections. A true nationalist would not disparage his own country's war heroes, or say in a television interview that "I think our country does plenty of killing also" as a justification for the brutality of a foreign leader.[17]

His populist style has long gone hand in hand with nationalism. But there is a fundamental difference between these two attitudes. Populism first asks who is on top and who is on the bottom. As Benjamin De Cleen has put it, it looks at the powerless-powerful dimension.[18] There is a public that has authentic and just demands, and an elite and political establishment that, so the populist alleges, seek to frustrate those demands.

In the West, "populism" has become a catch-all term to denote an entire range of phenomena. The result is that, by including everything, it means nothing. We are not witnessing the rise of a self-proclaimed populist party like that of the late nineteenth century, but rather the return of the concept as a descriptive and often derogatory term. "Populism" does not adequately describe the contemporary rise of nationalism, racism, fundamentalism, or radical left-wing ideas; importantly, and just as there are few self-proclaimed populists today, neither is there any coherent populist agenda. Scholars, pundits, journalists, and mainstream politicians sometimes prefer not to use more precise terminology, so they continue to refer to the ideology of the president and his supporters as Republican and to call the president a populist, even though the term he uses to describe himself is "nationalist." The fixation with calling today's insurgents "populists" is not confined to the United States. Europeans, too, are reluctant to apply the labels nationalist, fascist, or racist to far right-wing parties because of their association with the oppressive totalitarian regimes of the 1930s and the horrors of World War II. "Populism" is like a coat of many colors that looks benign and does not reek of malice or violence.

Nationalism, as opposed to populism, does not ask who rules a given society and who is ruled. Its view is not vertical but horizontal—it seeks to determine who is inside and who is outside its ideal national community, and who inside ought to be pushed outside.

The class and power relationships that interest the populist are of no appeal to the nationalist, who focuses on elements of identity that confirm people's inclusion or exclusion from the nation. "This is a country where we speak English, not Spanish," Trump said in one of the presidential debates in 2016. On another occasion he claimed that an American judge, born in the US to parents who had immigrated from Mexico, could not hear a suit against Trump because he had a conflict of interest.

Nationalism has another characteristic. It cannot long survive in symbiosis with other ideas. The appointments that Trump has made, the tax reform he has implemented, and the good relations he has developed with the Republican leadership have made it clear that he is no populist. Notably, Trump has not changed, and has not even sought to change, his country's basic power relations between the elite and everyone else. He has not handed power to the people, nor has he plied them with social benefits and government spending. On the contrary, all he has offered them is resentment, so that he can build his "big beautiful wall" and foreigners will not be able to "charge in."

It is the nature of the nationalist that he cannot hide his colors for long—there is no socialism in national socialism, and nationalist populism is ephemeral. Ethnic or racist nationalism can harmonize only with fascism, which it remarkably complements.

While there is an ongoing and fraught debate over whether Trump's rise has put America on the road to fascism, there is little evidence that Trump himself is a fascist with a concerted program to raze the institutions of democracy as part of a totalitarian project. The way in which Trump has begun to eat away at American democracy, especially after his acquittal in his impeachment trial, is not programmatic but rather personal and instinctive.

In Trump's world, no idea is greater than he himself is. But there is a direction, and it is nationalist. Trump's nationalism is crude. It centers on a sense that the nation is menaced, both from outside and from within, by forces that threaten the unique identity and success of the national community. Accordingly, the assumption that there are standards of truth and justice that stand independent of the nation threatens his view of the world. Socialism, conservatism, and democracy all address the civil community, but it will always be subsidiary for the nationalist. By its nature, nationalism always kills its host.

That's what has happened with Trump. He has not enacted any populist legislation, good or horrible, that fundamentally changes the ways elites rule, or which creates a new distribution of resources. His populism has been no more than a matter of style, a design accessory, a sentiment thundered at huge, scripted rallies.

Devoted nationalist that he is, Trump has turned a blind eye to the damage his trade wars have done to the American economy and workers. His sole focus is his determination that no other country should exploit America, without regard to the policy's actual implications for his country's standing and the welfare of its workers.

"Nationalism is power hunger tempered by self-deception," George Orwell wrote. "Although endlessly brooding on power, victory, defeat, revenge, the nationalist is often somewhat uninterested in what happens in the real world. What he wants is to *feel* that his own unit is getting the better of some other unit."[19]

For such an approach to be sustainable over time, and for politicians to be able to survive even if the consequences of their actions show them to be failures, indifference to objective truth must spread through society as a whole.

CHAPTER 19

The Implosion of Truth

Sean Spicer, our press secretary, gave alternative facts to that.

—KELLYANNE CONWAY, SENIOR COUNSELOR TO PRESIDENT TRUMP,[1] 2017

For the nationalists and fundamentalists of our age to prosper politically, and for the revolt to rage, it's not enough that the middle class feel economically and culturally disenfranchised. A growing indifference to facts needs to spread first, demoting ideas about objective truth and promoting sentiment instead. Social networks have contributed to that, but something even more profound has happened to our societies. It has more to do with the challenges that truth presents than to the usual talk of lies spreading online.

A FEW MONTHS AFTER DONALD TRUMP'S VICTORY, I TRAVELED to a different America—not to its neglected and forgotten hinterland but to Silicon Valley. The high-tech elite was in full panic mode. A haze hung over San Francisco Bay that morning, and it was drizzling. From my hotel south of Sausalito, not far from the Golden Gate Bridge, I could hear the honk of boats and make out bikers in windbreakers circling the green hills below. This part of California is lush, and rich. I felt the tickle of affluence as I drove the winding roads through small towns on my way to San Francisco.

The Bay Area is one of America's most economically disparate regions, where the gap between the poor and the rich is largest. Inequality has increased as the high-tech industry has boomed.[2] California is the most liberal state in the US. It has more homeless people than any other US state,[3] but the Bay Area is home to more billionaires than any other metropolitan area in the world.[4] In 2019, three of the five American companies with the largest market value were headquartered within fifteen miles of each other in Silicon Valley—Google, Apple, and Facebook.

"For the last seventeen years I've built a life in Silicon Valley," Apple CEO Tim Cook told George Washington University's graduating class in 2015. "It's a special place. The kind of place where there's no problem that can't be solved . . . A very sincere sort of optimism. Back in the '90s, Apple ran an advertising campaign we called 'Think Different.' It was pretty simple. Every ad was a photograph of one of our heroes . . . People like Gandhi and Jackie Robinson, Martha Graham and Albert Einstein, Amelia Earhart and Miles Davis . . . They remind us to live by our deepest values and reach for our highest aspirations."[5]

As Cook's speech purports, Silicon Valley executives do not see it merely as an aggregation of profit-seeking corporations. For them, it expounds an almost messianic message about improving the human condition. Or at least, that was what they were selling.

Then came November 2016. The social networks and communications media created in Silicon Valley, integrated and adapted into the smartphone universe, may have connected the world, but not as the tech wizards claimed they would. They produced closed communities of people who thought the same way, echo chambers that amplified the self-assurance of those inside. They were not exposed to a variety of opinions and thus had no impetus to check their facts. Russia exploited the system during the election campaign, disseminating, for example, doctored photos of Hillary Clinton with what were portrayed as her Muslim supporters, and by promoting posts aimed at suppressing the vote in black areas. One post, promoted by Russian agents, received 130,000 likes in the US: it depicted Donald Trump in the Oval Office, dressed in a Santa suit, with the caption "WE ARE GOING TO SAY MERRY CHRISTMAS AGAIN!" The post protested against people being "forced," out of political correctness, to say "happy holidays" instead.[6] That lie was relatively innocuous compared to other sinister conspiracy theories seeded online.

Search engines and mobile technology have made virtually limitless information instantly accessible. But the same thing has happened to disinformation. According to a *BuzzFeed* analysis, in the last three months of the presidential campaign of 2016, the twenty most popular fake stories received 8,711,000 shares, comments, and reactions. Stories widely disseminated on Facebook had headlines such as "Pope Francis Shocks World, Endorses Donald Trump," "Wikileaks Confirms: Hillary Clinton Sold Weapons to ISIS!" and "Just Read the Law: Hillary Is Disqualified from Holding Any Federal Office." By comparison, the twenty most popular stories on Facebook that came from established media networks such as the *Washington Post*, the *New York Times*, and NBC received 7,360,000 shares, reactions, and comments.[7] About two-thirds of Americans get their news from social media.[8] Post-election studies found that far more Americans than most observers expected were exposed to fake news sites. In October and November 2016, one out of every four Americans visited one. Facebook served as "a key vector of exposure to fake news."[9]

Wired magazine recounted how Boris (not his real name), a Macedonian teenager, began his career as a paid disseminator of false information. His first post on his website, Daily Interesting Things, linked to an article he had found online alleging that at a rally in North Carolina, Donald Trump had slapped a man in the audience for disagreeing with him. In fact, no such thing had happened. He shared the blog post on his personal Facebook page and shared it with groups devoted to American politics. *Wired* reported that it was shared some 800 times. "That month, in February 2016, Boris made more than $150 from the Google ads on his website. Deciding that this was the best possible use of his time, he stopped going to high school."[10] Boris's story is a tale of a teenager from a poor country in an undeveloped region who built up a profitable internet-based business. It is also a manifestation of the dark side of globalization.

In the winter of 2017, when I entered the sumptuously fashionable offices of a Silicon Valley company, with its generous kitchens and comfortable couches, it was clear that the usual calm pond of complacency had been replaced by a puddle of anxiety. "We understand the challenge here," one executive told me—off the record because he had not received permission to speak with journalists. "We realize what has happened to us, and we will take responsibility for this story and solve it." He looked exhausted. He had really believed that his job, with its stupendous salary and stock options, was connecting people; now he found himself in the middle of an especially horrifying episode of *Black Mirror*. He faced mounting and ominous challenges—people inciting and committing violence and using his platform to spread toxic stuff in real time. Two years after we spoke, the man who massacred people at mosques in New Zealand broadcast his attacks via Facebook Live.

This senior figure addressed the great long-term challenge. His platform needed a way to weed out lies, or at least to warn users against false stories. "In the near future there's no chance that we'll be able to do that using artificial intelligence. We'll need to do it with people. But if we start vetting and editing content, we will turn into the thing we replaced. We'll become traditional media." He was appalled by that prospect—the need to monitor a huge quantity of information, determine what is true, figure out how to measure truth, mitigate ongoing confrontations about content—in short, the work of editing a newspaper with many thousands of writers. When I asked him how his platform influenced societies outside the US, he seemed surprised. He indicated that they had not thought that through deeply, as their political problem was in the United States.

Since then, it has become clearer how YouTube, Google, Facebook, Twitter, and other such networks abet racists and hate-mongers around the world, who use these services to disseminate lies, undermining societies. In some places, ethnic relations are fragile to begin with.[11] In Myanmar, incitement propagated via Facebook Messenger was used to exacerbate tensions between the Bamar Buddhist majority and the Rohingya Muslim minority, who became the victims of genocidal violence.[12] A German study showed how Facebook posts by extreme rightist xenophobes tracked and presaged attacks on immigrants. The correlation was closest in areas in which the use of social media was highest.[13] A report by the global activism movement Avaaz claimed in 2020 that Facebook's algorithm is a "major threat to public health" in the context of battling COVID-19. According to the report, content from top websites spreading health misinformation had almost four times as many estimated views on Facebook as content from leading health institutions.[14]

As the large social media companies encountered increasing scrutiny, they tried to distance themselves from the fabrication industry and the radical rhetoric that flourished on their platforms. They put money into research, monitoring efforts, and the creation of mechanisms for filtering out problematic content. In an act of desperation, Facebook has reduced the frequency with which posts from business and news organizations of all kinds, as well as political and fan groups, pop up in the feeds of its users. Entrepreneurs have become much more involved in identifying and exposing lies as the public has become more and more concerned about the impact of misinformation. High-tech innovators in Tel Aviv often tell me about how they are developing tools that social media and websites can use to weed out disinformation, fact-check it in real time, or at least mark controversial information so that users will know that it needs to be looked into further. "If we can do that we can save the world," the vice president of an Israeli company told me, "or perhaps the truth."

The Medium Is the (Fake) Message

Behind the actions of the internet giants, and behind the public debate in general, lie several assumptions. The first is that the social networks and the internet as a whole played a critical role in Trump's victory and in the spread of false information of the kind promoted by the Russians. The second is that social networks can be repaired and turned into places where truth prevails, and that this can be done by means of enforcement, technological tools, and education. The third assumption is that people inadvertently take wrong turns into the forest of false information and are in need of protection and rescue.

Each one of these assumptions is problematic, or definitively false. First, the social networks were not the only or even the principal vectors for the spread of conspiracy theories; nor can Donald Trump's victory be attributed to the combination of Russian intervention and fake news. The effect was marginal in comparison with the usual factors that explain how elections are won—news stories, some false, reported by the mainstream media; party preference and turnout; the letter that FBI director James Comey sent Congress about the investigation of Hillary Clinton's emails, and the protest votes cast for Trump by white blue-collar workers.[15] In other words, more than the virtual world actually changed events. It reflected trends that were already in place, perhaps amplifying them to a more extreme level.

Silicon Valley's efforts to repair the system is at odds with the fundamental situation produced by the corporations operating there. Marshall McLuhan famously remarked that the medium is the message,[16] but perhaps he has not been taken seriously enough by Silicon Valley programmers. The new medium told its users that everyone has a voice and what matters is feeling good while interacting with other users of social networks, and to hell with facts or gatekeeper hierarchies that filter information. As the nouveau riche of Silicon Valley know all too well, the medium was created by multinational corporations to generate profits. Lies are well suited to social networks' algorithms, and the more outrageous the lies, the more response and engagement they get. They are built into it. It's not a bug, it's a feature.

The third assumption is the weakest. The view that the consumers of communications media are victims or fools who need protection has little empirical evidence to support it. There is little reason to believe that large parts of the public will trust the findings produced by new and improved fact-checking tools any more than they believe the news as reported by mainstream news organizations. Indeed, surveys indicate that almost half of Americans already see fact-checking sites as politically biased.[17] Many users are well aware that they consume, and at times spread, disinformation and fictitious stories. An international Ipsos study from 2018, covering twenty-seven countries, found that 65 percent of the 19,000 persons it polled agreed that most people live in their own "'bubble' on the internet . . . only connecting with people like themselves and looking for opinions they already agree with," while over a third said that applied to themselves. Six out of ten said that people don't care about facts anymore, they just "believe what they want"; the figure for the US was 68 percent, for Germany 62 percent.[18] Four out of ten of British social media users who responded to a poll conducted by Loughborough University's Online Civic Culture Centre (OCCC) admitted to sharing news reports that they later realized were false or inaccurate. One in six acknowledged deliberately spreading material known to be untrue. When asked why they shared political news on social networks at all, the most common answer was that they did it to express their feelings; almost a fifth said they wanted to upset others.[19]

Feelings are what is important now, much more so than rational and focused debate. In an interview during the Republican National Convention in 2016, former House Speaker Newt Gingrich defended Trump's lie that American cities were experiencing "soaring crime rates" by claiming that "people *feel* more threatened."[20] The CNN interviewer responded that he might be right about the feelings, but those feelings were not fact-based. "As a political candidate,"

Gingrich responded, "I'll go with how people feel and let you go with the theoreticians." Gingrich was not just displaying the cynicism of a politician. He was channeling what the public itself was saying about the centrality of its emotions.

A large-scale MIT study of 2018 looked at 126,000 news stories that were shared by 3 million Twitter users over ten years. The central conclusion was that truth was losing big. No matter what the subject of the story, from politics to entertainment, and by every measure, false or fake stories received more exposure, were more influential, and were spread much more rapidly than accurate content. False stories reached 1,500 people six times more rapidly than true ones. Most importantly, this happened not because of bots (automatic accounts) operated by interested parties but because of real people. The culture of lies flourishes "because humans, not robots, are more likely to spread it," researchers reported.[21] That, together with what people report about themselves, indicates that spreading lies online has become a kind of guilty pleasure.

Two out of every three people think that, fairly often or very often, the mainstream media maliciously print or broadcast fabricated stories.[22] Can they even tell when a story is true or false? Today, people are finding it harder and harder to make such distinctions. A Pew Research Institute survey from 2019 found that half of the Americans polled said that they have at times avoided conversations with someone out of fear that they might bring up made-up news.[23]

Something has changed profoundly in Western society. The key question is not why people fall for lies but why they've lost faith in traditional sources of authority, namely the government, scholarship, and the press. Countering false and misleading statements is today a favorite pastime of elites under siege, but what has caused the collapse of trust and the emergence of an ecosystem of lies?

"THERE IS A TRUST DEFICIT OUT THERE IN THE WORLD," Secretary-General António Guterres of the UN declared in 2019.[24] Trust and prosperity go hand in hand in societies; studies show a causal relationship between them—the more trust the members of a society have in each other, the more the society flourishes.[25] The Edelman Trust Barometer shows that people around the world, but especially in the West, are suffering a crisis of trust. Half of the people around the globe surveyed for the barometer voice a lack of confidence in government, the media, nongovernmental organizations, and businesses. Whether well-off and educated or poor without an academic degree, only one out of every five surveyed believes that the system is working for him or her. Seven out of ten Americans say that they have a "desire for change."[26] Only four out of every ten inhabitants of industrialized countries trust their governments.[27] Sixty percent of the people surveyed in the countries of the European Union said that they are not inclined to trust their national governments.[28]

Almost every institution in American society has seen its public trust decline significantly over the last forty years. Trust in Congress languishes in the single figures. Trust in the presidency has plummeted by 27 percent since the 1970s. Trust in the government as a whole stood at 53 percent in 1972; it has collapsed by more than two-thirds.[29] During these years the American economy, measured by GDP and productivity, has grown, the standard of living has gone up for all parts of the population, and the US has been one of two superpowers and now is the sole superpower. The later a person was born in the era of American supremacy, the more pessimistic she is likely to be about her country. Millennials have the least trust in elected officials, the military, religion, and the business world.[30]

But it is not just a matter of institutions. Interpersonal trust, measured by asking people whether they agree with a statement such as "most people can be trusted," has also declined; today, only one out of three Americans agrees.[31] On a scale of ten, Americans on the average rate their trust in others at 5.8; for Britons the average is 5.5, and for the French only 4.9.[32] Seven out of every ten young people maintain that most people will try to take advantage of you if they are given a chance. These are people who have enjoyed the best conditions humanity has ever known. Among Americans aged sixty-five and older, who grew up during the age of responsibility with a more modest standard of living, only four out of ten are so suspicious.[33] On the whole, older people trust foreigners twice as much as young people do.

Perhaps it has something to do with the kind of society one has grown up in. Researchers note a correlation, and suggest a causal relationship, between a lack of social equality and lack of trust.[34] It is those who feel betrayed who express the most mistrust: minorities, the poor, young people, those without a college education.[35] Trust declines dramatically in places that experience economic inequality, discrimination, and material want. The residents of Flint, Michigan, a long way away from Silicon Valley, know that better than anyone else.

Poisoned City

The sun is sparkling on the Flint River. Stepping off the path and approaching its muddy banks, I can see things in the water. Rusting iron rods, a sodden paper bag. Something oily floats. Over the last few years, this Michigan city has been cast in a leading role in a dark tale of today's America. Anyone who thinks that the collapse of trust and the culture of lies began in the digital world should go to a place like this and talk to people living here, people who lost their trust long before disinformation began to spread through social networks. It's a story about people who were lied to. About people to whom it has been permissible to lie.

I am standing close to a water purification plant by the side of the river. It's the most infamous water purification plant in the world. At its center is a glistening white water tower. Not far away is an open fire hydrant. Water is trickling out of it and forming a little pool, from which the water then flows back into the river. Local water is not in high demand in Flint.

Flint was on the trade route connecting Saginaw in the north with Detroit in the south. The whites pushed out the Sioux, who called the river Pawanunking, meaning "river of flint." African Americans relocated there from the South. Flint is the birthplace of General Motors and the entire range of brand-name automobiles that it produced.

Forty percent of the population in this majority black city lives under the poverty line. The city's financial problems led the state's governor to strip the mayor and city council of much of their power and, in 2011, to appoint emergency managers to run the city. These technocrats pushed an austerity regime that reduced the basic services provided to the city's vulnerable population. One of their efficiency initiatives involved saving money by redoing the water system, which piped water from Lake Huron. In the interim, water would be pumped from the Flint River—the same river into which much of the waste produced during the city's period of industrialization had been dumped.[36] The city concluded that it could save $5 million if the river's water was purified and channeled through the pipes to the city's inhabitants. But there was nothing to fear—technology would make the filthy water clean. The unelected executives signed off on the plan, declaring the water safe.

At a ceremony inaugurating the water purification plant, the mayor toasted "Here's to Flint!" and downed a glass of treated water.[37] But complaints began pouring in from people almost as soon as they opened their faucets. They complained that the water pumped into their homes was turbid and smelled bad. The technocrats reassured the public that the water was just fine. In January 2015, the water company found elevated levels of trihalomethanes (THMs), a suspected carcinogen, in the water. The mayor proposed hiring a water consultant. Legionnaires' disease broke out in the city, ultimately killing twelve, but public officials made no connection between the outbreak and the water supply. By June, the US Environmental Protection Agency had collected enough data to issue an emergency order, calling on the public not to drink the water.[38]

At the same time, people in Flint were sending the media photographs of brown sediment in their bathtubs. They reported skin rashes and sick children. Michigan's Department of Environmental Quality declared that "anyone who is concerned about lead in the drinking water in Flint can relax." But in response to a call he received from a city resident, Marc Edwards, a water engineer at Virginia Tech, took a team of scientists there to investigate. They discovered high levels of lead in the water supplied to much of Flint. In the homes of some families, the lead level was similar to that of untreated factory waste. The first reaction of the responsible officials was to insist that the water was safe. But this time there was scientific proof that could not be disregarded.

The facts quickly became public. It turned out that the water plant had not properly treated the water. It was so acidic that it caused massive corrosion in the already old and deteriorating water lines that supplied homes. Lead leached out of the pipes. The accumulation of high levels of lead in the body causes lifelong damage, especially to children. Lead exposure damages the brain, causes behavioral disturbances, delays puberty, and damages hearing, to name a few side effects. And the consequences are irreversible.

In 2017, three years after the tainted water began flowing through Flint's pipes, another study explained why Flint's birthrate had dropped after 2014. During the time the city was taking water from the river, fetal deaths rose by 58 percent. The birthrate declined by 12 percent.[39] All this had been permitted to happen because of collusion and misrepresentation, according to Michigan's state attorney. There "has been a fixation on finances and balance sheets," he declared. "This fixation has cost lives . . . It's all about numbers over people, money over health."[40]

Flint's travails are an egregious instance of the general deterioration of America's infrastructure. Unsafe levels of lead have been found in water supplies all over the country since, from Newark, New Jersey, to Houston County, Alabama.

A LARGE WHITE MOBILE PEDIATRIC CLINIC IN A MUNICIPAL public parking lot in Flint offers psychological evaluation and occupational therapy. The clinic came from New Orleans, where it had been used to care for victims of Hurricane Katrina. Renovated and repainted, it's ready to serve another group of Americans who were left behind. Inside there are a few well-worn dolls, along with arts and crafts supplies. Qiana Towns is one of the counselors who works with children here. She tells me that at her home the conversation is mostly about water: about staying away from the water, when to open the faucet, not to forget to drink purified water from the bottle or to open the filter. She has two daughters, seven and fourteen years old. She goes into panic mode when the younger one forgets to turn on the filter when she brushes her teeth, "because I don't know if this is the gulp that's going to make the difference." She is terrified that the lead will wreck their lives. "My seven-year-old is performing at an exceptional level," Towns says. "She is a very quick learner and a smart girl, and I think about her having drunk the water and whether that's ever going to affect her accelerated learning. That's when the guilt returns, when I think—if I just had paid attention . . ."

The word "guilt" comes up again and again in our conversation. She says she has "a tremendous amount of guilt . . . Because there were signs. I noticed the water turned the shower curtain to a particular color, I noticed there was a smell from the water I had not smelled before." Tears well up in her eyes. "I thought I could trust the people who were talking to me about the water being safe. So I didn't really react the way I generally do in keeping my kids safe."

When I hear that, I can't help telling her over and over again what she knows but does not feel: that it is not her fault. And I can't help thinking about what has gone lost here—trust. It's not much to expect that if you open a faucet and give your child a cup of water, it won't poison her. "I simply could not fathom being in a society, in, you know, the United States," she says, "where any municipality would allow the water to be in such bad shape."

But now she does fathom it, and her girls do, too. Karen Weaver, the impressive mayor voted in after the managers were kicked out, certainly knows that the people of her city are in the midst of a crisis of confidence. I meet her in her office. What happened here, she tells me, would not have happened if the majority of the population were not black and poor. No question about it. She says that the local GM plant asked to disconnect from the water supply coming from the river, because of the damage the water was doing to their metal products. The plant was promptly disconnected, while the populace continued to drink lead-laced water.

Among some of America's intellectual elite, there are romantic fantasies about the underprivileged marching to the polls to take revenge. In practice, when the weak are trampled and poisoned, when they lose trust in the system and themselves, they don't vote. In the 2016 presidential election, the turnout in Genesee County, where Flint is located, dropped by 3 to 4 percent,[41] and it was principally the members of the Democratic base in the poor cities who stayed home. Even more people in Flint stayed home, with the river's poison water behind their closed faucets. In 2012, the county had voted for Barack Obama by a margin of 57,000; Hillary Clinton won the county by only 19,000 votes four years later, and that helped to secure Trump's tiny margin of victory in Michigan. The Republicans did better than they had in the last twenty-eight years.

Flint's water system has since undergone intensive repair. Tests done by the EPA show that the water is safe, and that the lead levels are low. But as of 2019, five years after the crisis broke out, Mayor Weaver is still cautioning her city's inhabitants to drink only filtered or bottled water. Many of them are doing exactly as she says.

Flint was poisoned with water, and then with lies. The residue that lies leave behind is mistrust.

Lying to Grandma

My grandmother once asked me to buy a washing machine for her. The request was unprecedented—well into their eighties, my grandparents lived independently. Fastidious and diligent, as befits the generation of the era of responsibility, they always bought their goods at stores with good reputations, never on credit, and always after seeking and receiving recommendations about the items in question. Their request to me included precise instructions. I was to go to a particular store, there and nowhere else, and buy a particular model of washing machine that we had settled on. It was the store where they always bought their appliances. "We trust him," Grandma impressed on me.

When I went to the store, the price the proprietor gave me for the washing machine they wanted was very high—indeed, inflated. A minute's Google search showed that the same model, with the same warranty, could be purchased from another vendor not far away and delivered at a price that was a third less than what their store was demanding. When I told Grandma about the ridiculous price at "her" store, her eyes opened wide in astonishment. "Does that mean he cheated us all these years?" she asked.

I always check prices on Google before making a major purchase. And it was obvious to me why the store's price was higher. It was located in an upscale shopping center, where the customers were mostly older and largely set in their ways. But my grandmother took it personally. In her time, when there were no websites where you could compare prices, when people did not buy items of this kind by phone or mail, making such a purchase was an interpersonal transaction. It required time and investigation. It also required trust in the merchant, who would tell a story about the item. Grandma suddenly grasped that the store owner she had trusted for decades to give her a fair price had not been completely honest with her, perhaps over a period of many years.

In a way, we are all my Grandma—all the time. Imagine a person who, thirty years ago, hears a lecturer assert that, in a distant country, it is now customary to remove the kidneys of the dead before burial. According to the speaker, the practice began only a few years before. The listener thinks it's weird and implausible. He decides to check it out. But, as it's thirty years ago, he doesn't have many ways to do that. Library research won't be much help, as the books in the library are not likely to be sufficiently specialized or up-to-date. Encyclopedias won't do the job, either. Does he have to get on a plane and fly to the distant country? Wait for a scholarly journal to address the issue? Perhaps he could call the country's embassy, where someone on the staff could investigate and then confirm or disprove it. But it would be a major undertaking, requiring much time and considerable effort. Any correct answer would be conveyed to the researcher a few weeks or months down the line, long after he had practically forgotten about those strange foreigners and the kidneys of dead people.

Suppose that, in the meantime, the same person went for a medical checkup. His doctor might suggest replacing one of his regular medications with another one. Don't worry, the doctor tells him, this name-brand version is better than the generic one you were taking, and worth the extra money even if your health plan doesn't cover it. The patient will believe him. He really has no other choice. Theoretically, he could seek a second opinion, or read up on the subject in professional medical journals. But that's not likely to happen, because of the high cost involved, in money and time. Studies have consistently shown that many doctors are not open with their patients about their commercial conflicts of interest. According to one article, four out of every ten physicians did not inform their patients about their commercial relations with pharmaceutical companies or producers of other medical products.[42]

The same person turns on his TV at night. The secretary of the treasury cites economic data. Our victim—clearly, he is a victim—recalls different numbers. Perhaps he obsessively keeps the back issues of an economics journal and can check the figure. But that's not likely. He wants to check the number cited by the politician. How can he do it?

In the morning newspaper, he reads an astonishing report about the body of a diver discovered in a burned forest in Greece. How did it get there? The article cites experts who suggest that an air tanker plane employed in putting out the forest fire sucked the diver up with the seawater it filled its tanks with and cast him into the middle of the fire. Didn't he read a story strangely like this a couple of years ago? He thinks it might be an urban legend (which it is), but again, he will find it hard to disprove.

All these cases take place in a Google-less world, one with no good, fast, and inexpensive way of checking, comparing, confirming, verifying, and disproving stories and assertions.

And there's a lot that needs to be disproved. People tell lies each day. A classic 1977 work on the subject maintains that an average person encounters about two hundred lies a day, including white lies.[43] Another study finds that six out of every ten people cannot conduct a ten-minute conversation without lying two or three times. They themselves admit to telling the lies after hearing a recording of the conversation they took part in.[44] More optimistic studies claim that

Americans lie only once or twice a day on average. When the data was reexamined by other researchers, they found that the variation among people was very high. There are serial liars, and people who can go for days without telling a single lie.[45] It seems reasonable to assume that most people do not accurately report the number of times they lie, whether unintentionally or because they lie about this as well.

And how much do powerful people lie? It's an important question. They have considerable impact on the lives of others, and sometimes they can profit from lies. There is not enough research on just how honest people in power are, but researchers at Columbia Business School performed a fascinating study. Forty-seven people were assigned randomly to one of two groups, "leaders" and "subordinates." The leaders exercised power in the framework of complex role-playing in which they could make decisions regarding the economic and social standing of subordinates. Then, both leaders and subordinates were told to steal a hundred dollars and to try to persuade an interviewer (who was not informed about the nature of the experiment) that they had not committed the theft. Whoever could persuade the interviewer could keep the money.

The results showed that just a brief role-playing game turned the leaders into shameless liars. They displayed fewer cognitive and emotional markers of having lied than did the subordinates. And the difference was not just psychological—it presented itself biologically as well. Saliva samples were collected from each participant before and after the experiment to measure levels of cortisol, a hormone released under pressure. Significantly less cortisol appeared in the saliva of the leaders than in that of the subordinates. The latter also felt worse and suffered from cognitive impairments after they lied. The researchers concluded that "power acted as a buffer allowing the powerful to lie significantly more easily."[46]

Dishonesty Revealed

Here is a hypothesis: at first, it was not lies but rather information that multiplied, and therefore the ability to detect truths. It happened rapidly, with a technological lunge, and with the help of global interaction. The capacity to confirm, disprove, and check statements became so powerful that, abruptly, many lies were exposed, all at once. It became much easier to publicize refutations of false information. Truth was more accessible, threatening old power structures and conventions. Large cracks appeared in the trust people had previously placed in institutions and in the other members of their society. The impression it created was that everyone was lying to everyone, and that in and of itself encouraged people to lie.

The turning point came when the public at large realized just how profoundly and frequently they were being lied to by people in authority. Society, every society, grants some of its members formal or informal authority to provide truths about the world. These are the people who enjoy the public's legitimate expectation that they will tell real stories to large audiences or provide particularly crucial information to individuals about their lives.

Of course, these people—politicians, bankers, physicians, journalists, policemen, teachers, local community leaders, even the proprietor of my grandmother's appliance store—have nearly always lied, to varying degrees. They lie because, well, we all lie, and if the study cited above is correct, people in power lie better, and feel less guilty about it.

Until about thirty years ago, humanity lived behind a veil of dishonesty that provided those in power with a flexible space for imprecision and lies. That veil has not been entirely torn away, but it has rapidly become more transparent. More people today have the capacity to uncover lies, because it is easier to investigate exhaustively each element of a claim made by an authority figure.

As a result, many more lies are being called out. The significance of technology, or more precisely the significance of Google, is that we now have an unprecedentedly effective, easy, and inexpensive way of checking the validity of utterances about the world. We cannot uncover all lies, of course, but we can uncover many more than in the past. The extent of the truth and the timeliness of information on the internet will almost always be greater than the sources available in the predigital age, because the internet, unlike a printed book, is dynamic. To perform a search, a person needs access to the internet and digital literacy. As time passes, the number of people without these capabilities is dropping, and the percentage of the world's population that can confirm or disprove information is rising. And if you don't know how to check a particular fact, one of your Facebook friends or WhatsApp contacts will know.

There is an overload of information. Already in 2016, Google was coping with 4 million queries a minute, 2 trillion a year, many of them entirely new, questions that had never been asked of Google before.[47] This is an expansion of the truth, or the desire to attain truth, which continuously stretches and challenges society's conventions. Take, for example, physicians' hostility to "Dr. Google," with whom their patients second-guess diagnoses and prescriptions.

Is Dr. Google better for your children than a pediatrician? Probably not. But the shock of this new world, in which facts are available with a few keystrokes, and in which people encounter a flood of reports on the exposure of fabrications, has turned doubt into an argument in and of itself. One study found that search results that do not match what pediatricians say causes parents to lose trust in their children's doctor—even before they have checked the accuracy of the information brought up by the search.[48]

Hollow skepticism is the order of the day. Data on trust collected by Gallup over decades shows that the largest plunge in trust in American institutions happened between 1997 and 2007.[49] That is connected to the state of the middle class, economic inequality, and the 9/11 attacks. That is the soil from which the crisis of trust has sprouted. But there are other factors as well—the decade in question is the decade of the internet and Google, and many people have discovered good reasons not to rely anymore on people in power telling stories.

Millennials have grown up in this era. While their faith in most social institutions is extremely low, they display a high level of trust in three types of people, much higher than that granted to them by older generations: scientists, professors in academia, and journalists.[50] These are three professions that require fact-checking and comparison of sources as a matter of professional practice. Determining the truth of assertions is part of the job. For people born into the world of Google, that seems to engender more trust.

NOW WE COME TO LIES.

When the truth expanded so much and then imploded, it left a vacuum. Each time a deception or lie was perpetrated by a familiar social institution or a person of high reputation, it bolstered the power of the charlatans—those who peddle the supposed truth about vaccinations, about the conspiracy to hide the fact that the earth is flat, about the claim that George Soros controls the world. Many people discovered fabrications and became convinced they had found the truth.

Another thing happened to lies—they were no longer principally the preserve of those with power and authority, or of the street corner prophets crying "The end is nigh!" The ability to spread lies was privatized and decentralized, and passed into the hands of the masses. When that is the way things are, people make the perfectly rational choice of not trusting anyone.

The expansion of truth, the erosion of trust, and the ability to use Google to check facts made it easier to argue persuasively that "everybody lies," the title of Seth Stephens-Davidowitz's 2017 worldwide bestseller.[51] The book uses search engine data to show the fundamental hypocrisy of the way people lie to their friends, to researchers, and to themselves. He shows how the technological age expands information and thus reveals insincerity—people celebrate Martin Luther King Day at the same time that they seek racist jokes about black people on the internet, for example.

The belief that everybody lies gives people tacit justification for lying themselves, since they are just doing what everyone else is. Everybody lies, yes, but the big picture is that *everybody has been lied to*. That is the context of contemporary public discourse.

Two-thirds of Americans claim that they are being lied to by other people, according to one survey. That's 50 percent more than at the end of the 1980s.[52] Maybe they *feel* they are being lied to more. But, more significantly, they at least have the technological tools to easily uncover many of the lies. Never have people been able to win arguments so quickly or easily. They ask Siri and receive a pretty good answer via Google.

When truth imploded, people retreated to something they can control and about which they cannot be contradicted—their opinions, occasionally supported by convenient or manufactured facts. They say that facts are no longer so important in their social intercourse, or that they no longer want to take part in conversations that involve news and information. Whatever the case, more than half of Americans have trouble distinguishing factual utterances from expressions of opinion, according to one study.[53] Eighty-six percent of fifteen-year-olds could not tell the difference on PISA (Programme for International Student Assessment) tests. The question is whether people even want to make the distinction anymore.

In *The Origins of Totalitarianism*, Hannah Arendt writes about the apparatus of repression and its connection to the blurring of truth and falsehood, fact and opinion. "In an ever-changing, incomprehensible world," she argues, "the masses had reached the point where they would, at the same time, believe everything and nothing, think that everything was possible and that nothing was true. Mass propaganda discovered that its audience was ready at all times to believe the worst, no matter how absurd, and did not particularly object to being deceived because it held every statement to be a lie anyhow."[54]

The totalitarianism that preoccupied Arendt is, at least for now, dormant. Yet the utter cynicism about truth she described has again reared its head, fomenting violence. These days, it is not dictated by fascist elites but is rather a grassroots phenomenon, mobilized online. Where nothing is true and nothing false, progress itself is in danger.

CHAPTER 20

The Battle for Progress

I will show you fear in a handful of dust.

—T. S. ELIOT, *THE WASTELAND*

Humankind is writing a new chapter in its saga, in which we are more interconnected than ever before. The age of responsibility made wars between sovereign states less frequent. Industrialization and trade opened new opportunities for hundreds of millions of people in the global East and South who had been isolated from the prosperity of the West. On average, the human condition improved dramatically, but that does not tell the whole story. Entire communities lost their jobs, were pushed onto the social periphery, bore the brunt of climate change, and saw their towns or countries become exploitation hubs. Many sensed that their identities and way of life were under attack. That intuition is not misguided, and the understanding that globalization hurt many of them is no delusion.

The paradox at the heart of this promising yet frightening new world is that people seek the same things people have sought for centuries—security, a livelihood, love, faith, and, possibly, freedom. They are more interconnected now, enjoy the benefits of technological advances, and think of their culture in more supranational terms, but their basic needs and desires remain the same. Our world of meaning is primarily local. Globalization is a constant revolution fueled by this tension. The COVID-19 pandemic illustrated how your physical location, social environment, and the way your political community functions determine not only the risk to your health but also your economic and emotional well-being. When times are good one can live as a global avatar; when war is at the door, or a new pathogen, it's the neighbors and government who matter. Some people may find this consoling and reassuring, though not necessarily, and not with populist and nationalist leaders at the helm. Yet even regardless of the current leadership's shortcomings, a global crisis demands global bodies, solutions, and agreements. These are not in vogue.

"For much of the world, globalization as it has been managed seems like a pact with the devil," wrote the Nobel Prize–winning economist Joseph Stiglitz in 2006. "A few people in the country become wealthier; GDP statistics, for what they are worth, look better, but ways of life and basic values are threatened."[1] A year and a half later, the worst financial crisis since the Great Depression broke out. Since that crisis, the world has undergone rapid and frequent dramas—worldwide recession, government debt crisis, Brexit, Trump's accession to the presidency, the Syrian civil war, mass migration by desperate refugees, political instability, the rise of a violent extreme right wing, an avalanche of fake news, and a downward spiral of almost every measure of the environment and the global pandemic of COVID-19.

Resentment of globalization has intensified and spread. Its veteran opponents sniff impending victory. "I think that the elites have lived too long among themselves," taunts Marine Le Pen, who firmly opposes what the nationalist right calls "globalism." "They have acted like carnivores," she charges, "who used the world to enrich only themselves. And whether it's the election of Donald Trump, or Brexit, the elites have realized that the people have stopped listening to them, that the people want to determine their futures and, in a perfectly democratic framework, regain control of their destiny, and that panics them, because they are losing the power that they had given themselves."[2]

Le Pen disregards the fact that globalization is a product of democratic processes, and that it has saved many people from poverty and death. Indeed, it has enriched much of her constituency. But she is correct in saying that globalization has been led by the political elites whose power is weakening. As local traditions, communities, and businesses have struggled to survive the revolution of the information era, the elites have tuned out. They have sequestered themselves in wealthy communities, high-tech entrepreneurial islands, and the ivory towers of academe. Simultaneously, people are disengaging from the promises of the global village and have recast themselves as radicals. They represent large constituencies, mainly in the West, who feel increasingly marginalized and unmoored by a changing world.

LABELING THIS DISENGAGEMENT A "POPULIST WAVE" OR A "threat to democracy" is both simplistic and inadequate. The populism rubric leaves out the intensification of fundamentalist forces almost everywhere, from Myanmar to the Middle East and Europe. Fundamentalism is a force in contemporary politics, not a ghost from the ignorant past. For the weak and the poor, fundamentalism plays the same role that populism does for the middle class—it offers an ostensibly reasonable response to the fracture of the current order. And what seems at first to be populism can disguise the forces of racism or ultranationalism.

Another way of looking at the current crisis might be to think of the world as undergoing a deconsolidation of democracy, as suggested by Yascha Mounk and others. While democracy is indeed in crisis, authoritarian regimes also face increasing threats to their stability. China is cracking down on freedom of speech and fears growing resistance to the Communist Party. Iran faces growing resentment and protests. And Vladimir Putin is using all means at his disposal to repress, delegitimize, and liquidate his opponents. Each of these regimes is facing challenges, in the form of popular unrest that has its roots in the rise of global consciousness, the integrated world economy, and the spread of innovative technology.

Even in the West, young people's declining faith in democracy does not necessarily mean that they favor authoritarianism. They may say that democracy is not essential, as surveys show, but they are not marching in the streets demanding another form of government. Rather, many have lost confidence in their societies' institutions and power brokers, and are now experimenting with new ideas.[3] And while more are willing to fight for initiatives once considered to be on the margins of politics—basic income for all, non-policing ways of public safety ("defund the police"), a green new deal—most are just waiting; experiencing the limbo of an old age dying and another one just being born.

We are witnessing something more mercurial, motley, and messy than a simple us-them dichotomy. The latter might be appealing, but there no longer really is an "us," because identities have become more multilayered. There are myriad dividing lines today, not only those between urban elites and rural Americans without a college education, for example. Donald Trump did not win the 2016 election only because he got the votes of lower-middle-class whites but also because Americans earning more than $200,000 a year, the majority of whom live in the suburbs of large cities, preferred him to Hillary Clinton. Binary divisions were in vogue during the Cold War, but today's world is increasingly multipolar. Political paradigms are collapsing, and old labels are inadequate in the contemporary revolt. The world's most prominent liberal leader since 2016 has been German chancellor Angela Merkel, who leads her country's conservative party. The Republican Party of the United States, historically committed to the free market and free trade, has been taken over by advocates of tariffs and subsidies. Some in the radical left are displaying tolerance toward anti-Semitism, with Britain's Labour Party under Jeremy Corbyn only the latest example. Fundamentalism, populism, nationalism, and outliers of the radical left all deny inconvenient facts, dabble in conspiracy theories, or both. "Things fall apart; the centre cannot hold," as Yeats famously wrote.[4]

Neither coherent nor structured, the revolt rages on many fronts, taking different form on each. Its disparate facets are, like everything in the globalized world, connected. To view populism, fundamentalism, anti-scientism, and nationalist racism as mere bumps in the road on the way to the triumph of the liberal order, isolated events or temporary obstacles to the inevitable victory of Enlightenment values, is a perilous mistake.

What hangs in the balance today is not only peace, prosperity, and stability—the chief goals of the architects of the age of responsibility. Progress itself is under threat. Reason and scientific data are increasingly marginalized by the politics of sentiment, radicals who empower tribal politics of resentment or mainstream politicians who ride it. Increasingly, people with power "act contrary to the way reason points and enlightened self-interest suggests," to cite Barbara Tuchman's definition of folly.[5]

Newtown, Connecticut, Autumn 2016

I look at Dr. Jeremy Richman's face and decide at the last minute not to ask him about conspiracy theories. By that I mean the claims of those who maintain that no shots were fired at Sandy Hook Elementary School in Newtown, Connecticut, in 2012, and that six members of the staff and twenty first graders were not murdered there, among them Richman's daughter, Avielle Rose. Noah Pozner was six years old when he was killed; his family has had to move seven times since then, because they have been the targets of threats and provocations.

According to some hallucinatory narratives, the slain children were never even born. According to others, the children were indeed murdered, but the perpetrators were government agents. The advocates of these theories see a malevolent plot masterminded by the Obama administration to make a case for gun control.

I just cannot bear to ask the grieving Jeremy Richman about the most prominent purveyor of these falsehoods, the internet celebrity Alex Jones, whom Donald Trump thinks is "amazing."[6] Jones argued that "no one died" at Sandy Hook. (He was sued by the victims' families, retracted his claim, and was ordered to pay $100,000 in attorneys' fees and court costs.)

When I went to Newtown, the first falling leaves of September had begun to pile up. They slowly fluttered down from tall trees lining the park paths. A large sign next to a playground proclaimed: "Child protection area. We will call 911."

I met Kaitlin Roig-DeBellis. She was a first-grade teacher at Sandy Hook Elementary. On December 14, 2012, a fine, clear day, she was preparing to tell her pupils about holiday traditions like baking Christmas cookies, and Santa Claus. The killer entered the building at about 9:30 in the morning, after murdering his mother and taking her semiautomatic rifle and ten magazines of ammunition. He was clad in black and wore sunglasses and earplugs. He shot and killed members of the school administration. He then proceeded to the first-grade classroom of Lauren Rousseau, who had recently started as a full-time teacher. She and Rachel D'Avino, a therapist who was working with a pupil with special needs, tried to push their pupils into the bathroom to hide them, but to no avail. Both of them and fifteen of the first graders were shot dead.

Kaitlin is impressive and tough. When she heard the gunfire, she quickly pushed all her pupils into the bathroom. "Our bathroom was very small, so the fact that sixteen people including myself were going to fit in there seemed impossible," she recounts. "But it was either getting in there or not surviving." She warned the children to be very quiet.

On the other side of the wall was Victoria Soto's classroom. Victoria was murdered after pleading for the life of the children in her charge and then shielding them with her own body. Kaitlin and her pupils heard the screams and shots through the wall.

We sit on a small wooden bench in the park. "I thought I was going to die. I thought, I am going to watch each of my children be killed and then be killed," she recalls. The murderer never entered the bathroom where they were hiding, so they all survived. She tells me she had thought of the possibility that a shooter might attack the school even before it happened; she says that she was always ready for the worst-case scenario. A scenario that is most often American.

JEREMY RICHMAN AND JENNIFER HENSEL, AVIELLE'S PARents, established The Avielle Foundation in her memory. Jeremy had been a neuroscientist who investigated Alzheimer's disease; Jennifer is a microbiologist who has developed anticancer medications. The foundation's stated purpose is "to prevent violence and build compassion through neuroscience research, community engagement, and education."[7]

We sat on the lawn outside Dr. Richman's office in the center of Newtown. "Avielle was just one of those bright spirits, she could light a room by being in it. She had an infectious smile that everybody loved to have," he told me. "She wanted always to make people laugh and her best tool for that was telling a story. And she loved stories. She wanted to hear stories and as she got older she wanted to tell stories. Even if her day was mundane, she still wanted to make it a good read. 'I'm washing the dishes with Dad, how do I make it a good story?'"

We can hear the racket from the intermediate school that Avielle would have attended. "The kids laughing out there should include her. It kills me every day. I miss her a lot," he says. In the intervening years, he and Jennifer have had two more children, Imogen and Owen.

Dr. Richman spoke to me with a goal in mind—he wanted to talk about the scientific work that the foundation named after his daughter promotes. Its goal is to find out if science can provide a way for preventing violence in the US and the world. "We fund research that looks at the structure and chemistry of the brain, seeking a correlation between these and the behaviors we see in the real world. We are trying to bridge between biochemical and behavioral sciences," he told me. The accessibility of weapons in the home, he explained, is the leading risk factor for suicides and the murder of the gun owner him- or herself. He spoke at length about the data that proves it.

We met a few days after I accompanied an armed patrol of a militia in Michigan. Its members are determined never to lay down their arms, which they feel they must keep to defend themselves against impending tyranny. Some of them own dozens of weapons. "You know," I said to Dr. Richman, "when I told them about the statistics and the studies about the risk of owning guns, they said that it was Darwinism, that only idiots shoot themselves. Others said that if there had been guns in schools, maybe everything would have turned out differently. I heard Donald Trump saying that."

"Of course," Dr. Richman said bitterly. "They're not thinking, and they're not looking at data." He looked frustrated. "They are playing on fear and creating more fear, which worsens the problem and causes people to hold onto their contrived beliefs that they are safer with a firearm. Since it's so polarized, people don't want to hear the scientific data; they already have a conclusion and they're not listening."

Dr. Jeremy Richman was found dead near the Newtown city hall on March 25, 2019. He had died by suicide. Jennifer, his wife, issued a statement saying that "he succumbed to the grief he could not escape."

THE CONVERSATION WITH DR. RICHMAN HAS ECHOED IN MY mind in the years since, each time I reported on another mass shooting in the United States, on the hundreds of people who have been slaughtered because of the utter folly that is American gun policy.

I thought of him trying to address, scientifically and rationally, the horrific state of affairs that led to his daughter's murder. The American Medical Association considers gun violence a public health crisis. In the face of an inconceivable tragedy and the sadism denying the event even happened, Dr. Richman sought to promote scientific research to find a way to address that emergency, in the belief that things must and can change, and that the change will come about through an empirical and clearheaded examination of the human brain. That rational thinking, science, and notions of progress have improved the human condition and remain our best hope.

Thus, beyond the horrible personal tragedy involved, his death seems to be a warning. I hear his voice over and over in my head, telling me, "They are not listening."

Campo de' Fiori, Rome, 2018

It's evening. In this small piazza stands one of the oldest food markets in Europe. The boisterous end-of-day shouts of the produce hawkers ring through the air. At the piazza's center is a dark statue, slightly different from the classical sculptures that you see around the Italian capital. The figure gazes downward, with intense eyes, a monk's cowl over his head and a robe draping his body. His hands are crossed at his waist, holding a book. One leg steps forward, the other is ready to follow. A few wilted pink flowers lie at his feet.

The figure is Giordano Bruno. On February 17, 1600, a cheering Roman mob gathered at this site to see him burned at the stake, by order of the Inquisition. Before the fire was lit, he was tied to the stake upside down, because he had been convicted of heresy, and his tongue was nailed to the roof of his mouth so that he could not utter his sacrilegious views.

Bruno was a philosopher and a theologian. He openly accepted Copernicus's view that the earth circles the sun, asserted that there was no center to the universe, and accused the church of employing cheap tricks to gain favor with the masses. He wrote cosmological works that discussed the possibility that there were innumerable worlds and promoted mystical views at odds with the Catholicism of his time. Bruno was charged with blasphemy and heresy. At a trial that lasted seven years, he refused to recant his unorthodox claims.

Legend has it that his response to the judges who pronounced the death penalty was: “Perchance you who pronounce my sentence are in greater fear than I who receive it.”[8] His statue was erected in 1899, when the Industrial Revolution was proceeding apace. It has since become a fixture of the Roman landscape. His story is the standard one for an old hero. He was vilified, then forgotten, and later received posthumous fame. (The church has apologized for his execution.) Then, again, he was largely forgotten.

Today, scholars argue that his execution had more to do with his mystical-religious heresies, or his provocative personality, than with his views about cosmology. It no longer matters—Bruno was one of the first martyrs of science.

As I look further at the statue, I see very small scraps of paper taped to its sides. They are handwritten Italian texts criticizing Donald Trump, a direct continuation of the classical tradition of “talking sculptures,” wherein such monuments serve as places of discussion, often those that challenge the conventions of the ruling class. For these pamphleteers, Trump represents something diametrically opposed to Bruno’s devotion to the truth as he saw it, and his willingness to give his life for it.

When Bruno died, Europe was churning with a revolutionary idea—that there are objective facts that all people can agree on, no matter what their nation, tribe, or religion.

Under this new dispensation, multiple hypotheses needed to be proposed and investigated empirically. Faith in propositions handed down by religious authority or through the ages was replaced by the principle that a rational observation of the world, logic, and deduction would lead to more accurate results than would the purest of prayers. "The Scientific Revolution has not been a revolution of knowledge, it has been above all a revolution of ignorance," writes Yuval Noah Harari in his international bestseller *Sapiens*. "The great discovery that launched the Scientific Revolution was the discovery that humans do not know the answers to their most important questions."[9] It was the investigation of such questions that produced the Enlightenment, which sanctified the human.

The people who started asking such questions assumed that the world of yesterday would wither and die. As the Marquis de Condorcet wrote in 1793, "Then will arrive the moment in which the sun will observe in its course free nations only, acknowledging no other master than their reason . . . in which our only concern will be to lament their past victims and dupes, and, by the recollection of their horrid enormities, to exercise a vigilant circumspection, that we may be able instantly to recognise and effectually to stifle by the force of reason, the seeds of superstition and tyranny, should they ever presume again to make their appearance upon the earth."[10]

While the human condition improved significantly, the views of the Inquisition never quite disappeared as the Marquis de Condorcet had predicted; they only got shunted aside in the face of progress's accomplishments.

By "progress" I do not mean a deterministic notion of the linear improvement of human societies, but rather the values stemming from and informed by the Enlightenment: rational inquiry, human dignity, individual freedom, acceptance of science and technology as a means to improve the human condition, and liberalism, which protects these ideas through politics. Humans are the masters of their own fate and can, without divine intercession, achieve redemption. In the age of progress, redemption means happiness in the Aristotelian sense—achieving a state of human flourishing.

Historically, it is a tectonic shift. No longer do wisdom and knowledge flow from the ancients. No longer is history perceived as cyclical. Human knowledge, accumulated by scientific means, can be employed to help societies change, adapt, and improve. "It is pure illusion to think that an opinion that passes down from century to century, from generation to generation, may not be entirely false," wrote Pierre Bayle at the end of the seventeenth century.[11]

This constant drive for change and progress may lead to incredible cruelty. It can be committed by people who truly believe that the revolutionary nature of progress sometimes requires and justifies violence, or by those who make progress their banner while actually promoting greed, empire, or both. King Leopold II described his genocidal colonization of Congo as "a crusade worthy of this century of progress."[12] Indeed, every totalitarian regime of the previous century proclaimed itself to be at the forefront of progress, attempting to legitimize atrocities against those who stood in its way. Such actions sought to detach progress from Enlightenment values, leaving an empty and corrupted shell—better roads and factories, for instance—without a core. A country may have the brightest scientists and best-equipped hospitals for containing an epidemic, but if the ability to tell the truth about a spreading epidemic is silenced and freedom of speech oppressed, if physicians are threatened, technology may not be sufficient.

Battling Progress

The revolt against globalization is a platform for the real fight of our age—the fight over the ideas of progress themselves. Exploitation, inequality, catastrophic environmental threats, attempts to suppress or extinguish local identity, a sense of arbitrariness—all serve old foes of progress, among them fundamentalists and racist nationalists. Globalization itself, with the decentralized media and close interrelationships it has created, has armed those who disseminate rival ideas to its own liberal order. While the rivals of progress claim to represent the interests of the poor or the middle class, their real intent is more radical. Religious fundamentalists and ethnic nationalists, typically sworn enemies, have at least one thing in common—they are using the grievances of a globalized world to destroy the values that made it possible.

The revolt threatens progress in two main ways. The first is a sort of neo-Luddism. We have seen the effects of immigration and international trade, and the populist or nationalist claim that these have stolen jobs from the middle class. But the facts are clear—it is not trade or immigration that has agitated blue-collar workers in recent decades. Their jobs have largely been lost to industrialization and automation. By a conservative estimate, at least 8.5 percent of the world's industrial jobs will be eliminated by 2030, as workers will have been replaced by robots or software.[13] According to one report, that means that 20 million people will lose their jobs. No doubt they will be told that they can undergo vocational training and move into some other profession, but too often that hope is in vain.

The truth is that they will go the way of blacksmiths. In 1850, blacksmiths made up 2 percent of the American workforce. Their trade disappeared, of course.[14] As opposed to the Industrial Revolution, the information revolution and its follow-on, the artificial intelligence revolution, are not expected to create new jobs in the numbers needed to make up for those that they eliminate. Employment figures indicate as much.

In 1956, the big three automobile manufacturers—Chrysler, General Motors, and Ford—directly employed more than 900,000 workers. Today, the three Silicon Valley giants—Facebook, Apple, and Google—together employ fewer than 300,000 people, while having higher market values and revenues than the auto manufacturers ever did. And this in an America with double the population. The US Bureau of Labor Statistics estimates that a third of the country's jobs in word processing and typing as of 2018 will disappear by 2028, as will 28 percent of the jobs installing and repairing electronic equipment in motor vehicles and performing telephone operations, and 27 percent of postal worker jobs—and these are just a few examples.[15] New technologies also threaten traditions, communities, and religions. People may become neo-Luddites not only for materialist reasons; they can vilify or seek to destroy new technology out of fear that globalism is trampling local lifestyles, threatening power structures.

"In Trump's world and afterward I fear people with torches—and pitchforks," an Israeli-American Silicon Valley computer engineer told me in 2019. There is a truth behind his anxiety-fueled exaggeration. Only a small, select group has shared in the feast of burgeoning profits and salaries in the financial and technological sectors in recent decades. When you are invited to a party, you only rarely think of who was left out. These companies speak about one connected world even as they use their influence to promote weak regulation, to strangle competition from smaller entrepreneurs, and to aggressively pursue strategies for paying lower taxes. They have shirked responsibility for the incitement and verbal aggression published on their platforms while benefiting from the popularity of that violent discourse and selling advertising to appear alongside it. They have cannibalized literary, journalistic, and creative material while declining to invest significantly in the production of such content. For a single golden moment, these corporations flew under the radar of older politicians of the analog age, operating as unregulated monopolies, pursuing profit with no regard for the damage they were doing to society.

But that moment is past. As the voters and decision makers of the analog age die, their places are taken by a generation born into an online digital world. Younger people have nothing like the starry-eyed view of what older politicians still call "the internet." The backlash, already in progress, is vigorous. There are calls to break up huge tech corporations in order to maintain competition and protect democracy. Others talk about putting the brakes on technological development in general.

Fox News anchor Tucker Carlson, probably the most influential TV personality in the Trump camp, reflects this trend. He favors banning self-driving cars because "driving for a living is the single most common job for high school educated men in this country . . . the social cost of eliminating their jobs . . . is so high that it's not sustainable, so the greater good is protecting your citizens."[16]

He has a point. There are 3.5 million truck drivers in the US; they work long hours for relatively low pay and do a tough job that involves being alone for many hours on the road. They will soon face even tougher conditions because the tech giants are pulling out the stops in developing and implementing autonomous trucks that will drive themselves safely and cheaply on the highways, controlled by artificial intelligence.

Yet the same argument was made when people tried to block technological advances in the past, as when horses were replaced by tractors and when icemen went out of business with the advent of the electric refrigerator. The political phenomenon of middle-class revolt against stagnant salaries and the loss of job security is metamorphosing into specific policy proposals aimed at actually blocking technological progress.

Trump's vows to halt and even reverse the closures of coal-fired power plants are of this sort—they win him the support of the workers who have lost their jobs. But, if put into practice, such policies will impose costs on the rest of society, since the same energy can now be produced more efficiently, cheaply, and cleanly without coal.

The revolt will shift its focus from immigrants and trade to targeting technology, artificial intelligence, and robots. They, after all, are the real threat to many jobs, so a way will necessarily be found to demonize them and their creators.

THE SECOND WAY PROGRESS IS THREATENED IS MORE SUBstantive. The previous chapter showed how an overload of truth led to an implosion, actually producing more distrust and lies. The revolt creates an echo chamber in which insecurity is amplified, and in which radicals and conspiracy theorists promote their causes. Fundamentalism, an ancient enemy of progress on all accounts, has proven adept at exploiting mistrust, but so have anti-vaxxers, flat-earthers, racists, political liars, fascists of every creed, and tribalists, to name but a few. The unsustainability of the current globalized world, manifested in falling birthrates and environmental damage, is cited by some to show that progress itself has failed. Exploiting the growing distrust of social institutions, their agents can now recruit new constituencies to support their old goals. While the neo-Luddites focus on the material, these opponents of progress target consciousness.

The loss of trust in social institutions and the decline in a regard for facts have created fertile soil for extremist domestic politics in many countries, and have implications for international cooperation as well. The resolute way the world addressed the ozone depletion crisis in the 1980s contrasts starkly with the negligent and lethargic way it is addressing the current much more extreme and severe climate crisis. This march of folly is not the product of an unbending ideology that opposes Enlightenment values. Rather, it is political cynicism that seeks to avoid the electoral costs of responding to the warnings of scientists. When Trump disdains climate warming or mocks scientists who explain that some chemicals deplete ozone, he is not rebelling against the world order or trade agreements; he is challenging the assumption that science is a reliable way to explain the physical world, and he is associating scientists with left-wing politics. When Trump and his allies suggested that if the United States stops testing for the coronavirus "we'd have very few cases, if any," they were not only spouting relativistic nonsense, they were dismissing facts.

Both incendiary fundamentalism and a politics of cynicism are attacks on the tenet that reason must be the foundation of social discourse. Rational discourse itself is under attack when people say that facts are not important, and have difficulty distinguishing between real and fabricated information. It is under attack when Nazis and other racists manufacture bogus concepts of racial purity that they seek to restore, and when they and others say that what is important is not the substance of facts but the identity of the person who presents them. It is under attack when industrialists and governments disregard scientific data and continue to permit the unrestrained emission of greenhouse gases because of corporate lobbying and unhinged politicians, and when parents refuse to vaccinate their children on the basis of pseudo-information. When fundamentalists persuade people that the modern world has failed and that a fabricated piety is the answer, and when the earth barrels forward into an age of extinction that decision makers ignore, rational discourse and progress are under attack. The same is true when millions of people are trapped in the middle of an unprecedented experiment to block human movement, when they have no way to better their lives in their homelands, and when populists pursue failed economic policies that cause horrifying damage yet continue to be reelected.

What began as a justified revolt against globalization has mutated into a rejection of progress itself.

People chose modern over medieval medicine because it worked. They largely stopped using dowsing as a way to determine where to dig a well, because geologists were better at predicting a successful search for water. Agricultural researchers provided them with better seeds and fertilizers that increased yields and drastically reduced the danger of famine. Societies replaced absolute monarchy, and the belief that kings were anointed by God, with democracy. Democracy provided them with better government and happier lives. It is the prospect of increased revenue that impels a factory to replace a less efficient human production line with a more efficient robotic one—it simply works better. Science advances because it works, because societies, economies, and political structures have a vested interest in research and discovery.

Max Weber wrote that modernity "means that principally there are no mysterious incalculable forces that come into play, but rather that one can, in principle, master all things by calculation. This means that the world is disenchanted."[17]

But today's world is caught up in a different form of enchantment, of a human sort. Success is the powerful spell that progress casts. Success has engendered a belief that the world always progresses. Cities have been built on uncultivated land and swamps. Libraries have been built to serve previously illiterate populations. Our lives are easier than those of our grandparents. The world gets better—inexorably, inevitably.

In a broad sense, this myth is grounded in the facts of the last two centuries. The creation of wealth and the increase in longevity justify the feeling that progress is inexorable and that humankind will continue to move forward. Yes, we have told ourselves, from time to time violent storms have raged, like those of fascism and Nazism, but they have been defeated. Democracy, the form of government associated with progress, may not be perfect, but it is the most resilient, and will eventually triumph.

That too is a myth. We have no real reason to believe that because humanity has advanced so far in the last two hundred years it will continue to do so, or that it will do so at the same rate. Sometimes people, as communities, choose to return to a life that they think of as traditional, more efficient, racially pure, or pious, but which is retroactively exposed as irrational, or one that led to violent malevolence.

Success is the enchantment of our age, but when societies grapple with severe crises that call their fundamental values into question, the spell can be broken. It is replaced by an imagined past, a sense of superiority, or an addiction to lies. Think of a country like Iran prior to the Islamic revolution of 1979. The old regime was corrupt and oppressive, and Iranian society was hugely unequal, but the oppression of women was easing, the economy was growing, and the educational level of the public rising. Then came the revolution. The economy and civil society were subordinated to a rigid theocracy.

Turkey accomplished a shift from rule by a military autocracy to limited democracy. In the past decade, though, it has swung back and become an autocracy led by a ruling party with Islamist characteristics.

Life expectancy for males in Russia is lower by a decade than it is in developed countries. It has risen by only four years since the 1960s. There is widespread alcoholism, and the education and health systems are defective, compared to other industrial nations.

The fundamental malaise of American society demonstrates the extent to which leaders can drag communities backward, or even poison them, as happened in Flint, Michigan, and many other places. Americans traded their prosperous industrial state for a kind of financial oligarchy.

These are all examples of failures, yet the most acute one is environmental. Jared Diamond and Ronald Wright have portrayed vividly, in their books *Collapse*[18] and *A Short History of Progress*,[19] how the natural environment has been in the past the most important determinant of the stability and development of civilizations. The inference from this is that humanity's exploitation of natural resources today could lead to collapse and regression. Wright warns that population growth in earlier times led to a depletion of resources that occurred more quickly than nature could sustain. Diamond offers detailed accounts of civilizations that have collapsed because they grew unsustainably, exhausting their resources. The process could not be halted in time because of the "progress trap," a chain of successes that, when it reaches a certain level, leads to disaster and self-destruction.

Both of these writers maintain that the ever-closer bonds between countries in the Age of Globalization have produced a situation in which civilization itself is "worldwide," as Wright puts it. The corollary is that either *Homo sapiens* will succeed together or that we will experience a global collapse.[20] Glaciers are melting, species are going extinct, coral reefs are bleaching, sea level is rising, desertification is advancing. Half of the world's topsoil has disappeared in the past 150 years as a result of modern farming methods; we grow 95 percent of the world's food on what remains.[21] The denial of these facts is tantamount to turning our backs on the Scientific Revolution, precisely at the time when humankind desperately needs science.

Progress only seems like an unstoppable natural force whose bounty we enjoy. But it can only survive in a greenhouse of rational politics that respects natural and ecological resources and, no less important, rejects deliberate lies. What has been forgotten needs to be remembered again—that there is something fundamentally healthy about fearing that a great and devastating war could easily break out.

War is the ultimate danger that looms over a world that possesses the knowledge and the means to destroy itself. The Polish poet and Nobel laureate Czeslaw Milosz, witnessing the Warsaw Ghetto being burned by the Nazis in 1943, thought of another fire, that to which the Church consigned Giordano Bruno in Rome:

On this same square
they burned Giordano Bruno.
Henchmen kindled the pyre
close-pressed by the mob.
Before the flames had died
the taverns were full again,

baskets of olives and lemons
again on the vendors' shoulders.

I thought of the Campo dei Fiori
in Warsaw by the sky-carousel
one clear spring evening
to the strains of a carnival tune.
The bright melody drowned
the salvos from the ghetto wall,
and couples were flying
high in the cloudless sky.

At times wind from the burning
would drift dark kites along
and riders on the carousel
caught petals in midair.
That same hot wind
blew open the skirts of the girls
and the crowds were laughing
on that beautiful Warsaw Sunday.[22]

CHAPTER 21

A New Story

Before his tenure as secretary of defense, Gen. James Mattis commanded the First Marine Division in Iraq. One summer day in that country's western desert, soldiers under his command apprehended a young man as he was laying a roadside bomb that was intended to kill the general as his motorcade passed by. Hearing that the captive spoke English, Mattis went to speak with him. Over coffee and a cigarette, the would-be assassin explained that he wanted to kill the American commander and the marines accompanying him because they were foreign invaders on Iraqi soil. I understand how you feel, Mattis told the man, informing him that he would be sent to the infamous Abu Ghraib prison. "General," the man asked Mattis, "if I am a model prisoner, do you think someday I could emigrate to America?"[1]

The young Iraqi was fighting against infidel foreign invaders. The American general believed that he had been sent to Iraq to liberate it, while also protecting his own country against terrorists. The young man was being sent to prison, following a cup of coffee and a cigarette granted to him by his magnanimous captor. But the general's empire would eventually retreat and end its catastrophic occupation.

America is a dream for so many. The Iraqi was seeking not the US as it is but rather the shining city on a hill. That is still what the word "America" evokes in those of us who live outside it. The young man craved to live in a place of freedom, abundance, and opportunity. At present, the only model that has offered that in the entire history of human civilization is liberal democracy and the Enlightenment, and the progressive values on which both rest.

The young man who wanted to kill an American general but who also wanted to be American embodies the current radical moment. His wish evokes both how intrusive and alienating the liberal order is to local communities and power structures and how great its allure is.

Home

This book has been a journey to disparate places and an account of events over more than a decade. The people whose stories I have told—American workers, unemployed Greeks, Chinese entrepreneurs, Syrian refugees—walk a narrow path between hope and fear. I was a journalist who met them for a moment on their path, sometimes at a moment of disillusion, sometime in a moment of flight. Where is Lilan now, the seventeen-year-old Syrian refugee with the treble clef tattoo on her neck? What happened to Riyad, who fled ISIS and arrived in Europe with business cards at the ready, prepared in a heartbeat to start a job as a computer programmer, if he could find one? Have the Greek anarchists preparing Molotov cocktails returned to their erstwhile middle-class lives, or are they still holed up in an abandoned house somewhere, plotting another attack on the capitalist beast? What about the starving African penguins I saw being force-fed at the rescue clinic in Gansbaai—how many of them are free and back in the ocean?

At darker moments I think about the neo-Nazis and other racists I met in the trenches of the revolt. Their fantasies have started to come true. The world has become more divided and disconnected. Their propaganda is inspiring more mass murders. What are they planning now, as the West's sun sinks toward the horizon and the shadows grow longer?

In the past, these questions would have remained unanswered. In today's connected world, however, it's pretty easy to find out, via Facebook or WhatsApp. Those who wish to do so have continued to update me on their lives, the celebrations and the disasters. Riyad from Kobani now has a job in Germany as a programmer. He knows he is lucky to have found his place there, but also says he feels increasingly alienated by the society he lives in. Lilan resides in Holland and is pursuing her studies; she doesn't want to talk about her flight from Syria. Constantine Plevris, the writer of anti-Semitic screeds, is pleased as he watches men he has mentored take up positions in Greece's political leadership. Sampath Ekanayaka of Sri Lanka continues to promote electronic fences to block and save elephants, all the while remaining pessimistic about their ultimate fate.

Jeremiahs, prophets of doom, have not chalked up a good record over the last two centuries. Humanity has seen two world wars, the Holocaust, and other genocides, and yet optimists have again and again been more accurate in their predictions than have the doomsayers. Those who have predicted another Great Depression have been confounded. Those who trembled during the Cold War that the human race was about to destroy itself, and who said that the best approach was thus "eat, drink, and be merry, for tomorrow we die," discovered that the world not only survived but largely prospered in a new world order. The augurs of destruction who built nuclear shelters and hoarded gold paid a high price for their anxiety. It was the risk-takers, positive thinkers, and internationalists who prevailed.

Yet something similar happened to those who maintained that globalization and the values it disseminates would bring about a single and united world, eradicate poverty, or embed supranational universalism and Western democracy everywhere. We see globalization's destructive impact on the environment, on parts of the middle class, and on the culture, traditions, and stability of communities in many places. The optimists did not see that, for many, globalization felt violently arbitrary. Principally, whether because of a sense of superiority or out of fear, they have disregarded the importance of meaning and identity in people's lives. They were not as wrong as the fearmongers, but they were very wrong nonetheless. The revolt proves that.

An Age of Reform

These are fluid, volatile times. The sovereignty of nation-states is eroding, even as their ability to control, monitor, and manipulate their citizens grows. The world is more interconnected, while the barriers to migration are in some respects more effective than ever. It is a world built in and by an era of moderate politics, but the people born into this prosperity are increasingly voting for extremists. At a time when it has become easy, fast, and cheap to distinguish truth from lies, more and more people are certain that they are being lied to—and freely admit to spreading lies themselves. We are by every material measure better off than ever before, yet many people feel they have fewer opportunities. A society insisting on optimism, one that boasts of its achievements, nevertheless lives in fear of the collapse of civilization.

Industrialization, globalization, and the institutions that have emerged out of them have lifted a billion people from dire poverty, saved the lives of more than 100 million children, and fashioned international norms that have reduced conflicts between states to their lowest levels in modern times. Yet in many places, when global economic forces came, exploited, and left with no regard for local identities and grievances, they sowed the seeds of revolt. While largely advancing communities, globalization also created the social climate that strengthened those who seek to destroy it.

The world needs reforms on a global scale, a major rethinking of the global economy and the way countries manage their internal affairs. With the revolt against globalization morphing into a crusade against progress itself, the window of opportunity for such reforms will soon close.

All reforms must be predicated on the truism that the world and its people are interdependent, and that there is no going back. Even if nationalists will triumph and eradicate the liberal order, the world will remain connected—it will just lack effective institutions to manage the connections. Mutual responsibility is not a naive pipe dream—it is an essential need. "In a real sense we are all caught in an inescapable network of mutuality, tied in a single garment of destiny" said Martin Luther King Jr. "What affects one directly affects all indirectly."[2] The spread of the COVID-19 coronavirus at the beginning of 2020 has highlighted the extent of the globalization crisis. The viral aspects of the globalized world have suddenly become immediate, real, and lethal; closing borders became an argument for public health, not mere xenophobia or protectionism. The spread of the epidemic also highlighted the lack of international tools and rules to ensure safety and sustainability. In a world economy based on trade and massive air traffic, international institutions still lack the power to quickly intervene in sovereign states to monitor, investigate, and prevent the spread of dangerous pathogens. In 2008 it was a financial crisis, and in 2020 an aggressive pandemic that challenged the world's stability. Next time it could be a local conflict escalating into a global one. We live in an era of globalization but do not have the global responsibility and power necessary to manage it.

Many remedies and reforms have been proposed over the last two decades to tackle fundamental flaws in the current globalized order. This chapter cannot present them all; the point is to show that there are solutions for the taking.

The world needs to adopt binding protocols to address humanity's most daunting challenge, climate change. It is only logical that the West pay much more than it has up till now for the pollution and greenhouse gases it has emitted, and the way these caused disproportionate damage to the global South. Even more important, the world is in dire need of a new economic model. The concept of a circular economy, based on getting the maximum use out of resources through more reuse and less waste, offers important possibilities. Even if concerted measures are taken to counter the climate crisis, we will still need a coordinated international apparatus to address the needs of large numbers of people who may become climate refugees as a result of the expanding crisis. A global plan to address the loss of biodiversity and the extinction crisis is vital. It must adequately compensate poor countries for setting aside parts of their territories as essential ecological preserves.

Changes are urgently needed in international political bodies. Today, the Security Council, the UN's highest body, has five permanent members (the US, Russia, Britain, France, and China, representing the victorious powers of World War II) that wield veto power, giving them disproportionate control over the fifteen-member body. The Security Council needs reform, either by eliminating these special privileges or by granting representatives of Africa and Central and South America similar prerogatives. Countries on those continents should be able to exercise more influence over collective international decisions; it is no longer tenable for the UN to be organized on the basis of the outcome of the last world war.

America's retreat from its obligations and responsibilities as a superpower is as dangerous to the world as it is to itself. Isolation is not a wise choice for the world's biggest economy, especially given the US national debt and the country's interest in having the dollar continue to serve as the world's currency. The US has been weakening international institutions such as the United Nations and its organizations. These institutions, as well as the World Bank and the International Monetary Fund, were instrumental in ensuring global stability and prosperity, and are in dire need of increased American support.

An urgent reform in global trade rules is in order, one that will allow weak countries to export easily produced or agricultural goods to stronger countries. An attempt at such reform was made at the 2001–2008 Doha round of trade negotiations among the members of the World Trade Organization, but it failed.

Multinational corporations ought not to continue to enjoy moral and legal immunity, as manifested in their evasion both of taxes and of responsibility for the consequences of the sale of their products. The large internet companies are facing increasing criticism. Given their size and influence and the constant threat they pose to privacy and democracy, they should be broken up, or a legal structure should be put in place to address, on a case-by-case basis, the complex monopolies they have constructed.

Generally, if countries continue to engage in a race to the bottom on corporate and wealth tax in order to attract corporations and billionaires, their powers will continue to erode even as corporations become more important players on the world stage, taking the place of representative institutions. A global corporate and individual tax regime is needed to prevent this. The inequality in tax burden between the middle class and the top 1 percent is as malignant to capitalism as it is to an idea of a just society, and it is a petri dish for extremists of every ilk.

Global tax reform is needed also, because of what huge corporations are doing to societies beyond the prosperous global North. Many in Western countries remain oblivious to how huge corporations such as Facebook profit from local societies outside Europe and the United States, even as they enable a radicalization of the political discourse there in dangerous ways. They do so to the detriment of businesses employing local residents, while paying almost no taxes in these places. This cannot be corrected by case-by-case arrangements; in a world in which profits are made globally and by online transactions, taxes need to be paid to governments representing the people from which these revenues were earned.

Capitalism will need to adjust for societies with a shrinking population, and that is terra incognita. The migration crisis is an opportunity for the countries of the West. In a globalized world with a plummeting birthrate, countries need to develop and adopt an immigration ethos. It is a practical need. If the West nevertheless wants to keep immigration to a minimum, the best way to accomplish it is by making enormous investments in the countries the migrants are fleeing, in order to create more tolerable conditions for those populations. Even if the West is not prepared to accept responsibility for the damages and injustices caused by colonialism, it must, for its own good, invest larger sums in development and aid to poor countries.

These are only examples. None of them is particularly new or surprising, but they bear restating in the context of this book; they are largely common sense, but that does not make them any less necessary.

A New Narrative

Globalization remains unstable and defective, but not for lack of ideas about how to improve jurisdiction and authority on a global scale, reform trade policy, promote more effective development in the countries of the global South, and create a more environmentally sustainable economy. The fundamental problem is that there is little electoral will to fight for globalization of any sort.

Leaders and policy makers know that there are plans and initiatives that can help make the world better and more stable, but they are seldom prepared to risk their political capital by attempting to implement them. If they did, they would likely be swept out of office by a tidal wave of populism. They cannot place their hopes on the weakened mainstream. For the left, globalization and the practices it spreads are tainted by exploitation; for the right, these are a threat to communitarian values. The age of responsibility is over, and leaders, and the public behind them, were not witness to, and do not carry the scars of, a world war. Talk about international cooperation repels many; caution has been cast aside and replaced by adventurism. Enlightenment values are at best taken as a given; at worst, they are seen as an instrument of a de facto dictatorship, of an imagined deep state. Like Rome before its fall, globalization lacks not philosophers but warriors. In democracies, the warriors are voters.

Globalization itself is deeply unsustainable. It's difficult to transform it when the story that it offers is manifestly not credible: "We're all going to be winners. We're all a village!" People in Congolese coltan mines and in Detroit's worst neighborhoods already know that this sanguine description of trickle-down globalization is a fabrication, a fraud. The alternative narrative is cruel and spiritless. While industrialization and the laws of supply and demand are certainly better explanations of how our world functions, who will unite and fight under their banner? Only those in the topmost percentile.

Both of these narratives of globalization are deeply flawed. One is utterly false; the other cannot garner popular backing. The public does not trust the first and cannot support the second. And they both threaten all that is local. Communities suffer from insecure jobs, wage stagnation, pollution, from the erosion of their land and the rise of the sea. Religious and national identity are derided as passé. People increasingly believe that their institutions lack real power to act against the terrifying arbitrariness of global forces. Individuals are being expected to abandon the things that anchor their lives and adopt an alienating universal system dictated, for example, by the Federal Reserve's interest policy. The present global order, which is a political construct, presents itself as a force of nature to which citizens are expected to surrender, accepting whatever beneficence it might offer them. But people know that globalization and its values are neither preordained nor eternal, and that its institutions can be constrained and sometimes even dismantled. Industrialization and interconnectedness are not the wind, sun, and waves.

The challenge is clear, then. It means not only finding new and imaginative ways to reform a globalized world but also nourishing the motivation to do so. It means harnessing the energy of the revolt and directing it toward reform. The liberal order and the globalization to which it gave birth need a new narrative that is realistic and that is not afraid to refashion the mainstream in a radical way.

In this century, people will fight for their nations and religions. But will they also fight for the ideas of freedom, science, an interconnected world providing reciprocal benefits, and for universal values—and can they be persuaded that such things are worth fighting for?

The contours of this new story are already clear. Sovereign states are not the enemies of the global world, nor of universal values. The claim that patriotism is irreconcilable with universal interests presents a manifestly false dichotomy. It is a surrender to the nationalists. The international community must provide the means for nations to survive and flourish in a complex globalized world and avoid becoming failed states. The current international order has developed, to an extent, tools for intervening in conflicts between states, but it has no effective response to their internal collapse. When countries descend into civil war and mass murder, it has an effect on the entire world, and as global integration expands, so will the need for international intervention in conflicts that were once seen as local or regional.

Given the increasing rate of technological change, perhaps the most daunting problem that countries will face is unemployment. Bold decisions need to be made with regard to retirement age, pension systems, social security, and health.

The fundamental assumption ought to be that societies undergoing an accelerating technological revolution need to maintain better social safety nets. They need to do more, not less, to sustain civil cohesion. Higher tax rates will be necessary, along with new forms of taxes—for example, a progressive consumption tax. In a world of accelerating global interconnectedness, more solidarity is imperative.

Neither London nor Manhattan will crumble if a solidarity tax, to be used for infrastructure investment and education in rural areas, is levied on wealthy urbanites. If economic elites continue to ally themselves with nationalist or ultraconservative forces, if they do not allow these tax hikes and instead leave so many behind—the backlash will only intensify.

As a political idea, globalization must not dissipate identity, locality, and traditionalism. On the contrary, it should sanctify them. National sentiment is not the enemy; it can be a guarantor of liberalism. In a world with expanding homogenization, elites should embrace those who wave their national flags, not revile them as nationalists.

Communities cannot be pushed to accept immigrants without openly discussing an ethos of immigration; they cannot be totally at the mercy of international corporations seeking quick profits and an even quicker exit. When people feel sure that their identities will be respected, they can be inspired to fight for larger principles. Classic conservative liberalism is the sort of conservatism that rejects empire to protect the soul of the colonizing nation, a conservatism that promotes doing good in the name of traditional decency, and not only because of human rights. It has to be fused to the narrative of globalization.

Such a version, and vision, of globalization must not only do well on average. It must benefit and focus specifically on those communities in danger of being flattened by its force. Only a story focused on justice can unite and empower.

If we simply recite over and over the successes we have achieved through progress, we'll be back in November 2016 again. The Enlightenment's achievements are irrefutable, but reminding people of that will change nothing politically. Even worse, it further alienates the Enlightenment's skeptics. If people become convinced that the present state of affairs, in which they find themselves lost in a maze of alienation, is the epitome of progress, they are apt to conclude that opposing progress will gain them their freedom. They may reason that the way out of the maze is to go backward, not forward. Those who accept the ideas of the liberal order, left and right, are accustomed to being in the majority; it has eroded. They need to start thinking of themselves as an ambitious opposition. After years with Donald Trump as the most powerful man in the world, after the protests following the murder of George Floyd and others and the rise of grassroots environmental movements, the ground is beginning to shake.

"The conflict has been exciting, agitating, all-absorbing, and for the time being, putting all other tumults to silence. It must do this or it does nothing. If there is no struggle there is no progress," said Frederick Douglass not long before the American Civil War.[3] His words remind us that the way to guarantee progress is not by a nostalgic evocation of its benefits or sanctifying the status quo. Rather, it is by engaging in a struggle and demonstrating a willingness to change.

The revolt is everywhere. It is sweeping away the residue of the age of responsibility. The opportunity that this radical moment offers is greater than the dangers it portends—it is a chance to shape a more just and sustainable world. The aim is not to preserve the home that the previous age built but to replace it with a better, more viable one. It's on us.

—GANEY TIKVA, ISRAEL
PINAKATES, GREECE
2020

Acknowledgments

The volume before you is the product of journalistic work over some two decades, so I am indebted to the print and broadcast outlets where I work and have worked in the past. They allowed me enormous flexibility as I worked on the book while reporting and filing fast-moving news stories. Israel's News 10 and Reshet News 13 television, the newspapers *Yedioth Ahronoth* and *Ma'ariv*, and *Liberal* magazine all allowed me to use material from my columns and television reports, and I thank the editors at each of them. I am grateful to the two heads of Yediot Books publishers, Dov Eichenwald and Eyal Dadush, who were largely responsible for the success of the Hebrew version of this book. Deborah Harris is my literary agent but much more than that. Without her wisdom and determination, I would not have succeeded in turning my manuscript into a book.

My English translator, Haim Watzman, made astute comments and suggestions, and flagged errors. George S. Eltman made comments and corrections, and significantly improved the substance of this work with his brilliant perceptions. Inbal Asher checked facts and sources, but also made incisive comments on the text itself.

Dafna Maor, the editor of the Hebrew edition, who is also the international news editor for *Ha'aretz*, played a very important role in my work. Further investigative work was conducted by Noa Amiel Lavie and Inbar Golan. Many friends and experts assisted, added material, and corrected errors, among them Anshel Pfeffer, Prof. Liad Mudrik, Yair Assulin, Dr. Tomer Persico, Dr. Ori Katz, Prof. Uri Shanas, Dr. Sefy Hendler, Prof. Yoav Yair, Dr. Yuval Dror, Prof. Omer Moav, Saikat Datta, Dr. Jeremy Fogel, Ruti Koren, Hilik Sharir, Ariel Elgrabli, Emmanuelle Elbaz-Phelps, Gali Bartal, Orit Kopel, Antonia Yamin, David Agasi, Neta Livne, Dr. Noam Gidron, and Barak Ravid. Any errors found herein are mine alone.

The most important person is the one who gave me encouragement all along the way, my patient and loving wife, Tamar Ish Shalom. She was the first and most important reader of every word. I am grateful beyond words to her and our children, Zohar, Hilel, and Naomi, who sacrificed that most valuable resource on earth, family time.

Notes

INTRODUCTION: THE DEATH OF AN AGE

1. Howard J. Langer, ed., *World War II: An Encyclopedia of Quotations* (Abingdon, UK: Routledge, 2013), 39.
2. William A. Lydgate, "My Country, Right or Left?" *The Magazine of the Year*, 1947, http://www.oldmagazinearticles.com/cold_war_opinion_poll-pdf.
3. Ibid.
4. Edward T. Imparato, General MacArthur: Speeches and Reports, 1908–1964 (Nashville, TN: Turner, 2000), 192, 247; General Douglas MacArthur, radio broadcast from the battleship USS *Missouri*, September 2, 1945, *Missouri* battleship memorial, https://ussmissouri.org/learn-the-history/surrender/general-macarthurs-radio-address.
5. Martin W. Sandler, ed., *The Letters of John F. Kennedy* (New York: Bloomsbury, 2013), 230.
6. Roberto Stefan Foa and Yascha Mounk, "The Signs of Deconsolidation," *Journal of Democracy* 28, no. 1 (2017): 5–15.
7. "Trends in Armed Conflict, 1946–2017," Peace Research Institute Oslo (PRIO), May 2018, https://www.prio.org/utility/DownloadFile.ashx?id=1698&type=publicationfile.
8. Max Roser and Esteban Ortiz-Ospina, "Literacy," in "Our World in Data, 2019" (data sources: OECD; UNESCO), https://ourworldindata.org/literacy.
9. Tomas Hellebrandt and Paolo Mauro, "The Future of Worldwide Income Distribution," Peterson Institute for International Eco-

nomics, Working Paper Series 15–7, 2015 (data sources: OECD; Consensus Forecasts; IMF/World Bank; authors' forecasts for growth; United Nations for population projections; Luxembourg Income Study and World Bank for household survey data on income distribution).

10. Foa and Mounk, "The Signs of Deconsolidation."

CHAPTER 1: AN ATTACK ON A NEWSPAPER

1. "Poverty and Shared Prosperity 2018: Piecing Together the Poverty Puzzle," World Bank, 2018, https://openknowledge.worldbank.org/bitstream/handle/10986/30418/9781464813306.pdf.
2. Francisco Alcalá and Antonio Ciccone, "Trade and Productivity," *Quarterly Journal of Economics* 119, no. 2 (2004): 613–46; Steven N. Durlauf, Paul A. Johnson, and Jonathan R. W. Temple, "Growth Econometrics," *Handbook of Economic Growth* 1 (2005), pp. 555–677.
3. James C. Riley, "Estimates of Regional and Global Life Expectancy, 1800–2001," *Population and Development Review* 31, no. 3 (2005): 537–43; Richard A. Easterlin, "The Worldwide Standard of Living Since 1800," *Journal of Economic Perspectives* 14, no. 1 (2000): 7–26.
4. "How Has Life Expectancy Changed over Time?," Decennial Life Tables, Office for National Statistics (UK), September 9, 2015, https://www.ons.gov.uk/peoplepopulationandcommunity/birthsdeathsandmarriages/lifeexpectancies/articles/howhaslifeexpectancychangedovertime/2015-09-09; Max Roser, "Life Expectancy," in "Our World in Data, 2019" (data source: Human Mortality Database, University of California), https://ourworldindata.org/life-expectancy.
5. Max Roser and Esteban Ortiz-Ospina, "Global Extreme Poverty," in "Our World in Data, 2019" (data source: François Bourguignon and Christian Morrisson, 2002), https://ourworldindata.org/extreme-poverty.
6. Max Roser, "Child Mortality," in "Our World in Data, 2019" (data sources: Gapminder; World Bank), https://ourworldindata.org/child-mortality.
7. Martin Ravallion, "The Idea of Antipoverty Policy," Working Paper 19210, National Bureau of Economic Research (US), 2013, https://www.nber.org/papers/w19210.pdf.

8. Arthur Young, 1771, quoted in Edgar S. Furniss, *The Position of a Labourer in a System of Nationalism: A Study in the Labor Theories of the Later English Mercantilists* (Boston and New York: Houghton Mifflin, 1920), 118.
9. Bernard De Mandeville, "An Essay on Charity and Charity Schools," in *The Fable of the Bees: Or, Private Vices, Publick Benefits*, 3rd ed. (J. Tonson, 1724; reprinted from 1714), 328.
10. Philippe Hecquet, 1740, quoted in Daniel Roche, *The People of Paris: An Essay in Popular Culture in the Eighteenth Century* (Berkeley: University of California Press, 1987), 64.
11. Immanuel Kant, "What Is Enlightenment?" in *Eighteenth-Century Answers and Twentieth-Century Questions*, ed. James Schmidt (Berkeley: University of California Press, 1996), 58.
12. Karl Marx and Friedrich Engels, *The Communist Manifesto* (New York: Simon & Schuster, 2013; reprinted from 1848), 63.
13. Clayton Roberts, David F. Roberts, and Douglas R. Bisson, *A History of England, Volume 2: 1688 to the Present,* 6th ed. (Abingdon, UK: Routledge, 2016), 357.
14. David Mitch, "The Role of Education and Skill in the British Industrial Revolution," in Joel Mokyr, *The British Industrial Revolution: An Economic Perspective*, 2d ed. (Boulder, CO: Westview Press, 1998; reprinted from 1993), 241–79; Sascha O. Becker, Erik Hornung, and Ludger Woessmann, "Education and Catch-up in the Industrial Revolution," *American Economic Journal: Macroeconomics* 3, no. 3 (2011): 92–126.
15. Max Roser and Esteban Ortiz-Ospina, "Primary and Secondary Education," in "Our World in Data, 2019" (data sources: OECD and IIASA, 2016; Wittgenstein Centre for Demography and Global Human Capital, 2015), https://ourworldindata.org/primary-and-secondary-education.
16. Oded Galor and Omer Moav, "Das Human Kapital: A Theory of the Demise of the Class Structure," *Review of Economic Studies* 73 (2006): 85–117.
17. Roser and Ortiz-Ospina, "Global Extreme Poverty."
18. Voltaire, "Défense du Mondain ou l'apologie du luxe," 1736, in Theodore Besterman, *Voltaire's Notebooks* (Geneva: Voltaire Institute and Museum, 1952), 244.
19. Gregory Clark, "Introduction: The Sixteen-Page Economic History of the World," in *A Farewell to Alms: A Brief Economic*

History of the World (Princeton, NJ: Princeton University Press, 2007), 1.

20. Angus Maddison, *The World Economy* (Paris: OECD, 2003), 263.

CHAPTER 2: SHOWERING TWICE A MONTH

1. "GDP per Capita (current US$)—China," World Bank, 2019, https://data.worldbank.org/indicator/NY.GDP.PCAP.CD?locations=CN.
2. Hu Angang, Hu Linlin, and Chang Zhixiao, "China's Economic Growth and Poverty Reduction (1978–2002)," in *India's and China's Recent Experience with Reform and Growth*, eds. Wanda Tseng and David Cowen (Basingstoke, UK: Palgrave Macmillan, 2005), 59–90.
3. "China–Systematic Country Diagnostic," World Bank, 2017, http://documents.worldbank.org/curated/en/147231519162198351/pdf/China-SCD-publishing-version-final-for-submission-02142018.pdf, 20.
4. "Literacy Rate," Data for Sustainable Development Goals—China, UNESCO, 2018, http://uis.unesco.org/en/country/cn#slideoutmenu.
5. "Trends in Under-Five Mortality Rate," Key Demographic Indicators—China, UNICEF, 2018, https://data.unicef.org/country/chn/.
6. Quoted in Susan Whitfield, *Life Along the Silk Road* (Berkeley: University of California Press, 1999), 21.
7. Valerie Hansen, *The Silk Road: A New History* (New York: Oxford University Press, 2012), 9–10, 139.
8. Moahn Nair, "Understanding and Measuring the Value of Social Media," *Journal of Corporate Accounting & Finance* 22, no. 3 (2011): 45–51.
9. Richard Dobbs, James Manyika, and Jonathan Woetzel, "The Four Global Forces Breaking All the Trends," in *No Ordinary Disruption* (New York: Public Affairs and McKinsey Global Institute, 2015).
10. Paul Hirst and Grahame Thompson, "Global Myths and National Policies," in *Global Democracy: Key Debates*, ed. Barry Holden (Abingdon, UK: Routledge, 2000), 50.
11. Esteban Ortiz-Ospina, Diana Beltekian, and Max Roser, "Trade and Globalization," in "Our World in Data, 2018" (data source: Giovanni Federico and Antonio Tena-Junguito, 2016), https://

ourworldindata.org/trade-and-globalization#trade-has-grown-more-than-proportionately-with-gdp.

12. "Global Citizenship a Growing Sentiment Among Citizens of Emerging Economies: Global Poll," Globescan for BBC, April 27, 2016, https://globescan.com/wp-content/uploads/2016/04/BBC_GlobeScan_Identity_Season_Press_Release_April%2026.pdf, 1, 4.
13. Eric C. Marcus, Morton Deutsch, and Yangyang Liu, "A Study of Willingness to Participate in the Development of a Global Human Community," *Peace and Conflict: Journal of Peace Psychology*, 23, no. 1 (2017): 89–92.
14. Anthony Elliott, *Contemporary Social Theory: An Introduction* (Abingdon, UK: Routledge, 2014), 322–28.
15. Sugata Mitra, "Self-Organising Systems for Mass Computer Literacy: Findings from the 'Hole in the Wall' Experiments," *International Journal of Development Issues* 4, no. 1 (2005): 71–81.
16. Alvin Toffler, *Future Shock* (New York: Random House, 1970), 413–18.
17. Ian Johnson, "Chinese Activists Continue Calls for Protests," *New York Times*, February 25, 2011, https://www.nytimes.com/2011/02/26/world/asia/26china.html.
18. Andrew Jacobs and Jonathan Ansfield, "Catching Scent of Revolution, China Moves to Snip Jasmine," *New York Times*, May 10, 2011, http://www.nytimes.com/2011/05/11/world/asia/11jasmine.html.
19. *Life*, February 17, 1941, 65.
20. *The Bhagavad Gita*, trans. Juan Mascaro (New York: Penguin, 1962), 92.

CHAPTER 3: THE GLOBALIZATION WARS

1. Edward Wong, "In China, Breathing Becomes a Childhood Risk," *New York Times*, April 22, 2013, http://www.nytimes.com/2013/04/23/world/asia/pollution-is-radically-changing-childhood-in-chinas-cities.html.
2. Delin Fang et al., "Clean Air for Some: Unintended Spillover Effects of Regional Air Pollution Policies," *Science Advances* 5, no. 4 (2019), https://advances.sciencemag.org/content/5/4/eaav4707/tab-e-letters.
3. Celia Hatton, "Under the Dome: The Smog Film Taking China by Storm," BBC, March 2, 2015, http://www.bbc.com/news/blogs-china-blog-31689232.

4. "Air Pollution," World Health Organization, 2018, https://www.who.int/airpollution/en/.
5. "The Cost of a Polluted Environment: 1.7 Million Child Deaths a Year," World Health Organization, March 6, 2017, http://www.who.int/mediacentre/news/releases/2017/pollution-child-death/en/.
6. "9 out of 10 People Worldwide Breathe Polluted Air, but More Countries Are Taking Action," World Health Organization, May 2, 2018, https://www.who.int/news-room/detail/02–05–2018–9-out-of-10-people-worldwide-breathe-polluted-air-but-more-countries-are-taking-action.
7. Qiang Zhang et al., "Transboundary Health Impacts of Transported Global Air Pollution and International Trade," *Nature* 543 (2017): 705–9, https://doi.org/10.1038/nature21712.
8. Ibid., 708–9.
9. "The Air Quality Life Index," Energy Policy Institute at the University of Chicago (EPIC), https://aqli.epic.uchicago.edu/the-index/.
10. Immanuel Wallerstein's work on the dynamics of periphery and metropolitan areas in international trade and the international economy is the seminal work on these issues: Immanuel Wallerstein, *World-Systems Analysis: An Introduction* (Durham, NC: Duke University Press, 2004).
11. Jing Meng et al., "The Rise of South–South Trade and Its Effect on Global CO2 Emissions," *Nature Communications* 9, no. 1 (2018): 1–7, https://www.ncbi.nlm.nih.gov/pmc/articles/PMC5951843/.
12. Derek Thompson, "The Economic History of the Last 2,000 Years in 1 Little Graph," *The Atlantic*, June 19, 2012, https://www.theatlantic.com/business/archive/2012/06/the-economic-history-of-the-last-2–000-years-in-1-little-graph/258676/.
13. Gottfried Wilhelm Freiherr von Leibniz, *The Preface to Leibniz' Novissima Sinica: Commentary, Translation, Text*, ed. and trans. Donald Frederick Lach (Honolulu: University of Hawaii Press, 1957; original text published 1699), 69.
14. Emperor Qianlong, Letter to George III, 1793, in Harley Farnsworth MacNair, *Modern Chinese History: Selected Readings* (Shanghai: Commercial Press, 1923), 4–5.
15. Nick Robins, *The Corporation That Changed the World: How the East India Company Shaped the Modern Multinational* (London: Pluto Press, 2006), 152.
16. Hsin-pao Chang, *Commissioner Lin and the Opium War* (Cambridge, MA: Harvard University Press, 1964), 172–79.

17. Lin Zexu, "Letter to the Queen of England," in *The Chinese Repository*, Vol. 8 (Canton Press, 1840), https://books.google.com/books?id=ngMMAAAAYAAJ&printsec=frontcover&source=gbs_ge_summary_r&cad=0#v=onepage&q&f=false, 499.
18. Angus Maddison, *Contours of the World Economy 1–2030 AD: Essays in Macro-Economic History* (New York: Oxford University Press, 2007), 379.
19. Weimin Zhong, "The Roles of Tea and Opium in Early Economic Globalization: A Perspective on China's Crisis in the 19th Century," *Frontiers of History in China* 5, no. 1 (March 2010): 86–105.
20. Letter from Willaim Jardine to Dr. Charles Gutzlaff, 1832, quoted in Maurice Collis, *Foreign Mud: Being an Account of the Opium Imbroglio at Canton in the 1830s and the Anglo-Chinese War that Followed* (New York: New Directions, 2002; first published 1946), 82.
21. W. E. Gladstone, "War with China—Adjourned Debate," *Hansard Parliamentary Debates, House of Commons*, April 8, 1840, Vol. 53, cols. 817–18, https://api.parliament.uk/historic-hansard/commons/1840/apr/08/war-with-china-adjourned-debate#column_821.
22. Whitney Stewart, *Deng Xiaoping: Leader in a Changing China* (Minneapolis: Lerner, 2001), 23.
23. Zheng Bijian, "The Three Globalizations and China," *HuffPost*, November 26, 2014, https://www.huffpost.com/entry/globalization-and-china_b_4668216.
24. Paul Michael Linehan, *The Culture of Leadership in Contemporary China: Conflict, Values, and Perspectives for a New Generation* (Lanham, MD: Lexington Books, 2017), 107–22.
25. Yefu Gu et al., "Impacts of Sectoral Emissions in China and the Implications: Air Quality, Public Health, Crop Production, and Economic Costs," *Environmental Research Letters* 13, no. 8 (2018).
26. Library of Congress, "Federal Research Division Country Profile: Haiti, May 2006," https://www.loc.gov/rr/frd/cs/profiles/Haiti.pdf.
27. Malick W. Ghachem, "Prosecuting Torture: The Strategic Ethics of Slavery in Pre-revolutionary Saint-Domingue (Haiti)," *Law and History Review* 29, no. 4 (2011): 985–1029; Anthony Phillips, "Haiti, France and the Independence Debt of 1825," Canada Haiti Action Network, 2008, https://www.canadahaitiaction.ca/sites/default/files/Haiti%2C%20France%20and%20the%20Independence%20Debt%20of%201825_0.pdf.

28. Quoted in Carolyn E. Fick, *The Making of Haiti: The Saint Domingue Revolution from Below* (Knoxville: University of Tennessee Press, 1990), 19.
29. Ibid., 20.
30. David Geggus, *The Haitian Revolution: A Documentary History* (Indianapolis, IN: Hackett, 2014), 13.
31. James, *The Black Jacobins*, 74.
32. Ibid., 271.
33. Ibid., 78.
34. "Haitian Constitution of 1801," The Louverture Project, trans. Charmant Theodore, 2000, http://thelouvertureproject.org/index.php?title=Haitian_Constitution_of_1801_(English).
35. Tim Matthewson, "Jefferson and the Nonrecognition of Haiti," *American Philosophical Society* 140, no. 1 (1996): 22–48.
36. "Haiti's Troubled Path to Development," Council on Foreign Relations, March 12, 2018, https://www.cfr.org/backgrounder/haitis-troubled-path-development.
37. Herb Thompson, "The Economic Causes and Consequences of the Bougainville Crisis," *Resources Policy* 17, no. 1 (1991): 69–85.
38. "PNG Leader Apologises to Bougainville for Bloody 1990s Civil War," Australian Associated Press, January 29, 2014, https://www.theguardian.com/world/2014/jan/29/papua-new-guinea-apologises-bougainville-civil-war.
39. Daniel Flitton, "Rio Tinto's Billion-Dollar Mess: 'Unprincipled, Shameful and Evil,'" *Sydney Morning Herald*, August 19, 2016, http://www.smh.com.au/world/billiondollar-mess-a-major-disaster-the-people-do-not-deserve-to-have-20160817-gquzli.html.

CHAPTER 4: THE LAND OF THE LAST ELEPHANTS

1. Fred Kurt, Günther B. Hartl, and Ralph Tiedemann, "Tuskless Bulls in Asian Elephant *Elephas maximus*: History and Population Genetics of a Man-Made Phenomenon," *Acta Theriologica* 40 (1995): 125–43; Raman Sukumar, *The Living Elephants: Evolutionary Ecology, Behaviour and Conservation* (New York: Oxford University Press, 2003), 287.
2. Samuel White Baker, *The Rifle and the Hound in Ceylon* (London: Longman, Brown, Green, and Longmans, 1854), 9, 187, 373.
3. Monique Grooten and Rosamunde Almond, eds., "Living Planet Report–2018: Aiming Higher," WWF, 2018, https://c402277.ssl

.cf1.rackcdn.com/publications/1187/files/original/LPR2018_Full_Report_Spreads.pdf?1540487589.

4. Gerardo Ceballos, Paul R. Ehrlich, and Rodolfo Dirzo, "Biological Annihilation via the Ongoing Sixth Mass Extinction Signaled by Vertebrate Population Losses and Declines," *Proceedings of the National Academy of Sciences* 114, no. 30 (2017): e6089–96; "Global Assessment Report on Biodiversity and Ecosystem Services: Summary for Policymakers," IPBES, 2019, https://ipbes.net/system/tdf/inline/files/ipbes_global_assessment_report_summary_for_policymakers.pdf?file=1&type=node&id=36213, 12.
5. Vernon R. Booth and Kevin M. Dunham, "Elephant Poaching in Niassa Reserve, Mozambique: Population Impact Revealed by Combined Survey Trends for Live Elephants and Carcasses," *Oryx* 50, no. 1 (2016): 94–103.
6. Kenneth V. Rosenberg et al., "Decline of the North American Avifauna," *Science* 366, no. 6461 (2019): 120–24.
7. Caspar A. Hallmann et al., "More Than 75 Percent Decline over 27 Years in Total Flying Insect Biomass in Protected Areas," *PLOS ONE* 12, no. 10 (2017): e0185809.
8. Villy Christensen et al., "A Century of Fish Biomass Decline in the Ocean," *Marine Ecology Progress Series* 512 (2014): 155–66; Ransom A. Myers and Boris Worm, "Rapid Worldwide Depletion of Predatory Fish Communities," *Nature* 423 (2003): 280–83.
9. Boris Worm et al., "Global Catches, Exploitation Rates, and Rebuilding Options for Sharks," *Marine Policy* 40 (2013): 194–204.
10. "UN Report: Nature's Dangerous Decline 'Unprecedented'; Species Extinction Rates 'Accelerating,'" *Sustainable Development Goals* blog, May 6, 2019, https://www.un.org/sustainabledevelopment/blog/2019/05/nature-decline-unprecedented-report/.
11. Moses Maimonides, *The Guide for the Perplexed*, trans. Michael Friedländer (New York: E. P. Dutton & Co, 1904), Part 3, 274.
12. See Susan Scott's film *Stroop: Journey into the Rhino Horn War*, South Africa, 2018.

CHAPTER 5: "WE REFUSE TO DIE"

1. "Sri Lanka: Floods and Landslides Emergency Response Plan (June–October 2017)," UN, 2017, https://reliefweb.int/sites/reliefweb.int/files/resources/SriLanka_ResponsePlan_020617.pdf; "FAO/WFP Crop and Food Security Assessment Mission to Sri Lanka," Food and Agriculture Organization of the United

Nations and World Food Programme, June 22, 2017, http://www.fao.org/3/a-i7450e.pdf.

2. Noah S. Diffenbaugh and Marshall Burke, "Global Warming Has Increased Global Economic Inequality," *Proceedings of the National Academy of Sciences* 116, no. 20 (2019): 9808–13.
3. Stanford's School of Earth, Energy & Environmental Sciences. "Climate change has worsened global economic inequality," ScienceDaily, 2019. www.sciencedaily.com/releases/2019/04/190422151017.htm.
4. Marshall Burke, Solomon M. Hsiang, and Edward Miguel, "Global Non-Linear Effect of Temperature on Economic Production," *Nature* 527 (2015): 235.
5. Sebastian Bathiany et al., "Climate Models Predict Increasing Temperature Variability in Poor Countries," *Science Advances* 4, no. 5 (2018): eaar5809.
6. Martin Parry et al., "Climate Change and Hunger: Responding to the Challenge," World Food Programme, 2009, https://www.imperial.ac.uk/media/imperial-college/grantham-institute/public/publications/collaborative-publications/Climate-change-and-hunger-WFP.pdf; Terence P. Dawson, Anita H. Perryman, and Tom M. Osborne, "Modelling Impacts of Climate Change on Global Food Security," *Climatic Change* 134, no. 3 (2016): 429–40.
7. "Bangladesh: Reducing Poverty and Sharing Prosperity," World Bank, November 15, 2018, https://www.worldbank.org/en/results/2018/11/15/bangladesh-reducing-poverty-and-sharing-prosperity.
8. "Bangladesh Climate Change Strategy and Action Plan 2009," Government of the People's Republic of Bangladesh, September 2009, https://www.iucn.org/downloads/bangladesh_climate_change_strategy_and_action_plan_2009.pdf, 7–8.
9. Nellie Le Beau and Hugh Tuckfield, "The Change Luck City: Dhaka's Climate Refugees," *The Diplomat*, August 10, 2016, https://thediplomat.com/2016/08/the-change-luck-city-dhakas-climate-refugees/; Tim McDonnell, "Climate Change Creates a New Migration Crisis for Bangladesh," *National Geographic*, January 24, 2019, https://www.nationalgeographic.com/environment/2019/01/climate-change-drives-migration-crisis-in-bangladesh-from-dhaka-sundabans/?cjevent=92f17507352911e981a300f30a240612&utm_source=4003003&utm_medium=affiliates&utm_campaign=CJ/.

10. Kanta Kumari Rigaud et al., "Groundswell: Preparing for Internal Climate Migration," World Bank, 2018, 148; Scott A. Kulp and Benjamin H. Strauss, "New Elevation Data Triple Estimates of Global Vulnerability to Sea-Level Rise and Coastal Flooding," *Nature Communications* 10, no. 1 (2019): 1–12.
11. Julie Rozenberg and Stéphane Hallegatte, "The Impacts of Climate Change on Poverty in 2030 and the Potential from Rapid, Inclusive, and Climate-Informed Development," World Bank, November 8, 2015, http://documents.worldbank.org/curated/en/349001468197334987/pdf/WPS7483.pdf.
12. Mark Spalding, Corinna Ravilious, and Edmund Peter Green, *World Atlas of Coral Reefs* (Berkeley: University of California Press, 2001); Marjorie Mulhall, "Saving the Rainforests of the Sea: An Analysis of International Efforts to Conserve Coral Reefs," *Duke Environmental Law & Policy Forum* 19 (2009): 321–51.
13. Manfred Lenzen et al., "The Carbon Footprint of Global Tourism," *Nature Climate Change* 8, no. 6 (2018): 522–28.
14. Xavier Romero Frías, The Maldive Islanders: A Study of the Popular *Culture of an Ancient Ocean Kingdom (*Nova Ethnographia Indica, 1999), 443.
15. Joseph C. Farman, Brian G. Gardiner, and Jonathan D. Shanklin, "Large Losses of Total Ozone in Antarctica Reveal Seasonal ClOx/NOx Interaction," *Nature* 315 (1985): 207–10.
16. Robert Mackey, "Donald Trump's Hairspray Woes Inspire Climate Denial Riff," *The Intercept*, May 7, 2016, https://theintercept.com/2016/05/06/donald-trumps-got-hairspray-riff-hes-gonna-use/.
17. Douglas Adams, *The Hitchhiker's Guide to the Galaxy* (New York: Harmony Books, 1980; first published 1979), 35.
18. Alex Crawford, "Meet Dorsen, 8, Who Mines Cobalt to Make Your Smartphone Work," Sky News, February 28, 2017, https://news.sky.com/story/meet-dorsen-8-who-mines-cobalt-to-make-your-smartphone-work-10784120.
19. Naomi Klein, *This Changes Everything: Capitalism vs. the Climate* (New York: Simon & Schuster, 2014), 44.

CHAPTER 6: THE REBELLION'S HARBINGERS

1. "Terror in Mumbai," CNN Transcripts, December 12, 2009, http://transcripts.cnn.com/transcripts/0912/12/se.01.html.

2. Krishna Pokharel, "Investigators Trace Boat's Last Voyage," *Wall Street Journal*, December 2, 2008, https://www.wsj.com/articles/SB122816457079069941#.
3. Rahul Bedi, "India's Intelligence Services 'Failed to Act on Warnings of Attacks,'" *The Telegraph*, November 30, 2008, https://www.telegraph.co.uk/news/worldnews/asia/india/3537279/Indias-intelligence-services-failed-to-act-on-warnings-of-attacks.html.
4. "Terror in Mumbai," CNN Transcripts, December 12, 2009.
5. Nadav Eyal, "Darkness and Terror in Mumbai," *Ma'ariv Daily*, November 30, 2008.
6. Guillaume Lavallée, "'Banned' Group Thrives in Pakistan," AFP, UCA News, February 10, 2015, https://www.ucanews.com/news/banned-group-thrives-in-pakistan/72963.
7. John C. M. Calvert, "The Striving Shaykh: Abdullah Azzam and the Revival of Jihad," in Ronald A. Simkins, ed., "The Contexts of Religion and Violence," *Journal of Religion & Society*, Supplement Series 2 (2007): 83–102.
8. Shaykh Abdullah Azzam, "Join the Caravan," 1987, https://archive.org/stream/JoinTheCaravan/JoinTheCaravan_djvu.txt, 24.
9. Ibid., 10.
10. Zbigniew Brzezinski to the Mujahideen, "Your cause is right and God is on your side!," YouTube, September 4, 2014, https://www.youtube.com/watch?v=A9RCFZnWGE0.
11. Peter L. Bergen, *Holy War, Inc.: Inside the Secret World of Osama bin Laden* (New York: Simon & Schuster, 2002), 56.
12. Andrew McGregor, "'Jihad and the Rifle Alone': 'Abdullah 'Azzam and the Islamist Revolution," *Journal of Conflict Studies* 23, no. 2 (2003): 92–113, https://journals.lib.unb.ca/index.php/jcs/article/view/219/377.
13. "US Embassy Cables: Lashkar-e-Taiba Terrorists Raise Funds in Saudi Arabia," *The Guardian*, December 5, 2010, https://www.theguardian.com/world/us-embassy-cables-documents/220186.
14. John Rollins, Liana Sun Wyler, and Seth Rosen, "International Terrorism and Transnational Crime: Security Threats, US Policy, and Considerations for Congress," Congressional Research Service, January 5, 2010, https://fas.org/sgp/crs/terror/R41004-2010.pdf, 15.

15. "Lashkar-E-Tayyiba," United Nations Security Council, 2010, https://www.un.org/securitycouncil/sanctions/1267/aq_sanctions_list/summaries/entity/lashkar-e-tayyiba.
16. Rituparna Chatterje, "Dawood Ibrahim's Wife Tells TV Channel World's Most Wanted Terrorist Is in Karachi, Sleeping at the Moment," *HuffPost*, August 22, 2015, https://www.huffingtonpost.in/2015/08/22/dawood-ibrahim_n_8024254.html.
17. "Al-Mourabitoun," Counter Extremism Project, March 28, 2019, https://www.counterextremism.com/threat/al-mourabitoun; "Mali: Group Merges with Al Qaeda," Associated Press/*New York Times*, December 4, 2015, https://www.nytimes.com/2015/12/05/world/africa/mali-group-merges-with-al-qaeda.html.
18. Simon Usborne, "Dead or Alive? Why the World's Most-Wanted Terrorist Has Been Killed at Least Three Times," *The Guardian*, November 28, 2016, https://www.theguardian.com/world/shortcuts/2016/nov/28/dead-or-alive-mokhtar-belmokhtar-most-wanted-terrorist-killed-three-times.
19. Ishaan Tharoor, "Paris Terror Suspect Is 'a Little Jerk,' His Lawyer Says," *Washington Post*, April 27, 2016, https://www.washingtonpost.com/news/worldviews/wp/2016/04/27/paris-terror-suspect-is-a-little-jerk-his-lawyer-says/?noredirect=on&utm_term=.0d0887cd1bb2.
20. Paul Tassi, "ISIS Uses 'GTA 5' in New Teen Recruitment Video," *Forbes*, September 20, 2014, https://www.forbes.com/sites/insertcoin/2014/09/20/isis-uses-gta-5-in-new-teen-recruitment-video/#59240edb681f.
21. Andrew K. Przybylski and Netta Weinstein, "Violent Video Game Engagement Is Not Associated with Adolescents' Aggressive Behaviour: Evidence from a Registered Report," *Royal Society Open Science* 6, no. 2 (2019), https://royalsocietypublishing.org/doi/10.1098/rsos.171474.
22. Jean Baudrillard, *Simulacra and Simulation*, trans. Sheila Faria Glaser (Ann Arbor: University of Michigan Press, 1994; first published 1981), 84.
23. "For What It's Worth," Buffalo Springfield, 1966, https://genius.com/Buffalo-springfield-for-what-its-worth-lyrics.
24. Abdullah Azzam, "So That the Islamic Nation Does Not Die an Eternal Death," *al-Jihad* 63 (1990): 29.

25. Asaf Maliach, "Abdullah Azzam, al-Qaeda, and Hamas: Concepts of Jihad and Istishhad," *Military and Strategic Affairs* 2, no. 2 (2010): 80.
26. Bernard Lewis and Buntzie Ellis Churchill, *Islam: The Religion and the People* (Upper Saddle River, NJ: Pearson Prentice Hall, 2008), 153.
27. Robert Allen Denemark and Mary Ann Tétreault, *Gods, Guns, and Globalization: Religious Radicalism and International Political Economy* (Boulder, CO: Lynne Rienner, 2004).
28. Ibid., 1–3.
29. Michael J. Stevens, "The Unanticipated Consequences of Globalization: Contextualizing Terrorism," in *The Psychology of Terrorism: Theoretical Understandings and Perspectives*, Vol. 3, ed. Chris E. Stout (Westport, CT: Greenwood Publishing Group, 2002), 31–56.

CHAPTER 7: TALKING WITH NATIONALISTS

1. Nicholas Cronk, *Voltaire: A Very Short Introduction* (New York: Oxford University Press, 2017), 37.
2. Tom Baldwin and Fiona Hamilton, "Times Interview with Nick Griffin: The BBC Is Stupid to Let Me Appear," *The Times*, October 22, 2009, https://www.thetimes.co.uk/article/times-interview-with-nick-griffin-the-bbc-is-stupid-to-let-me-appear-lkqvlv6r6vk.
3. "Barack Obama's Speech in Independence, Mo.," *New York Times*, June 30, 2008, https://www.nytimes.com/2008/06/30/us/politics/30text-obama.html?mtrref=www.google.com
4. David Nakamura, "Obama: Biggest Mistake Was Failing to 'Tell a Story' to American Public," *Washington Post*, July 12, 2012, https://www.washingtonpost.com/blogs/election-2012/post/obama-biggest-mistake-was-failing-to-tell-a-story-to-american-public/2012/07/12/gJQANHBFgW_blog.html?noredirect=on&utm_term=.547a520e6035.
5. "Countering Violent Extremism," US Government Accountability Office, April 2017, https://www.gao.gov/assets/690/683984.pdf, 0, 4.
6. "Timothy McVeigh: The Path to Death Row," CNN Transcripts, June 9, 2001, http://edition.cnn.com/transcripts/0106/09/pitn.00.html.
7. Angelique Chrisafis, "Jean-Marie Le Pen Convicted of Contesting Crimes against Humanity," *The Guardian*, February 16, 2012,

https://www.theguardian.com/world/2012/feb/16/jean-marie-le-pen-convicted.

8. Jeremy Diamond, "Trump Embraces 'Nationalist' Title at Texas Rally," CNN, October 23, 2018, https://edition.cnn.com/2018/10/22/politics/ted-cruz-election-2018-president-trump-campaign-rival-opponent/index.html.
9. Sigmund Freud, *The Future of an Illusion*, trans. and ed. James Strachey (New York: W. W. Norton & Company, 1961; first published 1927), 12.
10. Noam Gidron and Jonathan J. B. Mijs, "Do Changes in Material Circumstances Drive Support for Populist Radical Parties? Panel Data Evidence from the Netherlands During the Great Recession 2007–2015," *European Sociological Review* 35, no. 5 (2019): 637–50.
11. Carlo Bastasin, "Secular Divergence: Explaining Nationalism in Europe," Brookings Institution, May 2019, https://www.brookings.edu/wp-content/uploads/2019/05/FP_20190516_secular_divergence_bastasin.pdf.
12. Ronald F. Inglehart and Pippa Norris, "Trump, Brexit, and the Rise of Populism: Economic Have-Nots and Cultural Backlash," Harvard JFK School of Government Faculty Working Paper Series No. RWP16–026, August 2016, 1–52.

CHAPTER 8: A NAZI REVIVAL

1. Nadav Eyal, "Hatred: A Journey to the Heart of Antisemitism," Channel 10, Israel, October 7, 2014, https://www.youtube.com/watch?v=helC1_cog0A.
2. Thomas Rogers, "Heil Hipster: The Young Neo-Nazis Trying to Put a Stylish Face on Hate," *Rolling Stone*, June 23, 2014, https://www.rollingstone.com/culture/culture-news/heil-hipster-the-young-neo-nazis-trying-to-put-a-stylish-face-on-hate-64736/.
3. Conrad Hackett, "5 Facts About the Muslim Population in Europe," Pew Research Center, November 29, 2017, https://www.pewresearch.org/fact-tank/2017/11/29/5-facts-about-the-muslim-population-in-europe.
4. "Europe's Growing Muslim Population," Pew Research Center, November 29, 2017, https://www.pewforum.org/2017/11/29/europes-growing-muslim-population/.
5. J. D. Hunter, "Fundamentalism in Its Global Contours," in *The Fundamentalist Phenomenon: A View from Within; A Response*

from Without, ed. Norman J. Cohen (Grand Rapids, MI: William B. Eerdmans, 1990), 59.

6. Alon Confino, *A World Without Jews: The Nazi Imagination from Persecution to Genocide* (New Haven, CT: Yale University Press, 2014); Avner Shapira, "The Nazi Narrative: How a Fantasy of Ethnic Purity Led to Genocide," *Ha'aretz*, April 23, 2017 (Hebrew), https://www.haaretz.co.il/gallery/literature/.premium-1.4039220.
7. "International Military Trials—Nurnberg," in Office of United States Chief of Counsel for Prosecution of Axis Criminality, *Nazi Conspiracy and Aggression*, Volume 4 (Washington, DC: US Government Printing Office, 1946), http://www.loc.gov/rr/frd/Military_Law/pdf/NT_Nazi_Vol-IV.pdf, 563.
8. Michael B. Salzman, "Globalization, Religious Fundamentalism and the Need for Meaning," *International Journal of Intercultural Relations* 32, no. 4 (2008): 319.
9. Garry Wills, *Under God: Religion and American Politics* (New York: Simon & Schuster, 1990), 15–16.
10. General Social Survey Data (GSS), NORC at the University of Chicago, 2018, http://www.norc.org/Research/Projects/Pages/general-social-survey.aspx.

CHAPTER 9: THE MIDDLE-CLASS MUTINIES

1. "Wall Street and the Financial Crisis: The Role of Investment Banks," Hearing Before the Permanent Subcommittee on Investigations of the Committee on Homeland Security and Governmental Affairs, United States Senate, 111th Congress, second session, vol. 4 of 5, April 27, 2010, https://www.govinfo.gov/content/pkg/CHRG-111shrg57322/pdf/CHRG-111shrg57322.pdf.
2. Bruce Horovitz, "Shoppers Splurge for Their Country," *USA Today*, October 3, 2001, http://usatoday30.usatoday.com/money/retail/2001–10–03-patriotic-shopper.htm.
3. "Defence Expenditure of NATO Countries (2010–2017)," NATO Public Diplomacy Division, March 15, 2018, https://www.nato.int/nato_static_fl2014/assets/pdf/pdf_2018_03/20180315_180315-pr2018–16-en.pdf; "Defense Budget Overview," United States Department of Defense, Fiscal Year 2020 Budget Request, March 5, 2019, https://comptroller.defense.gov/Portals/45/Documents

/defbudget/fy2020/fy2020_Budget_Request_Overview_Book .pdf.

4. Moritz Kuhn, Moritz Schularick, and Ulrike Steins, "Asset Prices and Wealth Inequality," VOX CEPR Policy Portal, August 9, 2018, https://voxeu.org/article/asset-prices-and-wealth -inequality.
5. Andrew G. Haldane and Piergiorgio Alessandri, "Banking on the State," Bank of England, September 25, 2009, https://www .bis.org/review/r091111e.pdf; Andrew G. Haldane, "The Contribution of the Financial Sector: Miracle or Mirage?" Bank of England, July 14, 2010, https://www.bis.org/review/r100716g.pdf.
6. Julia Finch and Katie Allen, "What Do Bankers Spend Their Bonuses On?," *The Guardian*, December 14, 2007, https://www .theguardian.com/business/2007/dec/14/banking.
7. "Northern Rock Besieged by Savers," BBC, September 17, 2007, http://newsvote.bbc.co.uk/2/hi/business/6997765.stm#story.
8. Jonny Greatrex, "West Midlands Men Planning Credit Crunch Full Monty," *Birmingham Mail*, April 19, 2009, https://www .birminghammail.co.uk/news/local-news/west-midlands-men -planning-credit-239734.
9. William Boston, "Financial Casualty: Why Adolf Merckle Killed Himself," *Time*, January 6, 2009, http://content.time.com/time /business/article/0,8599,1870007,00.html.
10. Nic Allen and Aislinn Simpson, "City Banker Spent £43,000 on Champagne," *The Telegraph*, February 20, 2009, https://www .telegraph.co.uk/news/newstopics/howaboutthat/4700148/City -banker-spent-43000-on-champagne.html.
11. Rebecca Smithers, "Au ATM: UK's First Gold Vending Machine Unveiled," *The Guardian*, July 1, 2011, https://www.theguardian .com/money/2011/jul/01/au-atm-gold-vending-machine; Wei Xu, "Gold ATM Activated, but Not for Long," *China Daily*, September 27, 2011, http://www.chinadaily.com.cn/business/2011–09/27 /content_13801006.htm; Associated Press, "Gold-Dispensing ATM Makes U.S. Debut in Fla.," CBS News, December 17, 2010, https://www.cbsnews.com/news/gold-dispensing-atm-makes-us -debut-in-fla/.
12. "Report of the Study Group on the Role of Public Finance in European Integration," Vols. 1 and 2, Commission of the European Communities, European Union, April 1977, https://www

.cvce.eu/content/publication/2012/5/31/91882415–8b25–4f01-b18c-4b6123a597f3/publishable_en.pdf; https://www.cvce.eu/content/publication/2012/5/31/c475e949-ed28–490b-81ae-a33ce9860d09/publishable_en.pdf.

13. "Why Europe Can't Afford the Euro," *The Times*, November 19, 1997, from *The Collected Works of Milton Friedman*, eds. Robert Leeson and Charles G. Palm, https://miltonfriedman.hoover.org/friedman_images/Collections/2016c21/1997novtimesWhyEurope.pdf.

CHAPTER 10: ANARCHISTS WITH FERRARIS

1. Serge Berstein and Jean-François Sirinelli, eds., *Les années Giscard: Valéry Giscard d'Estaing et l'Europe, 1974–1981* (Paris: Armand Colin, 2007; first published 2005), 135.
2. "'Seventeen Countries Were Far Too Many,'" *Der Spiegel*, September 11, 2012, https://www.spiegel.de/international/europe/spiegel-interview-with-helmut-schmidt-and-valery-giscard-d-estaing-a-855127.html.
3. "Taking responsibility for the arson of yachts on 30/3," April 3, 2009, https://bellumperpetuum.blogspot.com/2009/04/303_03.html.
4. Henry Miller, *The Colossus of Maroussi*, 2nd ed. (New York: New Directions, 2010; first published 1941), 14.
5. Joergen Oerstroem Moeller, "The Greek Crisis Explained," *Huffington Post*, June 22, 2015, https://www.huffingtonpost.com/joergen-oerstroem-moeller/the-greek-crisis-explaine_b_7634564.html.
6. "Europe Balks at Greece's Retire-at-50 Rules," AP, May 17, 2010, https://www.cbsnews.com/news/europe-balks-at-greeces-retire-at-50-rules; "Pensions at a Glance 2013: OECD and G20 Indicators," OECD, 2013, http://dx.doi.org/10.1787/pension_glance-2013-en.
7. "Greece 10 Years Ahead: Defining Greece's New Growth Model and Strategy," McKinsey, June 1, 2012, https://www.mckinsey.com/featured-insights/europe/greece-10-years-ahead.
8. Suzanne Daley, "Greek Wealth Is Everywhere but Tax Forms," *New York Times*, May 1, 2010, http://www.nytimes.com/2010/05/02/world/europe/02evasion.html?th&emc=th&mtrref=undefined&gwh=C3F3DF2E8C5C22D2A667A933C80604C9&gwt=pay.
9. Elisabeth Oltheten et al., "Greece in the Eurozone: Lessons from a Decade of Experience," *Quarterly Review of Economics and Finance* 53, no. 4 (2013): 317–35; Beat Balzli, "How Gold-

man Sachs Helped Greece to Mask Its True Debt," *Der Spiegel*, February 2, 2010, http://www.spiegel.de/international/europe/greek-debt-crisis-how-goldman-sachs-helped-greece-to-mask-its-true-debt-a-676634.html.

10. Nikos Roussanoglou, "Thousands of Empty Properties Face the Prospect of Demolition," *Kathimerini*, March 19, 2017, http://www.ekathimerini.com/216998/article/ekathimerini/business/thousands-of-empty-properties-face-the-prospect-of-demolition.
11. "Youth Unemployment Rate," OECD data, 2019, https://data.oecd.org/unemp/youth-unemployment-rate.htm.
12. "Severely Materially Deprived People," Eurostat, July 2019, https://ec.europa.eu/eurostat/databrowser/view/tipslc30/default/table?lang=en.
13. Nicole Itano, "In Greece, Education Isn't the Answer," Public Radio International, May 14, 2009, https://www.pri.org/stories/2009–05–14/greece-education-isnt-answer.
14. Alyssa Rosenberg, "'Girls' Was About the Path—and Costs—to Being 'A Voice of a Generation,'" *Washington Post*, April 14, 2017, https://www.washingtonpost.com/news/act-four/wp/2017/04/14/girls-was-about-the-path-and-costs-to-being-a-voice-of-a-generation/?utm_term=.0f1a5526c60e.
15. J. Rocholl and A. Stahmer, "Where Did the Greek Bailout Money Go?," ESMT White Paper No. WP–16–02, 2016, http://static.esmt.org/publications/whitepapers/WP-16–02.pdf.
16. Susanne Kraatz, "Youth Unemployment in Greece: Situation Before the Government Change," European Parliament, 2015, http://www.europarl.europa.eu/RegData/etudes/BRIE/2015/542220/IPOL_BRI(2015)542220_EN.pdf.
17. Karolina Tagaris, "After Seven Years of Bailouts, Greeks Sink Yet Deeper in Poverty," Reuters, February 20, 2017, https://www.reuters.com/article/us-eurozone-greece-poverty/after-seven-years-of-bailouts-greeks-sink-yet-deeper-in-poverty-idUSKBN15Z1NM.
18. "Fertility Rates," "Population," OECD data, 2019, https://data.oecd.org; Lois Labrianidis and Manolis Pratsinakis, "Outward Migration from Greece during the Crisis," LSE for the National Bank of Greece, 2015, https://www.lse.ac.uk/europeanInstitute/research/hellenicObservatory/CMS%20pdf/Research/NBG_2014_-Research_Call/Final-Report-Outward-migration-from-Greece-during-the-crisis-revised-on-1–6–2016.pdf.

19. David Molloy, "End of Greek Bailouts Offers Little Hope to Young," BBC, August 19, 2018, https://www.bbc.com/news/world-europe-45207092.
20. Dunja Mijatović, "Report of the Commissioner for Human Rights of the Council of Europe," Council of Europe, 2018, https://rm.coe.int/report-on-the-visit-to-greece-from-25-to-29-june-2018-by-dunja-mijatov/16808ea5bd.
21. Marina Economou et al., "Enduring Financial Crisis in Greece: Prevalence and Correlates of Major Depression and Suicidality," *Social Psychiatry and Psychiatric Epidemiology* 51, no. 7 (2016): 1015–24.
22. Ibid., 1020.
23. Herb Keinon, "Greek Minister Distances Himself from Past Associations with Neo-Nazi Groups," *Jerusalem Post*, July 15, 2019, https://www.jpost.com/Diaspora/Antisemitism/Greek-Minister-distances-himself-from-past-associations-with-neo-Nazi-groups-595623.
24. "How Some Made Millions Betting Against the Market," National Public Radio, May 2, 2011, https://www.npr.org/2011/05/02/135846486/how-some-made-millions-betting-against-the-market.
25. *Dune*, David Lynch, dir., 1984.
26. "Flashback: Elizabeth Warren (Basically) Predicts the Great Recession," *Moyers on Democracy*, June 25, 2004, https://billmoyers.com/segment/flashback-elizabeth-warren-basically-predicts-the-great-recession/.
27. Thomas Philippon, "Has the US Finance Industry Become Less Efficient? On the Theory and Measurement of Financial Intermediation," *American Economic Review* 105, no. 4 (2015): 1408–38.
28. Sameer Khatiwada, "Did the Financial Sector Profit at the Expense of the Rest of the Economy? Evidence from the United States," International Institute for Labor Studies, Cornell University and International Labor Organization, 2010, https://digitalcommons.ilr.cornell.edu/cgi/viewcontent.cgi?article=1101&context=intl.
29. "Household Debt, Loans and Debt Securities Percent of GDP," IMF, 2018, https://www.imf.org/external/datamapper/HH_LS@GDD/CAN/ITA/USA; "How Has the Percentage of Consumer Debt Compared to Household Income Changed over the Last Few Decades? What Is Driving These Changes?," Federal Reserve Bank of San Francisco, 2009, https://www.frbsf.org/education

/publications/doctor-econ/2009/july/consumer-debt-household-income/.

30. "Household Debt and Credit Report (Q1 2019)," Federal Reserve Bank of New York, 2019, https://www.newyorkfed.org/medialibrary/interactives/householdcredit/data/pdf/hhdc_2019q1.pdf.
31. Martin Wolf, "Bank of England's Mark Carney Places a Bet on Big Finance," *Financial Times*, October 29, 2013, https://www.ft.com/content/08dea9d4–4002–11e3–8882–00144feabdc0.
32. Report on the Economic Well-Being of U.S. Households (SHED), Federal Reserve Board's Division of Consumer and Community Affairs (DCCA), 2018, https://www.federalreserve.gov/publications/report-economic-well-being-us-households.htm.

CHAPTER 11: DISAPPEARING CHILDREN

1. "Mobile Population Survey (November)," population by municipality, Gunma Prefecture statistical information, November 2019, https://toukei.pref.gunma.jp/idj/idj201911.htm.
2. Kiyoshi Takenaka and Ami Miyazaki, "'Vanishing Village' Looks to Japan's LDP for Survival," Reuters, October 17, 2018, https://www.reuters.com/article/us-japan-election-ageing/vanishing-village-looks-to-japans-ldp-for-survival-idUSKBN1CM0VM.
3. Ben Dooley, "Japan Shrinks by 500,000 People as Births Fall to Lowest Number Since 1874," *New York Times*, December 24, 2019,https://www.nytimes.com/2019/12/24/world/asia/japan-birthrate-shrink.html.
4. Charlotte Edmond, "Elderly People Make up a Third of Japan's Population—and It's Reshaping the Country," World Economic Forum, September 17, 2019, https://www.weforum.org/agenda/2019/09/elderly-oldest-population-world-japan/; "Population Projections for Japan (2016–2065)," National Institute of Population and Social Security Research (Japan), April 2017, http://www.ipss.go.jp/pp-zenkoku/e/zenkoku_e2017/pp_zenkoku2017e.asp; "2019 Revision of World Population Prospects," United Nations, 2019, https://population.un.org/wpp/.
5. "Family Database: The Structure of Families," OECD statistics, 2015, https://stats.oecd.org/Index.aspx?DataSetCode=FAMILY/.
6. Alana Semuels, "Japan Is No Place for Single Mothers," *The Atlantic*, September 7, 2017, https://www.theatlantic.com/business/archive/2017/09/japan-is-no-place-for-single-mothers/538743/;

"Child poverty," OECD Social Policy Division, November 2019, https://www.oecd.org/els/CO_2_2_Child_Poverty.pdf.

7. "Declining Birthrate White Paper, 2018," Cabinet Office (Japan), 2018, https://www8.cao.go.jp/shoushi/shoushika/whitepaper/measures/english/w-2018/index.html.
8. Mizuho Aoki, "In Sexless Japan, Almost Half of Single Young Men and Women are Virgins," *Japan Times*, September 16, 2016, https://www.japantimes.co.jp/news/2016/09/16/national/social-issues/sexless-japan-almost-half-young-men-women-virgins-survey/#.WmxosqiWY2x.
9. Abigail Haworth, "Why Have Young People in Japan Stopped Having Sex?" *The Guardian*, October 20, 2013, https://www.theguardian.com/world/2013/oct/20/young-people-japan-stopped-having-sex; "The Fifteenth Japanese National Fertility Survey in 2015, Marriage Process and Fertility of Married Couples, Attitudes toward Marriage and Family Among Japanese Singles," National Institute of Population and Social Security Research, March 2017, http://www.ipss.go.jp/ps-doukou/e/doukou15/Nfs15R_points_eng.pdf.
10. Cyrus Ghaznavi et al., "Trends in Heterosexual Inexperience Among Young Adults in Japan: Analysis of National Surveys, 1987–2015," *BMC Public Health* 19, no. 355 (2019), https://bmcpublichealth.biomedcentral.com/articles/10.1186/s12889–019–6677–5.
11. Léna Mauger, *The Vanished: The "Evaporated People" of Japan in Stories and Photographs*, trans. Brian Phalen, with photographs by Stéphane Remael (New York: Skyhorse, 2016).
12. Frank Baldwin and Anne Allison, eds., *Japan: The Precarious Future* (New York: NYU Press, 2015).
13. Justin McCurry, "Japanese Minister Wants 'Birth-Giving Machines', aka Women, to Have More Babies," *The Guardian*, January 29, 2007, https://www.theguardian.com/world/2007/jan/29/japan.justinmccurry.
14. Baldwin and Allison, *Japan*, 58–59.
15. "OECD Economic Surveys: Japan 2017," OECD, April 13, 2017, https://www.oecd-ilibrary.org/economics/oecd-economic-surveys-japan-2017/the-wage-gap-between-regular-and-non-regular-workers-is-large_eco_surveys-jpn-2017-graph28-en; Koji Takahashi, "Regular/Non-Regular Wage Gap Between and Within Japanese Firms," Japan Institute for Labour Policy and Train-

ing, 2016, https://www.jil.go.jp/profile/documents/ktaka/asa14_proceeding_721357.pdf.

16. "Employed Persons by Age Group and Employee by Age Group and Type of Employment," Historical data (9), Japan Statistics Bureau, 2019, https://www.stat.go.jp/english/data/roudou/lngindex.html.
17. Kathy Matsui, Hiromi Suzukib, and Kazunori Tatebe, "Womenomics 5.0," Portfolio Strategy Research, Goldman Sachs, April 18, 2019, https://www.goldmansachs.com/insights/pages/womenomics-5.0/multimedia/womenomics-5.0-report.pdf, 14.
18. "The Global Gender Gap Report 2018," World Economic Forum, 2018, http://www3.weforum.org/docs/WEF_GGGR_2018.pdf, p. 8.
19. "Record Low of 16,772 Children on Day Care Waiting Lists in Japan, Welfare Ministry Says," *Japan Times*, September 6, 2019, https://www.japantimes.co.jp/news/2019/09/06/national/japan-day-care-waiting-record-low/#.XfyXbOgzY2w; "Report on the Status Related to Day-Care Centers," Ministry of Health, Labor and Welfare (Japan), April 1, 2019, https://www.mhlw.go.jp/stf/houdou/0000176137_00009.html.
20. Justin McCurry, "Japanese Women Suffer Widespread 'Maternity Harassment' at Work," November 18, 2015, https://www.theguardian.com/world/2015/nov/18/japanese-women-suffer-widespread-maternity-harassment-at-work.
21. Matsui, Suzukib, and Tatebe, "Womenomics 5.0."
22. Mary Brinton, "Gender Equity and Low Fertility in Postindustrial Societies," Lecture at the Radcliffe Institute for Advanced Study, Harvard University, April 9, 2014, https://www.youtube.com/watch?v=XiKYU07QqPI.
23. Ibid.
24. "Employees Working Very Long Hours," Better Life Index, OECD, 2019, http://www.oecdbetterlifeindex.org/topics/work-life-balance/.
25. "White Paper on Measures to Prevent Karoshi," Ministry of Health, Labour and Welfare (Japan), 2017, https://fpcj.jp/wp/wp-content/uploads/2017/11/8f513ff4e9662ac515de9e646f63d8b5.pdf.
26. "Japan's State-Owned Version of Tinder," *The Economist*, October 3, 2019, https://www.economist.com/asia/2019/10/03/japans-state-owned-version-of-tinder.
27. Chizuko Ueno, "The Declining Birth Rate: Whose Problem?," *Review of Population and Social Policy* 7 (1998): 103–28.

CHAPTER 12: "HUMANKIND IS THE *TITANIC*"

1. "Fertility Rate, Total (Births per Woman)," World Bank, 2019 (data source: United Nations Population Division, World Population Prospects: 2019 Revision), https://data.worldbank.org/indicator/SP.DYN.TFRT.IN.
2. Christopher J. L. Murray et al., "Population and Fertility by Age and Sex for 195 Countries and Territories, 1950–2017: A Systematic Analysis for the Global Burden of Disease Study 2017," *The Lancet* 392, no. 10159 (2018): 1995–2051.
3. Anthony Cilluffo and Neil G. Ruiz, "World's Population Is Projected to Nearly Stop Growing by the End of the Century," Pew Research Center, June 17, 2019, https://www.pewresearch.org/fact-tank/2019/06/17/worlds-population-is-projected-to-nearly-stop-growing-by-the-end-of-the-century/; Max Roser, "Future Population Growth," in "Our World in Data, 2019," https://ourworldindata.org/future-population-growth.
4. "Vital Statistics Rapid Release, Births: Provisional Data for 2018," Report No. 7, National Center for Health Statistics (US), May 2019, https://www.cdc.gov/nchs/data/vsrr/vsrr-007–508.pdf.
5. "Fertility Rate, Total (Births per Woman)," World Bank, 2019, https://data.worldbank.org/indicator/sp.dyn.tfrt.in.
6. "Population Growth (Annual %)," World Bank, 2019, https://data.worldbank.org/indicator/SP.POP.GROW?locations=ES-PT.
7. Rachel Chaundler, "Looking for a Place in the Sun? How About an Abandoned Spanish Village," *New York Times*, April 9, 2019, https://www.nytimes.com/2019/04/09/realestate/spain-abandoned-villages-for-sale.html.
8. J. C. Caldwell, *Demographic Transition Theory* (Dordrecht, The Netherlands: Springer, 2006), 249.
9. "Life Expectancy," World Health Organization, 2020, http://www.who.int/gho/mortality_burden_disease/life_tables/situation_trends_text/en/.
10. "Mapped: The Median Age of the Population on Every Continent," World Economic Forum, February 20, 2019 (data source: The World Factbook, CIA, 2018), https://www.weforum.org/agenda/2019/02/mapped-the-median-age-of-the-population-on-every-continent/; Charles Goodhart and Manoj Pradhan, "Demographics Will Reverse Three Multi-Decade Global Trends," Bank of International Settlements, Working Paper No. 656, 2017, https://www.bis.org/publ/work656.pdf, 21.

11. Jay Winter and Michael Teitelbaum, *Population, Fear, and Uncertainty: The Global Spread of Fertility Decline* (New Haven, CT: Yale University Press, 2013).
12. "Germany's Population by 2060, Results of the 13th Coordinated Population Projection," Federal Statistical Office of Germany, 2015, https://www.destatis.de/GPStatistik/servlets/MCRFileNodeServlet/DEMonografie_derivate_00001523/5124206159004.pdf;jsessionid=0EDFA73EBE669FB229AAED0566265526, 6, 20.
13. "The Labor Market Will Need More Immigration from Non-EU Countries in the Future," Bertelsmann Stiftung, 2015, https://www.bertelsmann-stiftung.de/en/topics/aktuelle-meldungen/2015/maerz/immigration-from-non-eu-countries/.
14. Lorenzo Fontana and Ettore Gotti Tedeschi, *La culla vuota della civiltà: All'origine della crisi* (Verona: Gondolin, 2018).
15. George Alter and Gregory Clark, "The Demographic Transition and Human Capital," in *The Cambridge Economic History of Modern Europe: Volume 1, 1700–1870*, ed. Stephen Broadberry and Kevin H. O'Rourke (Cambridge: Cambridge University Press, 2010), 64.
16. John Bingham, "Falling Birth Rates Could Spell End of the West—Lord Sacks," *The Telegraph*, June 6, 2016, https://www.telegraph.co.uk/news/2016/06/06/falling-birth-rates-could-spell-end-of-the-west–lord-sacks/.
17. Fabrice Murtin, "Long-Term Determinants of the Demographic Transition, 1870–2000," *Review of Economics and Statistics* 95, no. 2 (2013): 617–31.
18. Una Okonkwo Osili and Bridget Terry Long, "Does Female Schooling Reduce Fertility? Evidence from Nigeria," *Journal of Development Economics* 87, no. 1 (2008): 57–75.
19. Amartya Sen, *Development as Freedom* (New York: Oxford University Press, 2001; first published 1999), 153; Max Roser, "Fertility Rate" (under "Empowerment of Women"), in "Our World in Data, 2017," https://ourworldindata.org/fertility-rate.
20. Gary S. Becker, *A Treatise on the Family* (Cambridge, MA: Harvard University Press, 1981); Gary S. Becker, "An Economic Analysis of Fertility," in *Demographic and Economic Change in Developed Countries*, ed. Gary S. Becker (New York: Columbia University Press, 1960), 209–40.
21. Luis Angeles, "Demographic Transitions: Analyzing the Effects of Mortality on Fertility," *Journal of Population Economics* 23, no. 1 (2010): 99–120.

22. Hagai Levine et al., "Temporal Trends in Sperm Count: A Systematic Review and Meta-Regression Analysis," *Human Reproduction Update* 23, no. 6 (2017): 646–59.
23. Chuan Huang et al., "Decline in Semen Quality Among 30,636 Young Chinese Men from 2001 to 2015," *Fertility and Sterility* 107, no. 1 (2017): 83–88; Priyanka Mishra et al., "Decline in Seminal Quality in Indian Men over the Last 37 Years," *Reproductive Biology and Endocrinology* 16, no. 1 (2018), article 103.
24. Conversation with the author, October 2019.
25. Netta Ahituv, "Western Men's Free-Falling Sperm Count Is a 'Titanic Moment for the Human Species,'" *Ha'aretz*, November 17, 2017, https://www.haaretz.com/science-and-health/.premium.MAGAZINE-western-men-s-dropping-sperm-count-is-a-titanic-moment-for-humans-1.5466078.
26. "South Korea's Fertility Rate Falls to a Record Low," *The Economist*, August 30, 2019, https://www.economist.com/graphic-detail/2019/08/30/south-koreas-fertility-rate-falls-to-a-record-low.
27. Joori Roh, "Not a Baby Factory: South Korea Tries to Fix Demographic Crisis with More Gender Equality," Reuters, January 4, 2019, https://www.reuters.com/article/us-southkorea-economy-birthrate-analysis/not-a-baby-factory-south-korea-tries-to-fix-demographic-crisis-with-more-gender-equality-idUSKCN1OY023.
28. A. M. Devine, "The Low Birth-Rate in Ancient Rome: A Possible Contributing Factor," *Rheinisches Museum für Philologie* 128, nos. 3–4 (1985): 313–17.
29. Goran Therbon, *Between Sex and Power: Family in the World, 1900–2000* (Abingdon, UK: Routledge, 2004), 255; Kate Bissell, "Nazi Past Haunts 'Aryan' Children," BBC, May 13, 2005, http://news.bbc.co.uk/2/hi/europe/4080822.stm8.
30. Wang Feng, Yong Cai, and Baochang Gu, "Population, Policy, and Politics: How Will History Judge China's One-Child Policy?," *Population and Development Review* 38, Supplement 1 (2013): S115–29; Stuart Gietel-Basten, Xuehui Han, and Yuan Cheng. "Assessing the Impact of the 'One-Child Policy,' in "China: A Synthetic Control Approach," *PLOS ONE* 14, no. 11 (2019).
31. James Renshaw, *In Search of the Romans,* 2nd ed. (London: Bloomsbury, 2019), 244.
32. "China," *The World Factbook*, Central Intelligence Agency, 2018, https://www.cia.gov/library/publications/the-world-factbook/geos/ch.html; Simon Denyer and Annie Gowen, "Too Many Men,"

Washington Post, April 18, 2018, https://www.washingtonpost.com/graphics/2018/world/too-many-men/.

33. Valerie M. Hudson and Andrea M. den Boer, *Bare Branches: The Security Implications of Asia's Surplus Male Population* (Cambridge, MA: MIT Press, 2004).
34. "World Population Prospects," UN, 2019, https://population.un.org/wpp/DataQuery/.

CHAPTER 13: FACES OF EXODUS

1. "Refugee Data Finder," Refugee Statistics, UNHCR, https://www.unhcr.org/refugee-statistics/.
2. "Forced Displacement in 2015," Global Trends, UNHCR, June 20, 2016, https://www.unhcr.org/576408cd7.
3. "Refugee Data Finder," Refugee Statistics, UNHCR, https://www.unhcr.org/refugee-statistics/.
4. "Syria Refugee Crisis," UNHCR, 2019, https://www.unrefugees.org/emergencies/syria/.
5. Max Roser, "War and Peace After 1945," in "Our World in Data, 2019" (data sources: UCDP; PRIO), https://ourworldindata.org/war-and-peace#war-and-peace-after-1945.
6. UNHCR's Populations of Concern, UNHCR Statistics, 2019, http://popstats.unhcr.org.
7. Mary Kaldor, *New and Old Wars: Organized Violence in a Global Era* (Cambridge, UK: Polity Press, 1999).
8. "UN and Partners Call for Solidarity, as Venezuelans on the Move Reach 4.5 million," UN News, October 23, 2019, https://news.un.org/en/story/2019/10/1049871.
9. Ted Enamorado et al., "Income Inequality and Violent Crime: Evidence from Mexico's Drug War," Latin America and the Caribbean Region, Poverty Reduction and Economic Management Unit, World Bank, June 1, 2014.
10. Kimberly Heinle, Octavio Rodríguez Ferreira, and David A. Shirk, "Drug Violence in Mexico," Department of Political Science & International Relations, University of San Diego, March 2017, https://justiceinmexico.org/wp-content/uploads/2017/03/2017_DrugViolenceinMexico.pdf.
11. "Refugee Data Finder," Refugee Statistics, UNHCR, https://www.unhcr.org/refugee-statistics/.
12. Hugh Naylor, "Desperate for Soldiers, Assad's Government Imposes Harsh Recruitment Measures," *Washington Post*, December 28, 2014,

https://www.washingtonpost.com/world/middle_east/desperate-for-soldiers-assads-government-imposes-harsh-recruitment-measures/2014/12/28/62f99194–6d1d-4bd6-a862-b3ab46c6b33b_story.html; Erin Kilbride, “Forced to Fight: Syrian Men Risk All to Escape Army Snatch Squads,” *Middle East Eye*, April 3, 2016, http://www.middleeasteye.net/news/escape-assads-army-373201818.

13. “Gen. Breedlove’s Hearing with the House Armed Services Committee,” United States European Command Library, February 25, 2016, https://www.eucom.mil/media-library/transcript/35355/gen-breedloves-hearing-with-the-house-armed-services-committee.

CHAPTER 14: AN EXPERIMENT AND ITS COSTS

1. Universal Declaration of Human Rights, Article 13, Clause 2, UN, https://www.ohchr.org/EN/UDHR/Documents/UDHR_Translations/eng.pdf.
2. Joseph de Veitia Linage, *Norte de la contratacion de las Indias Occidentales* (Sevilla: por Juan Francisco de Blas, 1672), cited in Bernard Moses, *The Casa de Contratacion of Seville*, 1896, 113. https://books.google.com/books?id=JyTDJEsXMqUC&printsec=frontcover&source=gbs_ge_summary_r&cad=0#v=onepage&q&f=false.
3. Prudentius, “The Divinity of Christ,” in *Prudentius*, trans. H. J. Thomson, Vol. 1 (London: William Heinemann and Harvard University Press, 1949), 161.
4. Haim Beinart, *The Expulsion of the Jews from Spain*, trans. Jeffrey M. Green (Oxford: Littman Library of Jewish Civilization, 2001), 285.
5. Francois Soyer, “King John II of Portugal ‘O Principe Perfeito’ and the Jews (1481–1495),” *Sefarad* 69, no. 1 (2009): 75–99.
6. Moises Orfali and Tom Tov Assis, eds., *Portuguese Jewry at the Stake: Studies on Jews and Crypto-Jews* [Hebrew] (Jerusalem: Magnes, 2009), 30.
7. Richard Zimler, “Identified as the Enemy: Being a Portuguese New Christian at the Time of the Last Kabbalist of Lisbon,” *European Judaism* 33, no. 1 (2000): 32–42.
8. Rachel Zelnick-Abramovitz, *Not Wholly Free: The Concept of Manumission and the Status of Manumitted Slaves in the Ancient Greek World* (New York: Brill, 2005).

9. John C. Torpey, *The Invention of the Passport: Surveillance, Citizenship and the State*, 2nd ed. (Cambridge: Cambridge University Press, 2018; first published 2000), 27.
10. Alan Dowty, *Closed Borders: The Contemporary Assault on the Freedom of Movement* (New Haven, CT: Yale University Press, 1987); Bonnie Berkowitz, Shelly Tanand, and Kevin Uhrmacher, "Beyond the Wall: Dogs, Blimps and Other Things Used to Secure the Border," *Washington Post*, February 8, 2019, https://www.washingtonpost.com/graphics/2019/national/what-is-border-security/?utm_term=.cd9d7eb58313.
11. Mae M. Ngai, "Nationalism, Immigration Control, and the Ethnoracial Remapping of America in the 1920s," *OAH Magazine of History* 21, no. 3 (2007): 11–15.
12. Daniel C. Turack, "Freedom of Movement and the International Regime of Passports," *Osgoode Hall Law Journal* 6, no. 2 (1968): 230.
13. Richard Plender, *International Migration Law* (Leiden, The Netherlands: Martinus Nijhoff, 1988); Martin Lloyd, *The Passport: The History of Man's Most Travelled Document* (Canterbury, UK: Queen Anne's Fan, 2008; first published 2003), 95–115.
14. Ibid.
15. Mae M. Ngai, *Impossible Subjects: Illegal Aliens and the Making of Modern America* (Princeton, NJ: Princeton University Press, 2014).
16. "Immigration Timeline," The Statue of Liberty–Ellis Island Foundation, https://www.libertyellisfoundation.org/immigration-timeline.
17. "Russell Brand: Messiah Complex (2013)—Full Transcript," *Scraps from the Loft*, November 7, 2017, https://scrapsfromtheloft.com/2017/11/07/russell-brand-messiah-complex-2013-full-transcript/.
18. Theresa May, "Theresa May's Conference Speech in Full," *The Telegraph*, October 5, 2016, http://www.telegraph.co.uk/news/2016/10/05/theresa-mays-conference-speech-in-full/.
19. Irene Skovgaard-Smith and Flemming Poulfelt, "Imagining 'Non-Nationality': Cosmopolitanism as a Source of Identity and Belonging," *Human Relations*, 71, no. 2 (2018): 129–54; Pnina Werbner, ed., *Anthropology and the New Cosmopolitanism: Rooted, Feminist and Vernacular Perspectives* (New York: Berg, 2008); Kwame Anthony Appiah, "Cosmopolitan Patriots," *Critical Inquiry* 23, no. 3 (1997): 617–39.

20. World Economic Forum, Global Shapers Annual Survey 2017, http://www.shaperssurvey2017.org/.
21. "Global Citizenship a Growing Sentiment Among Citizens of Emerging Economies: Global Poll," Globescan for BBC, April 27, 2016, https://globescan.com/wp-content/uploads/2016/04/BBC_GlobeScan_Identity_Season_Press_Release_April%2026.pdf.
22. Brittany Blizzard and Jeanne Batalova, "Refugees and Asylees in the United States," Migration Policy Institute, June 13, 2019, https://www.migrationpolicy.org/article/refugees-and-asylees-united-states.
23. Ronald Reagan, January 19, 1989, in *Public Papers of the Presidents of the United States: Ronald Reagan, 1988–1989* (Washington, DC: US Government Printing Office, 1990), https://www.reaganlibrary.gov/research/speeches/011989b, 1752.

CHAPTER 15: RIVERS OF BLOOD

1. The conversations with the Syrian refugees were part of a documentary for Channel 10 TV, Israel.
2. "Global Views on Immigration and the Refugee Crisis," Ipsos, September 13, 2017, https://www.ipsos.com/sites/default/files/ct/news/documents/2017–09/ipsos-global-advisor-immigration-refugee-crisis-slides_0.pdf.
3. Florence Jaumotte, Ksenia Koloskova, and Sweta Chaman Saxena, "Impact of Migration on Income Levels in Advanced Economies," International Monetary Fund, 2016, https://www.imf.org/en/Publications/Spillover-Notes/Issues/2016/12/31/Impact-of-Migration-on-Income-Levels-in-Advanced-Economies-44343.
4. Lena Groeger, "The Immigration Effect," *ProPublica*, July 19, 2017, https://projects.propublica.org/graphics/gdp.
5. "Second-Generation Americans: A Portrait of the Adult Children of Immigrants," Pew Research Center, February 7, 2013, https://www.pewsocialtrends.org/2013/02/07/second-generation-americans/.
6. "The Progressive Case for Immigration," *The Economist*, March 18, 2017, https://www.economist.com/news/finance-and-economics/21718873-whatever-politicians-say-world-needs-more-immigration-not-less?fsrc=scn/tw/te/bl/ed/.
7. Ryan Edwards and Francesc Ortega, "The Economic Contribution of Unauthorized Workers: An Industry Analysis," *Regional Science and Urban Economics* 67 (2017): 119–34.

8. Francine D. Blau and Christopher Mackie, *The Economic and Fiscal Consequences of Immigration* (Washington, DC: National Academies Press, 2017).
9. George J. Borjas, "Among Many Other Things, That Current Policy Creates a Large Wealth Transfer from Workers to Firms," *National Review*, September 22, 2016, http://www.nationalreview.com/article/440334/national-academies-sciences-immigration-study-what-it-really-says, accessed January 29, 2018.
10. George J. Borjas, "The Labor Demand Curve Is Downward Sloping: Reexamining the Impact of Immigration on the Labor Market," *Quarterly Journal of Economics* 118, no. 4 (2003): 1335–74.
11. Borjas, "Among Many Other Things."
12. Christian Dustmann, Uta Schönberg, and Jan Stuhler, "Labor Supply Shocks, Native Wages, and the Adjustment of Local Employment," *Quarterly Journal of Economics* 132, no. 1 (2017): 435–83.
13. Jynnah Radford, "Key Findings About U.S. Immigrants," Pew Research Center, June 17, 2019 (data sources: US Census Bureau; American Community Survey [IPUMS]), https://www.pewresearch.org/fact-tank/2019/06/17/key-findings-about-u-s-immigrants/.
14. "Proportion of Resident Population Born Abroad, England and Wales; 1951–2011," Office for National Statistics (UK), 2013, http://www.ons.gov.uk/ons/rel/census/2011-census-analysis/immigration-patterns-and-characteristics-of-non-uk-born-population-groups-in-england-and-wales/chd-figure-1.xls; "Population of the UK by Country of Birth and Nationality: 2018," Office for National Statistics (UK), May 24, 2019, https://www.ons.gov.uk/peoplepopulationandcommunity/populationandmigration/internationalmigration/bulletins/ukpopulationbycountryofbirthandnationality/2018.
15. Jens Manuel Krogstad, Jeffrey S. Passel, and D'vera Cohn, "5 Facts about Illegal Immigration in the U.S.," Pew Research Center, June 12, 2019, https://www.pewresearch.org/fact-tank/2019/06/12/5-facts-about-illegal-immigration-in-the-u-s/.
16. "An Edgy Inquiry," *The Economist*, April 4, 2015 (data source: Insee–National Institute of Statistics and Economic Studies [France], France strategie), https://www.economist.com/news/europe/21647638-taboo-studying-immigrant-families-performance-fraying-edgy-inquiry.

17. "Settling In 2018: Indicators of Immigrant Integration," OECD, 2018, https://www.oecd.org/publications/indicators-of-immigrant-integration-2018–9789264307216-en.htm.
18. Rick Noack, "Some French Wanted to Find Out How Racist Their Country Is. They Might Get Sued for It," *Washington Post*, February 4, 2016, https://www.washingtonpost.com/news/worldviews/wp/2016/02/04/why-it-can-be-illegal-to-ask-people-about-their-religion-or-ethnicity-in-france/?utm_term=.21f58814349b.
19. "Timeline: Deadly Attacks in Western Europe," Reuters, August 17, 2017, https://www.reuters.com/article/us-europe-attacks-timeline-idUSKCN1AX2EV; David Batty, "Timeline: 20 Years of Terror That Shook the West," *The Guardian*, November 14, 2015, https://www.theguardian.com/world/2015/nov/14/paris-attacks-timeline-20-years-of-terror.
20. "The Perils of Perception 2018," Ipsos MORI, 2018, https://www.ipsos.com/ipsos-mori/en-uk/perils-perception-2018.
21. Hackett, "5 Facts about the Muslim Population in Europe."
22. "The Perils of Perception 2018," Ipsos MORI, 2018; Pamela Duncan, "Europeans Greatly Overestimate Muslim Population, Poll Shows," *The Guardian*, December 13, 2016, https://www.theguardian.com/society/datablog/2016/dec/13/europeans-massively-overestimate-muslim-population-poll-shows.
23. The conversation was conducted by Inbar Golan, the researcher who worked with me on a series of reports called "Exodus" for Israel's Channel 10, and on this book. It was during the preparation of these reports that we met the Aboudan family.
24. "German Spy Agency Says ISIS Sending Fighters Disguised as Refugees," Reuters, February 5, 2016, https://www.reuters.com/article/us-germany-security-idUSKCN0VE0XL; Anthony Faiola and Souad Mekhennet, "Tracing the Path of Four Terrorists Sent to Europe by the Islamic State," *Washington Post*, April 22, 2016, https://www.washingtonpost.com/world/national-security/how-europes-migrant-crisis-became-an-opportunity-for-isis/2016/04/21/ec8a7231–062d-4185-bb27-cc7295d35415_story.html?utm_term=.7cf4615e01c9.
25. Alan Travis, "Net Immigration to UK Nears Peak as Fewer Britons Emigrate," *The Guardian*, May 26, 2016, https://www.theguardian.com/uk-news/2016/may/26/net-migration-to-uk-nears-peak-fewer-britons-emigrate.

26. Heather Stewart and Rowena Mason, "Nigel Farage's Anti-Migrant Poster Reported to Police," *The Guardian*, June 16, 2016, https://www.theguardian.com/politics/2016/jun/16/nigel-farage-defends-ukip-breaking-point-poster-queue-of-migrants.
27. "The Vote to Leave the EU," in *British Social Attitudes* 34, The National Centre for Social Research, 2017, http://www.bsa.natcen.ac.uk/media/39149/bsa34_brexit_final.pdf; Daniel Boffey, "Poll Gives Brexit Campaign Lead of Three Percentage Points," *The Observer (The Guardian)*, June 5, 2016, https://www.theguardian.com/politics/2016/jun/04/poll-eu-brexit-lead-opinium.
28. Rose Meleady, Charles R. Seger, and Marieke Vermue, "Examining the Role of Positive and Negative Intergroup Contact and Anti-Immigrant Prejudice in Brexit," *British Journal of Social Psychology* 56, no. 4 (2017): 799–808.
29. Yago Zayed, "Hate Crimes: What Do the Stats Show?" House of Commons Library, April 8, 2019 (data source: Home Office, Office for National Statistics), https://commonslibrary.parliament.uk/home-affairs/justice/hate-crimes-what-do-the-stats-show/.
30. Hannah Corcoran and Kevin Smith, "Hate Crime, England and Wales, 2016/16," Home Office (UK), October 13, 2016, https://assets.publishing.service.gov.uk/government/uploads/system/uploads/attachment_data/file/559319/hate-crime-1516-hosb1116.pdf.
31. Qasim Peracha, "How Hate Crimes Have Spiked in London since the Brexit Referendum," *My London*, May 3, 2019 (data source: London Metropolitan Police), https://www.mylondon.news/news/zone-1-news/how-hate-crimes-spiked-london-16217897; "Hate Crime or Special Crime Dashboard," London Metropolitan Police, 2019, https://www.met.police.uk/sd/stats-and-data/met/hate-crime-dashboard/.
32. Robert Booth, "Racism Rising since Brexit Vote, Nationwide Study Reveals," *The Guardian*, May 20, 2019 (data source: opinion survey, 2014–16), https://www.theguardian.com/world/2019/may/20/racism-on-the-rise-since-brexit-vote-nationwide-study-reveals.
33. Franz Solms-Laubach, "Mehr als 6 Millionen Flüchtlinge auf dem Weg nach Europa," *Bild*, May 23, 2017, https://www.bild.de/politik/ausland/fluechtlinge/6-millionen-warten-auf-reise-nach-europa-51858926.bild.html.

34. Peter Heather, *Empires and Barbarians: Migration, Development and the Birth of Europe* (London: Macmillan, 2009).
35. "Two Americas: Immigration," August 18, 2016, YouTube, https://youtu.be/3mKzYPt0Bu4.
36. "Transcript of the Second Debate," *New York Times*, October 10, 2016, https://www.nytimes.com/2016/10/10/us/politics/transcript-second-debate.html.
37. "Exit Polls," CNN, November 23, 2016, http://edition.cnn.com/election/results/exit-polls/national/president; Philip Bump, "In Nearly Every Swing State, Voters Preferred Hillary Clinton on the Economy," *Washington Post*, December 2, 2016, https://www.washingtonpost.com/news/the-fix/wp/2016/12/02/in-nearly-every-swing-state-voters-preferred-hillary-clinton-on-the-economy/?utm_term=.cf8fbdc0763f.

CHAPTER 16: A SUBJECT OF THE EMPIRE SPEAKS

1. Jon Wiener, "Relax, Donald Trump Can't Win," *The Nation*, June 21, 2016, https://www.thenation.com/article/trump-cant-win/.
2. Jonathan Chait, "Why Hillary Clinton Is Probably Going to Win the 2016 Election," *New York Magazine*, April 12, 2015, http://nymag.com/daily/intelligencer/2015/04/why-hillary-clinton-is-probably-going-to-win.html.
3. The interviews presented here were conducted for a series of reports on Israel's Channel 10 in advance of the 2016 elections.
4. Thomas Jefferson to James Madison, April 27, 1809, National Archives, https://founders.archives.gov/documents/Jefferson/03–01–02–0140.
5. Jonathan McClory, "The Soft Power 30, A Global Ranking of Soft Power, 2018," Portland and USC Center on Public Diplomacy, July 2018, https://www.uscpublicdiplomacy.org/sites/uscpublicdiplomacy.org/files/useruploads/u39301/The%20Soft%20Power%2030%20Report%202018.pdf.
6. Iliana Olivié and Manuel Gracia, "Elcano Global Presence Report 2018," Elcano Royal Institute, 2018, http://www.realinstitutoelcano.org/wps/wcm/connect/897b80cc-47fa-4130–9c3d-24e16c7f0a66/Global_Presence_2018.pdf?MOD=AJPERES&CACHEID=897b80cc-47fa-4130–9c3d-24e16c7f0a66.
7. Michael Scherer, "Obama Too Is an American Exceptionalist," *Time*, April 4, 2009, https://swampland.time.com/2009/04/04/obama-too-is-an-american-exceptionalist/.

8. Virgil, *The Aeneid*, Book Six, trans. David Ferry (Chicago: University of Chicago Press, 2017), 201.
9. Reinhold Niebuhr, *The Irony of American History* (Chicago: University of Chicago Press, 2008), 74.
10. Alan P. Dobson and Steve Marsh, *US Foreign Policy since 1945 (The Making of the Contemporary World)* (London and New York: Routledge, 2006), 55.
11. Charles L. Mee Jr., *The Marshall Plan* (New York: Simon & Schuster, 1985), 99–100.
12. John T. Bethell, "How the Press Missed 'Mr. Marshall's Hint,'" *Washington Post*, May 25, 1997, http://www.washingtonpost.com/wp-srv/inatl/longterm/marshall/bethell.htm.
13. Bruce D. Jones, ed., *The Marshall Plan and the Shaping of American Strategy* (Washington, DC: Brookings Institution Press, 2017).
14. Niall Ferguson, *Empire: The Rise and Demise of the British World Order and the Lessons for Global Power* (New York: Basic Books, 2003).
15. Michael Ignatieff, "American Empire (Get Used to It)," *New York Times Magazine*, January 5, 2003, https://www.nytimes.com/2003/01/05/magazine/the-american-empire-the-burden.html.
16. Richard H. Immerman, *Empire for Liberty: A History of American Imperialism from Benjamin Franklin to Paul Wolfowitz* (Princeton, NJ: Princeton University Press, 2010), 3.
17. Molly Ivins, "Cheney's Card: The Empire Writes Back," *Washington Post*, December 30, 2003, https://www.washingtonpost.com/archive/opinions/2003/12/30/cheneys-card-the-empire-writes-back/18317ced-c7d4–4ea2-a788-d9a67cd72f86/.
18. Amy Belasco, "The Cost of Iraq, Afghanistan, and Other Global War on Terror Operations Since 9/11," Congressional Research Service, Report RL33110, 2014, https://fas.org/sgp/crs/natsec/RL33110.pdf.
19. Joseph Stiglitz and Linda J. Bilmes, *The Three Trillion Dollar War* (New York: W. W. Norton & Company, 2008).
20. Neta C. Crawford, "United States Budgetary Costs of the Post-9/11 Wars Through FY2019: $5.9 Trillion Spent and Obligated," Brown University, November 14, 2018, https://watson.brown.edu/costsofwar/files/cow/imce/papers/2018/Crawford_Costs%20of%20War%20Estimates%20Through%20FY2019.pdf.

21. "Israeli Journalist Mines a Story in Marianna," *Observer-Reporter* [Washington, PA], July 30, 2016, updated December 5, 2017, https://observer-reporter.com/news/localnews/israeli-journalist-mines-a-story-in-marianna/article_923b8bbb-e3a8–54c8–992b-90e65404d987.html.
22. The conversations in Marianna were for a documentary aired on Israel's Channel 10 in July 2016.

CHAPTER 17: "MY MOTHER WAS MURDERED HERE"

1. "Hutchins Intermediate School," 1922, http://detroiturbex.com/content/schools/hutchins/index.html.
2. Detroit, Michigan, Quick Facts, United States Census Bureau, 2018, https://www.census.gov/quickfacts/fact/table/detroitcitymichigan/PST045218.
3. "1950 Census of Population, Population of Michigan by Counties," United States Census Bureau, April 1, 1950, https://www2.census.gov/library/publications/decennial/1950/pc-02/pc-2–36.pdf.
4. The conversation is documented in my series of reports "The Battle for America" on Israel's Channel 10 News, broadcast in October 2016.
5. Ed Mazza, "Ron Baity, Baptist Preacher, Claims God Will Send Something Worse Than Ebola as Punishment for Gay Marriage," *HuffPost*, October 15, 2014, https://www.huffpost.com/entry/ron-baity-ebola-gay-marriage_n_5987210.
6. Anna North and Catherine Kim, "The 'Heartbeat' Bills That Could Ban Almost All Abortions, Explained," Vox, June 28, 2019, https://www.vox.com/policy-and-politics/2019/4/19/18412384/abortion-heartbeat-bill-georgia-louisiana-ohio-2019; Jacob Gershman and Arian Campo-Flores, "Antiabortion Movement Begins to Crack, After Decades of Unity," *Wall Street Journal*, July 17, 2019, https://www.wsj.com/articles/antiabortion-movement-begins-to-crack-after-decades-of-unity-11563384713.
7. "All Employees: Total Nonfarm Payrolls," Federal Reserve Bank of St. Louis, 2019 (data source: US Bureau of Labor Statistics), https://fred.stlouisfed.org/graph/?g=4EKm.
8. Lee E. Ohanian, "Competition and the Decline of the Rust Belt," Economic Policy Paper No. 14–6, Federal Reserve Bank of Minneapolis, 2014.
9. David H. Autor, David Dorn, and Gordon H. Hanson, "The China Shock: Learning from Labor-Market Adjustment to

Large Changes in Trade," *Annual Review of Economics* 8 (2016): 205–40.

10. Zeeshan Aleem, "Another Kick in the Teeth: A Top Economist on How Trade with China Helped Elect Trump," Vox, March 29, 2017, https://www.vox.com/new-money/2017/3/29/15035498/autor-trump-china-trade-election.
11. Anne Case and Angus Deaton, "Mortality and Morbidity in the 21st Century," Brookings Papers on Economic Activity, Vol. 1, 2017, https://www.brookings.edu/wp-content/uploads/2017/08/casetextsp17bpea.pdf, 397–476.
12. Andrew Buncombe, "Donald Trump's Detroit Speech: Read the Full Transcript," *The Independent*, August 8, 2016, http://www.independent.co.uk/news/world/americas/us-elections/donald-trumps-detroit-speech-read-the-full-transcript-a7179421.html.
13. Michael J. Hicks and Srikant Devaraj, "The Myth and the Reality of Manufacturing in America," Center for Business and Economic Research, Ball State University, 2015 https://conexus.cberdata.org/files/MfgReality.pdf.
14. Ryan A. Decker et al., "Where Has All the Skewness Gone? The Decline in High-Growth (Young) Firms in the US," *European Economic Review* 86 (2016): 4–23 (data source: US Census Bureau).
15. Ronald S. Jarmin, Shawn D. Klimek, and Javier Miranda, "The Role of Retail Chains: National, Regional and Industry Results," in *Producer Dynamics: New Evidence from Micro Data*, ed. Tim Dunne (Chicago: University of Chicago Press, 2009), 237–62.
16. Neela Banerjee, Lisa Song, and David Hasemyer, "Exxon's Own Research Confirmed Fossil Fuels' Role in Global Warming Decades Ago," Inside Climate News, September 16, 2015, https://insideclimatenews.org/news/15092015/Exxons-own-research-confirmed-fossil-fuels-role-in-global-warming.
17. Geoffrey Supran and Naomi Oreskes, "Assessing ExxonMobil's Climate Change Communications (1977–2014)," *Environmental Research Letters* 12, no. 8 (2017): 084019.
18. Art Van Zee, "The Promotion and Marketing of Oxycontin: Commercial Triumph, Public Health Tragedy," *American Journal of Public Health* 99, no. 2 (2009): 221–27, doi 10.2105/AJPH.2007.131714.
19. Thomas Piketty, Emmanuel Saez, and Gabriel Zucman, "Distributional National Accounts: Methods and Estimates for the

United States," *Quarterly Journal of Economics* 133, no. 2 (2017): 553–609, doi 10.3386/w22945.

20. Facundo Alvaredo et al., "World Inequality Report, 2018," World Inequality Lab, 2018, https://wir2018.wid.world/files/download/wir2018-full-report-english.pdf, 82.
21. Bruce Sacerdote, "Fifty Years of Growth in American Consumption, Income, and Wages," Working Paper No. 23292, National Bureau of Economic Research, 2017; Michael R. Strain, "The Link between Wages and Productivity Is Strong," American Enterprise Institute, 2019, https://www.aei.org/wp-content/uploads/2019/02/The-Link-Between-Wages-and-Productivity-is-Strong.pdf.
22. "Average Weekly Earnings of Production and Nonsupervisory Employees, 1982–84 Dollars, Total Private, Seasonally Adjusted," https://data.bls.gov/pdq/SurveyOutputServletEmployment; "Hours, and Earnings from the Current Employment Statistics Survey (National)," Bureau of Labor Statistics (US), 2019, https://www.bls.gov/webapps/legacy/cesbtab8.htm.
23. Drew DeSilver, "For Most U.S. Workers, Real Wages Have Barely Budged in Decades," Pew Research Center, August 7, 2018, https://www.pewresearch.org/fact-tank/2018/08/07/for-most-us-workers-real-wages-have-barely-budged-for-decades/.
24. "The Distribution of Household Income, 2016," Congressional Budget Office, July 2019, https://www.cbo.gov/publication/55413.
25. David Leonhardt, "Our Broken Economy, in One Simple Chart," *New York Times*, August 7, 2017, https://www.nytimes.com/interactive/2017/08/07/opinion/leonhardt-income-inequality.html?smid=tw-share.
26. Raj Chetty et al., "The Fading American Dream: Trends in Absolute Income Mobility Since 1940," *Science* 356, no. 6336 (2017): 398–406.
27. Alvaredo et al., "World Inequality Report, 2018," 45.
28. Raquel Meyer Alexander, Stephen W. Mazza, and Susan Scholz, "Measuring Rates of Return for Lobbying Expenditures: An Empirical Case Study of Tax Breaks for Multinational Corporations," *Journal of Law and Politics* 25, no. 401 (2009): 401–58.
29. David Autor et al., "Importing Political Polarization? The Electoral Consequences of Rising Trade Exposure," National Bureau of Economic Research, Working Paper No. w22637 (2016): 936–53, doi 10.3386/w22637.

30. Data USA, 2017 (data source: US Census Bureau), https://datausa.io/profile/geo/waynesburg-pa/?compare=pennsylvania#about.
31. Julian Turner, "Lean and Clean: Why Modern Coal-Fired Power Plants are Better by Design," June 21, 2016, https://www.power-technology.com/features/featurelean-and-clean-why-modern-coal-fired-power-plants-are-better-by-design-4892873/.
32. Bryan Walsh, "How the Sierra Club Took Millions from the Natural Gas Industry—And Why They Stopped," *Time*, February 2, 2012, http://science.time.com/2012/02/02/exclusive-how-the-sierra-club-took-millions-from-the-natural-gas-industry-and-why-they-stopped/.
33. Neil Irwin, "How Are American Families Doing? A Guided Tour of Our Financial Well-Being," *New York Times*, September 8, 2014, https://www.nytimes.com/2014/09/09/upshot/how-are-american-families-doing-a-guided-tour-of-our-financial-well-being.html?module=inline.
34. *Terminator 2: Judgment Day*, James Cameron, dir., 1991.

CHAPTER 18: THE ANTI-GLOBALIZER

1. Hanoch Levin, *Schitz*, trans. Naaman Tammuz, in *Selected Plays One (1975–1983)* (London: Oberon, 2020), 110.
2. The conversation with the Quigley family was conducted by my researcher, Inbar Golan, in July 2019.
3. Winston S. Churchill, *The World Crisis: The Aftermath* (New York: Scribner, 1929), 63.
4. Marc Fisher and Will Hobson, "Donald Trump Masqueraded as Publicist to Brag About Himself," *Washington Post*, May 13, 2016, https://www.washingtonpost.com/politics/donald-trump-alter-ego-barron/2016/05/12/02ac99ec-16fe-11e6-aa55–670cabef46e0_story.html.
5. Chris Cillizza, "Donald Trump's 'John Miller' Interview Is Even Crazier Than You Think," *Washington Post*, May 16, 2016, https://www.washingtonpost.com/news/the-fix/wp/2016/05/16/donald-trumps-john-miller-interview-is-even-crazier-than-you-think/.
6. Patrick Radden Keefe, "How Mark Burnett Resurrected Donald Trump as an Icon of American Success," *New Yorker*, December 27, 2018, https://www.newyorker.com/magazine/2019/01/07/how-mark-burnett-resurrected-donald-trump-as-an-icon-of-american-success.

7. David A. Fahrenthold, "Trump Recorded Having Extremely Lewd Conversation About Women in 2005," *Washington Post*, October 8, 2016, https://www.washingtonpost.com/politics/trump-recorded-having-extremely-lewd-conversation-about-women-in-2005/2016/10/07/3b9ce776–8cb4–11e6-bf8a-3d26847eeed4_story.html.
8. "Transcript of Mitt Romney's Speech on Donald Trump," *New York Times*, March 3, 2016, https://www.nytimes.com/2016/03/04/us/politics/mitt-romney-speech.html.
9. Brad Plumer, "Full Transcript of Donald Trump's Acceptance Speech at the RNC," Vox, July 21, 2016, https://www.vox.com/2016/7/21/12253426/donald-trump-acceptance-speech-transcript-republican-nomination-transcript.
10. Craig Timberg and Tony Romm, "New Report on Russian Disinformation, Prepared for the Senate, Shows the Operation's Scale and Sweep," *Washington Post*, December 17, 2018, https://www.washingtonpost.com/technology/2018/12/16/new-report-russian-disinformation-prepared-senate-shows-operations-scale-sweep/; Philip N. Howard, Bharath Ganesh, and Dimitra Liotsiou, "The IRA, Social Media and Political Polarization in the United States, 2012–2018," University of Oxford, 2018, https://comprop.oii.ox.ac.uk/wp-content/uploads/sites/93/2018/12/The-IRA-Social-Media-and-Political-Polarization.pdf.
11. David E. Sanger and Catie Edmondson, "Russia Targeted Election Systems in All 50 States, Report Finds," *New York Times*, July 25, 2019, https://www.nytimes.com/2019/07/25/us/politics/russian-hacking-elections.html; "Report of the Select Committee on Intelligence, United States Senate, on Russian Active Measures Campaigns and Interference in the 2016 U.S. Election, Volume 1: Russian Efforts Against Election Infrastructure," US Senate, July 25, 2019, https://www.intelligence.senate.gov/sites/default/files/documents/Report_Volume1.pdf, 21–28.
12. "Transcript: Donald Trump's Foreign Policy Speech," *New York Times*, April 27, 2016, https://www.nytimes.com/2016/04/28/us/politics/transcript-trump-foreign-policy.html.
13. "Speech: Donald Trump Holds a Political Rally in Houston, Texas," Factbase, October 22, 2018, https://factba.se/transcript/donald-trump-speech-maga-rally-houston-tx-october-22–2018.
14. Salena Zito, "Taking Trump Seriously, Not Literally," *The Atlantic*, September 23, 2016, https://www.theatlantic.com

/politics/archive/2016/09/trump-makes-his-case-in-pittsburgh/501335/.

15. Rosie Gray, "Trump Defends White-Nationalist Protesters: 'Some Very Fine People on Both Sides,'" *The Atlantic*, August 15, 2017, https://www.theatlantic.com/politics/archive/2017/08/trump-defends-white-nationalist-protesters-some-very-fine-people-on-both-sides/537012/.
16. Katie Rogers and Nicholas Fandos, "Trump Tells Congresswomen to 'Go Back' to the Countries They Came From," *New York Times*, July 14, 2019, https://www.nytimes.com/2019/07/14/us/politics/trump-twitter-squad-congress.html.
17. Paul Waldman, "Trump Sucks up to Putin, Embarrassing Us Yet Again," *Washington Post*, June 28, 2019, https://www.washingtonpost.com/opinions/2019/06/28/trump-sucks-up-putin-embarrassing-us-yet-again/.
18. Benjamin De Cleen, "Populism and Nationalism," in *The Oxford Handbook of Populism*, ed. Cristóbal Kaltwasser Rovira et al. (New York: Oxford University Press, 2017), 342–62.
19. George Orwell, "Notes on Nationalism," *Polemic* 1 (October 1945), paragraphs 4, 15.

CHAPTER 19: THE IMPLOSION OF TRUTH

1. Alexandra Jaffe, "Kellyanne Conway: WH Spokesman Gave 'Alternative Facts' on Inauguration Crowd," NBC, January 22, 2017, https://www.nbcnews.com/storyline/meet-the-press-70-years/wh-spokesman-gave-alternative-facts-inauguration-crowd-n710466.
2. "Income Inequality in the San Francisco Bay Area," Silicon Valley Institute for Regional Studies, June 2015, https://jointventure.org/images/stories/pdf/income-inequality-2015–06.pdf.
3. "California Homelessness Statistics," United States Interagency Council on Homelessness, 2018, https://www.usich.gov/homelessness-statistics/ca.
4. Theodore Schleifer, "One Out of Every 11,600 People in San Francisco Is a Billionaire," Vox, May 9, 2019, https://www.vox.com/recode/2019/5/9/18537122/billionaire-study-wealthx-san-francisco; "The Wealth-X Billionaire Census 2019," Wealth-X, May 9, 2019, https://www.wealthx.com/report/the-wealth-x-billionaire-census-2019/?utm_campaign=bc-2019&utm_source=broadcast&utm_medium=referral&utm_term=bc-2019-press&utm_source=broadcast&utm_medium=referral.

5. Tim Cook, "Tim Cook to Grads: This Is Your World to Change," *Time*, May 18, 2015, http://time.com/collection-post/3882479/tim-cook-graduation-speech-gwu/.
6. Mike Isaac and Scott Shane, "Facebook's Russia-Linked Ads Came in Many Disguises," *New York Times*, October 2, 2017, https://www.nytimes.com/2017/10/02/technology/facebook-russia-ads-.html?rref=collection%2Fbyline%2Fmike-isaac&action=click&contentCollection=undefined%C2%AEion=stream&module=stream_unit&version=latest&contentPlacement=5&pgtype=collection.
7. Craig Silverman, "This Analysis Shows How Viral Fake Election News Stories Outperformed Real News on Facebook," *BuzzFeed*, November 17, 2016, https://www.buzzfeed.com/craigsilverman/viral-fake-election-news-outperformed-real-news-on-facebook?utm_term=.uyRyVedQ2P#.hj5KkW1nXJ.
8. Kurt Wagner, "Two-Thirds of Americans Are Now Getting News from Social Media," Vox, September 7, 2017, https://www.vox.com/2017/9/7/16270900/social-media-news-americans-facebook-twitter; "In 2017 Two-Thirds of U.S. Adults Get News from Social Media," Pew Research Center, September 5, 2017, https://www.journalism.org/2017/09/07/news-use-across-social-media-platforms-2017/pi_17–08–23_socialmediaupdate_0–01/.
9. Andrew Guess, Brendan Nyhan, and Jason Reifler, "Selective Exposure to Misinformation: Evidence from the Consumption of Fake News During the 2016 US Presidential Campaign," European Research Council, January 9, 2018, http://www.dartmouth.edu/~nyhan/fake-news-2016.pdf.
10. Samanth Subramanian, "The Macedonian Teens Who Mastered Fake News," *Wired*, February 15, 2017, https://www.wired.com/2017/02/veles-macedonia-fake-news/.
11. Thomas Fuller, "Internet Unshackled, Burmese Aim Venom at Ethnic Minority," *New York Times*, June 15, 2012, https://www.nytimes.com/2012/06/16/world/asia/new-freedom-in-myanmar-lets-burmese-air-venom-toward-rohingya-muslim-group.html?searchResultPosition=8&module=inline.
12. "Report of the Independent International Fact-Finding Mission on Myanmar," Human Rights Council, UN, September 17, 2018, https://www.ohchr.org/EN/HRBodies/HRC/Pages/NewsDetail.aspx?NewsID=23575&LangID=E.

13. Karsten Müller and Carlo Schwarz, "Fanning the Flames of Hate: Social Media and Hate Crime," 2018, http://dx.doi.org/10.2139/ssrn.3082972.
14. "Facebook's Algorithm: A Major Threat to Public Health," AVAAZ, August 19, 2020, https://avaazimages.avaaz.org/facebook_threat_health.pdf.
15. Alan I. Abramowitz, "Did Russian Interference Affect the 2016 Election Results?," Sabato's Crystal Ball, University of Virginia Center for Politics, August 8, 2019, http://crystalball.centerforpolitics.org/crystalball/articles/did-russian-interference-affect-the-2016-election-results/; Morgan Marietta, "Did Russian Interference Change Votes in 2016?" *Psychology Today*, August 15, 2019, https://www.psychologytoday.com/us/blog/inconvenient-facts/201908/did-russian-interference-change-votes-in-2016; Yochai Benkler, Robert Faris, and Hal Roberts, *Network Propaganda: Manipulation, Disinformation, and Radicalization in American Politics* (New York: Oxford University Press, 2018), 235–68.
16. Herbert Marshall McLuhan, *Understanding Media: The Extensions of Man* (New York: McGraw-Hill, 1964).
17. Mason Walker and Jeffrey Gottfried, "Republicans Far More Likely than Democrats to Say Fact-Checkers Tend to Favor One Side," Pew Research Center, June 27, 2019, https://www.pewresearch.org/fact-tank/2019/06/27/republicans-far-more-likely-than-democrats-to-say-fact-checkers-tend-to-favor-one-side/.
18. "Fake News, Filter Bubbles, Post-Truth and Trust," Ipsos, September 2018, https://www.ipsos.com/sites/default/files/ct/news/documents/2018–09/fake-news-filter-bubbles-post-truth-and-trust.pdf.
19. Andrew Chadwick and Cristian Vaccari, "News Sharing on UK Social Media Misinformation, Disinformation, and Correction," Online Civic Culture Centre, Loughborough University, May 2, 2019, https://www.lboro.ac.uk/media/media/research/o3c/Chadwick%20Vaccari%20O3C-1%20News%20Sharing%20on%20UK%20Social%20Media.pdf.
20. Francine Prose, "Truth Is Evaporating Before Our Eyes," *The Guardian*, December 19, 2016, https://www.theguardian.com/commentisfree/2016/dec/19/truth-is-evaporating-before-our-eyes.
21. Soroush Vosoughi, Deb Roy, and Sinan Aral, "The Spread of True and False News Online," *Science* 359, no. 6380 (2018): 1146–51.

22. "Fake News," Ipsos, September 2018.
23. Galen Stocking, "Political Leaders, Activists Viewed as Prolific Creators of Made-Up News; Journalists Seen as the Ones to Fix It," Pew Research Center, June 5, 2019, https://www.journalism.org/2019/06/05/political-leaders-activists-viewed-as-prolific-creators-of-made-up-news-journalists-seen-as-the-ones-to-fix-it/.
24. António Guterres, "Secretary-General's Remarks to UNA-USA Global Engagement Summit," United Nations, Secretary-General, UN, February 22, 2019, https://www.un.org/sg/en/content/sg/statement/2019–02–22/secretary-generals-remarks-una-usa-global-engagement-summit-delivered.
25. Yann Algan and Pierre Cahuc, "Inherited Trust and Growth," *American Economic Review* 100, no. 5 (2010): 2060–92; Oguzhan C. Dincer and Eric M. Uslaner, "Trust and Growth," *Public Choice* 142 (2010): 59–67.
26. "2019 Edelman Trust Barometer, Global Report," Edelman, 2019, https://www.edelman.com/sites/g/files/aatuss191/files/2019–02/2019_Edelman_Trust_Barometer_Global_Report.pdf.
27. "Trust in Government," Directorate for Public Governance, OECD, 2019, https://www.oecd.org/gov/trust-in-government.htm.
28. Public Opinion, Eurobarometer Interactive, European Commission, June 2019, https://ec.europa.eu/commfrontoffice/public opinion/index.cfm/Chart/getChart/themeKy/18/groupKy/98; "Standard Eurobarometer 89 Spring 2018," European Commission, 2018, https://ec.europa.eu/commfrontoffice/publicopinion/index.cfm/ResultDoc/download/DocumentKy/83548.
29. "Confidence in Institutions," Gallup, 2019, https://news.gallup.com/poll/1597/confidence-institutions.aspx; "Public Trust in Government: 1958–2019," Pew Research Center, April 11, 2019, https://www.people-press.org/2019/04/11/public-trust-in-government-1958–2019/.
30. John Gramlich, "Young Americans Are Less Trusting of Other People—and Key Institutions—Than Their Elders," Pew Research Center, August 6, 2019, https://www.pewresearch.org/fact-tank/2019/08/06/young-americans-are-less-trusting-of-other-people-and-key-institutions-than-their-elders/.
31. Esteban Ortiz-Ospina and Max Roser, "Trust," in "Our World in Data, 2019" (data source: US General Survey Data, 2014), https://ourworldindata.org/trust#in-the-us-people-trust-each

-other-less-now-than-40-years-ago; US General Survey Data, 2018, https://gssdataexplorer.norc.org/variables/441/vshow.

32. Trustlab, OECD, 2019 (data source: OECD survey, 2016–18), https://www.oecd.org/sdd/trustlab.htm.
33. Gramlich, "Young Americans."
34. Alberto Alesina and Eliana La Ferrara, "The Determinants of Trust," National Bureau of Economic Research, Working Paper No. 7621, 2000; Henrik Jordahl, "Economic Inequality," in *Handbook of Social Capital*, ed. Gert Tinggaard Svendsen and Gunnar Lind Haase Svendsen (Cheltenham, UK: Edward Elgar, 2009), 323–36.
35. Lee Rainie and Andrew Perrin, "Key Findings about Americans' Declining Trust in Government and Each Other," Pew Research Center, July 22, 2019, https://www.pewresearch.org/fact-tank/2019/07/22/key-findings-about-americans-declining-trust-in-government-and-each-other/.
36. Susan J. Masten, Simon H. Davies, and Shawn P. McElmurry, "Flint Water Crisis: What Happened and Why?" *Journal of the American Water Works Association* 108, no. 12 (2016): 22–34.
37. Mitch Smith, Julie Bosman, and Monica Davey, "Flint's Water Crisis Started 5 Years Ago. It's Not Over," *New York Times*, April 25, 2019, https://www.nytimes.com/2019/04/25/us/flint-water-crisis.html.
38. "High Lead Levels in Flint, Michigan: Interim Report," United States Environmental Protection Agency, June 24, 2015, http://flintwaterstudy.org/wp-content/uploads/2015/11/Miguels-Memo.pdf.
39. Daniel S. Grossman and David J. G. Slutsky, "The Effect of an Increase in Lead in the Water System on Fertility and Birth Outcomes: The Case of Flint, Michigan," University of West Virginia and University of Kansas, 2017.
40. James Salzman, *Drinking Water: A History*, rev. ed. (New York: Abrams, 2017), 149–50.
41. Lauren Gibbons, "See How Voter Turnout Changed in Every Michigan County from 2012 to 2016," Michigan Live, November 11, 2016, https://www.mlive.com/news/2016/11/see_how_every_michigan_county.html.
42. Sowmya R. Rao et al., "Survey Shows That at Least Some Physicians Are Not Always Open or Honest with Patients," *Health Affairs* 31, no. 2 (2012): 383–91; Marcia Frellick, "Physicians, Nurses Draw Different Lines for When Lying Is OK," Med-

scape, January 31, 2019, https://www.medscape.com/viewarticle/908418.

43. Jerald M. Jellison, *I'm Sorry, I Didn't Mean To, and Other Lies We Love To Tell* (Chicago: Chatham Square Press, 1977).
44. Robert S. Feldman, James A. Forrest, and Benjamin R. Happ, "Self-Presentation and Verbal Deception: Do Self-Presenters Lie More?" *Basic and Applied Social Psychology* 24, no. 2 (2002): 163–70, https://doi.org/10.1207/S15324834BASP2402_8.
45. Kim B. Serota, Timothy R. Levine, and Franklin J. Boster, "The Prevalence of Lying in America: Three Studies of Self-Reported Lies," *Human Communication Research* 36, no. 1 (2010): 2–25.
46. Dana Carney et al., "People with Power Are Better Liars," Columbia Business School, 2017, https://www0.gsb.columbia.edu/mygsb/faculty/research/pubfiles/3510/Power.Lying.pdf; D. R. Carney, "People with Power Are Better Liars" (October 2009), presented at the Person Memory Interest Group, Boothbay Harbor, ME. 2.
47. Danny Sullivan, "Google Now Handles at Least 2 Trillion Searches per Year," Search Engine Land, May 24, 2016, https://searchengineland.com/google-now-handles-2–999-trillion-searches-per-year-250247.
48. Nikita Sood et al., "Paging Dr. Google: The Effect of Online Health Information on Trust in Pediatricians' Diagnoses," *Clinical Pediatrics* 58,no. 8 (2019): 889–96.
49. Reid Wilson, "Fury Fuels the Modern Political Climate in US," *The Hill*, September 20, 2019 (data source: Gallup), https://thehill.com/homenews/state-watch/351432-fury-fuels-the-modern-political-climate-in-us; Frank Newport, "Americans' Confidence in Institutions Edges Up," Gallup, June 26, 2017, https://news.gallup.com/poll/212840/americans-confidence-institutions-edges.aspx.
50. Gramlich, "Young Americans."
51. Seth Stephens-Davidowitz, *Everybody Lies: Big Data, New Data, and What the Internet Can Tell Us About Who We Really Are* (New York: Harper Collins, 2017).
52. Exclusive Third Rail with OZY–The Marist Poll, September 2017, https://www.pbs.org/wgbh/third-rail/episodes/episode-1-is-truth-overrated/americans-value-ideal-truth-american-society/.

53. Amy Mitchell et al., "Distinguishing Between Factual and Opinion Statements in the News," Pew Research Center, June 18, 2018, https://www.journalism.org/2018/06/18/distinguishing-between-factual-and-opinion-statements-in-the-news/.
54. Hannah Arendt, *Totalitarianism: Part Three of The Origins of Totalitarianism* (New York: Harcourt Brace and Company, 1973; originally published 1951), 382.

CHAPTER 20: THE BATTLE FOR PROGRESS

1. Joseph E. Stiglitz, *Making Globalization Work* (New York: W. W. Norton & Company, 2006), 292.
2. "CNBC Transcript: French Presidential Candidate & National Front Party Leader Marine Le Pen Speaks with CNBC's Michelle Caruso-Cabrera Today," CNBC, November 21, 2016, https://www.cnbc.com/2016/11/21/cnbc-transcript-french-presidential-candidate-national-front-party-leader-marine-le-pen-speaks-with-cnbcs-michelle-caruso-cabrera-today.html.
3. Roberto Stefan Foa and Yascha Mounk, "The Danger of Deconsolidation: The Democratic Disconnect," *Journal of Democracy* 27, no. 3 (2016): 5–17.
4. W. B. Yeats, "The Second Coming," 1919, https://www.poetryfoundation.org/poems/43290/the-second-coming.
5. Barbara W. Tuchman, *The March of Folly* (New York: Knopf, 1984), 5.
6. Eric Bradner, "Trump Praises 9/11 Truther's 'Amazing' Reputation," CNN, December 2, 2015, https://edition.cnn.com/2015/12/02/politics/donald-trump-praises-9–11-truther-alex-jones/index.html.
7. The Avielle Foundation, 2019, https://aviellefoundation.org/about-the-foundation/welcome-message/.
8. Dorothea Waley Singer, *Giordano Bruno: His Life and Thought* (New York: Henry Schuman, 1950), 179.
9. Yuval Noah Harari, *Sapiens: A Brief History of Humankind* (London: Harvill Secker, 2014), 215.
10. Marie Jean-Antoine-Nicolas de Caritat, Marquis de Condorcet, *Outlines of an Historical View of the Progress of the Human Mind*, trans. from the French (London: Printed for J. Johnson, 1795), 327.
11. Pierre Bayle, *Various Thoughts on the Occasion of a Comet*, 1682, trans. Robert C. Bartlett (Albany: SUNY Press, 2000), 130.

12. Quoted in Mark Twain, "King Leopold's Soliloquy" (New Delhi: LeftWord Books, 1970; first published 1905), 12.
13. "How Robots Change the World," Oxford Economics, June 2019, https://www.oxfordeconomics.com/recent-releases/how-robots-change-the-world;.
14. Quoctrung Bui, "How Machines Destroy (And Create!) Jobs, in 4 Graphs," National Public Radio, May 18, 2015 (data source: IPUMS-USA, University of Minnesota), https://www.npr.org/sections/money/2015/05/18/404991483/how-machines-destroy-and-create-jobs-in-4-graphs.
15. "Fastest Declining Occupations, 2018 and Projected 2028," Bureau of Labor Statistics, United States Department of Labor, September 4, 2019, https://www.bls.gov/emp/tables/fastest-declining-occupations.htm.
16. "Ben Shapiro and Tucker Carlson Debate the Impact of Driverless Cars," YouTube, November 4, 2018, https://www.youtube.com/watch?v=o5zPKxpPHFk.
17. Max Weber, "Science as a Vocation," *From Max Weber: Essays in Sociology*, translated, edited, and with an introduction by H. H Gerth and C. Wright Mills (Abingdon, UK: Routledge, 1971), 139.
18. Jared Diamond, *Collapse: How Societies Choose to Fail or Succeed* (New York: Viking, 2005).
19. Ronald Wright, *A Short History of Progress* (Toronto: House of Anansi Press, 2004).
20. Wright, *A Short History*, 64; Diamond, *Collapse*, 64, 118–19.
21. Susan Cosier, "The World Needs Topsoil to Grow 95% of Its Food—But It's Rapidly Disappearing," *The Guardian*, May 30, 2019, https://www.theguardian.com/us-news/2019/may/30/topsoil-farming-agriculture-food-toxic-america.
22. Czeslaw Milosz, "Campo dei Fiori," 1943, trans. David Brooks and Louis Iribarne, Poetry Foundation, https://www.poetryfoundation.org/poems/49751/campo-dei-fiori, from *The Collected Poems: 1931–1987* (New York: Ecco, 1988), 33–35.

CHAPTER 21: A NEW STORY

1. Robert Burns, "Away from Washington, a More Personal Mattis Reveals Himself," Associated Press, January 9, 2018, https://www.apnews.com/667bd4c51217464487e44948ccf6b631.

2. Martin Luther King Jr., Methodist Student Leadership Conference Address, Lincoln, Nebraska, 1964, American Rhetoric Online Speech Bank, https://americanrhetoric.com/speeches/mlkmethodistyouthconference.htm.
3. Douglass, "West India Emancipation Speech."

Index